PRAISE FOR THE *MANGA GUIDE* SERIES

"I love, love, love this idea! EduManga is a great concept."
—The Otaku Librarian

"These books will make studying fun for teens."
—CONNECTINGWITHYOURTEENS.BLOGSPOT.COM

"The art is nicely clean and filled with just the right amount of detail, and the characters are quite expressive and given to funny asides."
—Warren Peace

"This is a perfect addition to a home school curriculum."
—Parenting Squad

"The series' shoujo style and spunky schoolgirl heroines make the books a good potential fit for those manga readers who've made *Fruits Basket* a hit."
—Blogcritics.org

PRAISE FOR *THE MANGA GUIDE TO STATISTICS*

"This is really what a good math text should be like. Unlike the majority of books on subjects like statistics, it doesn't just present the material as a dry series of pointless-seeming formulas. It presents statistics as something fun, and something enlightening."
—Good Math, Bad Math

"*The Manga Guide to Statistics* is a solid introduction to the world of statistical analysis done in a fun and accessible way."
—Active Anime

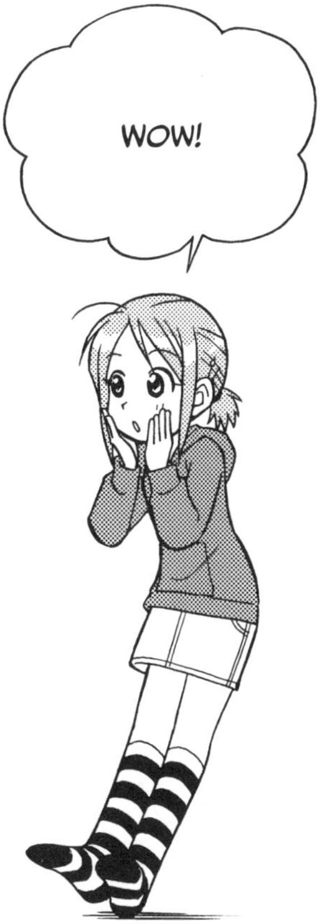
WOW!

"Inoue's art is clean, cute and simplified, and it works with machine-like efficiency—the artist not only knows and speaks the language of manga, but does so fluently."
—Newsarama

"For those who've been taking math classes in high school or college, I think this book will be very helpful in making statistics easier to understand."
—Astronerdboy's Anime & Manga Blog

"This book is an excellent introductory book on statistics."
—Golden Triangle PC Club

"This manga succeeds in presenting important statistical concepts in an accessible and entertaining way. The charm of the book lies in the real-world examples it uses."
—Curled Up with a Good Book

THE MANGA GUIDE™ TO ELECTRICITY

THE MANGA GUIDE™ TO
ELECTRICITY

KAZUHIRO FUJITAKI
MATSUDA
TREND-PRO CO., LTD.

The Manga Guide to Electricity is a translation of the Japanese original, *Manga de Wakaru Denki*, published by Ohmsha, Ltd. of Tokyo, Japan, © 2006 by Kazuhiro Fujitaki and TREND-PRO Co., Ltd.

This English edition is co-published by No Starch Press, Inc. and Ohmsha, Ltd.

13 12 11 10 09   1 2 3 4 5 6 7 8 9

ISBN-10: 1-59327-197-2
ISBN-13: 978-1-59327-197-8

FIBER USED IN THIS PRODUCT LINE MEETS THE SOURCING REQUIREMENTS OF THE SFI PROGRAM
SFI CERTIFIED SOURCING
WWW.SFIPROGRAM.ORG

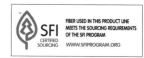

Publisher: William Pollock
Author: Kazuhiro Fujitaki
Illustrator: Matsuda
Producer: TREND-PRO Co., Ltd.
Production Editor: Megan Dunchak
Developmental Editor: Tyler Ortman
Translator: Arnie Rusoff
Technical Reviewers: Dave Issadore and Keith Brown
Compositor: Riley Hoffman
Proofreader: Cristina Chan
Indexer: Sarah Schott

For information on book distributors or translations, please contact No Starch Press, Inc. directly:
No Starch Press, Inc.
555 De Haro Street, Suite 250, San Francisco, CA 94107
phone: 415.863.9900; fax: 415.863.9950; info@nostarch.com; http://www.nostarch.com/

Library of Congress Cataloging-in-Publication Data

Fujitaki, Kazuhiro.
[Manga de wakaru denki. English]
The manga guide to electricity / Kazuhiro Fujitaki, Matsuda, and Trend-pro Co., Ltd.
p. cm.
Includes index.
ISBN-13: 978-1-59327-197-8
ISBN-10: 1-59327-197-2
1. Electricity--Comic books, strips, etc. 2. Electricity--Popular works. I. Trend-pro Co. II. Title.
QC527.F8513 2009
537--dc22

2008054602

CONTENTS

5
HOW CAN YOU CONVENIENTLY USE ELECTRICITY? 155

PREFACE

Our current lifestyle necessitates the use of electricity. Electric current is often explained by comparing it to flowing water, but since electricity cannot be seen by the naked eye, this metaphor can be difficult to understand. What can we do to better understand electricity?

Electricity is very helpful in almost every facet of our lives; it produces light, heat, and power. Even though we can see these benefits, we are usually unaware of electricity itself. However, if we simply learn the basics, we can get a clear picture of how electricity works.

This book explains fundamental electrical concepts using a story told through manga followed by further explanations in written text. There are no complicated explanations—readers simply listen along with the heroine Rereko as her teacher Hikaru explains concepts. Even people who have had a hard time understanding electricity will find Hikaru's explanations easy to comprehend.

I am extremely grateful to Matsuda, who provided the artwork, and to everyone at TREND-PRO, who produced the book. I would also like to give my sincere thanks to Professor Masaaki Mitani for checking my work. I am also very thankful to Ohmsha, Ltd. for giving me the opportunity to write this book.

I hope that in reading this book you will learn about electricity and gain a familiarity with it.

KAZUHIRO FUJITAKI
DECEMBER 2006

PROLOGUE:
FROM ELECTOPIA,
THE LAND OF ELECTRICITY

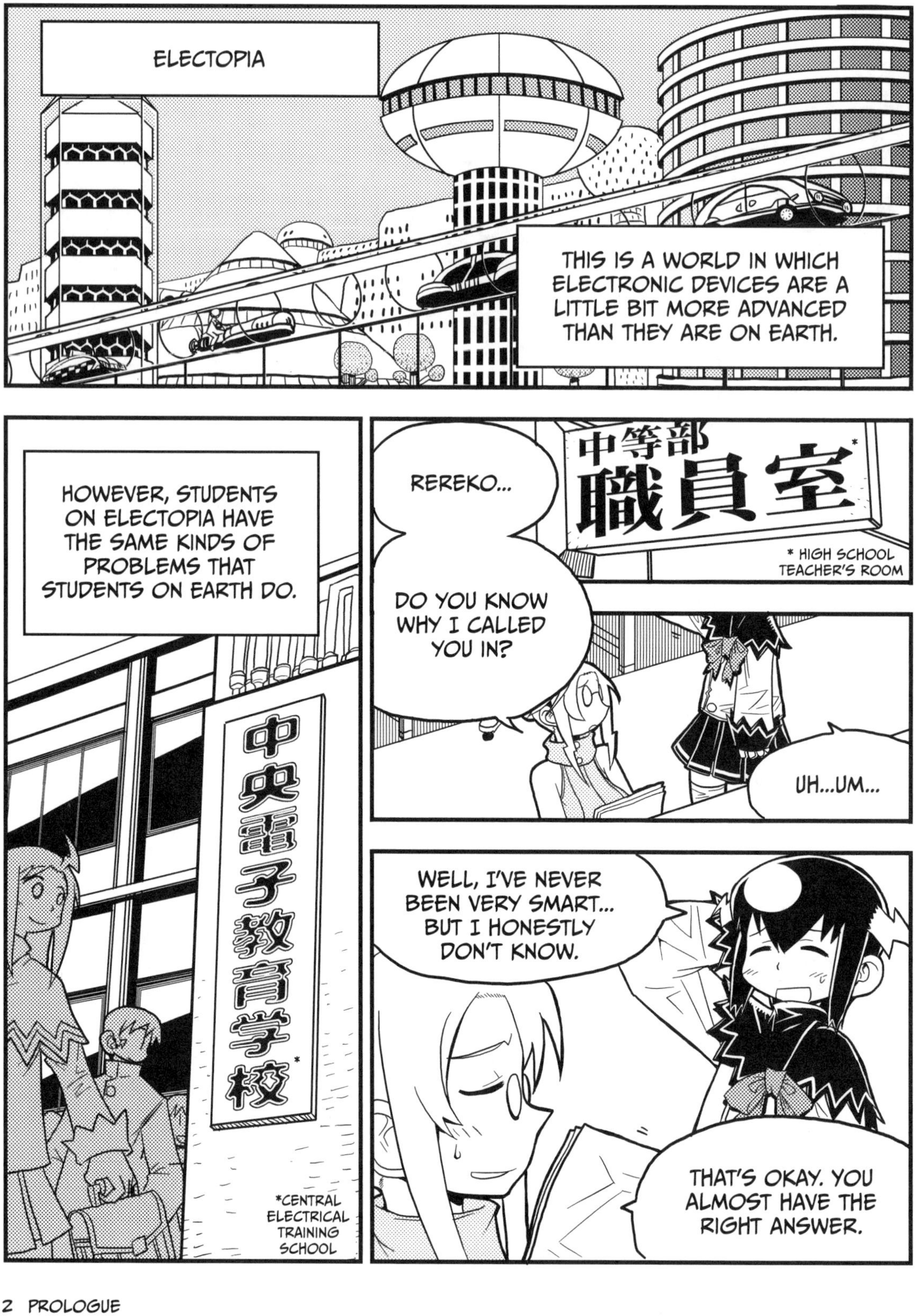
ELECTOPIA
THIS IS A WORLD IN WHICH ELECTRONIC DEVICES ARE A LITTLE BIT MORE ADVANCED THAN THEY ARE ON EARTH.
HOWEVER, STUDENTS ON ELECTOPIA HAVE THE SAME KINDS OF PROBLEMS THAT STUDENTS ON EARTH DO.
中央電子教育学校*
*CENTRAL ELECTRICAL TRAINING SCHOOL
REREKO...
中等部 職員室*
* HIGH SCHOOL TEACHER'S ROOM
DO YOU KNOW WHY I CALLED YOU IN?
UH...UM...
WELL, I'VE NEVER BEEN VERY SMART... BUT I HONESTLY DON'T KNOW.
THAT'S OKAY. YOU ALMOST HAVE THE RIGHT ANSWER.

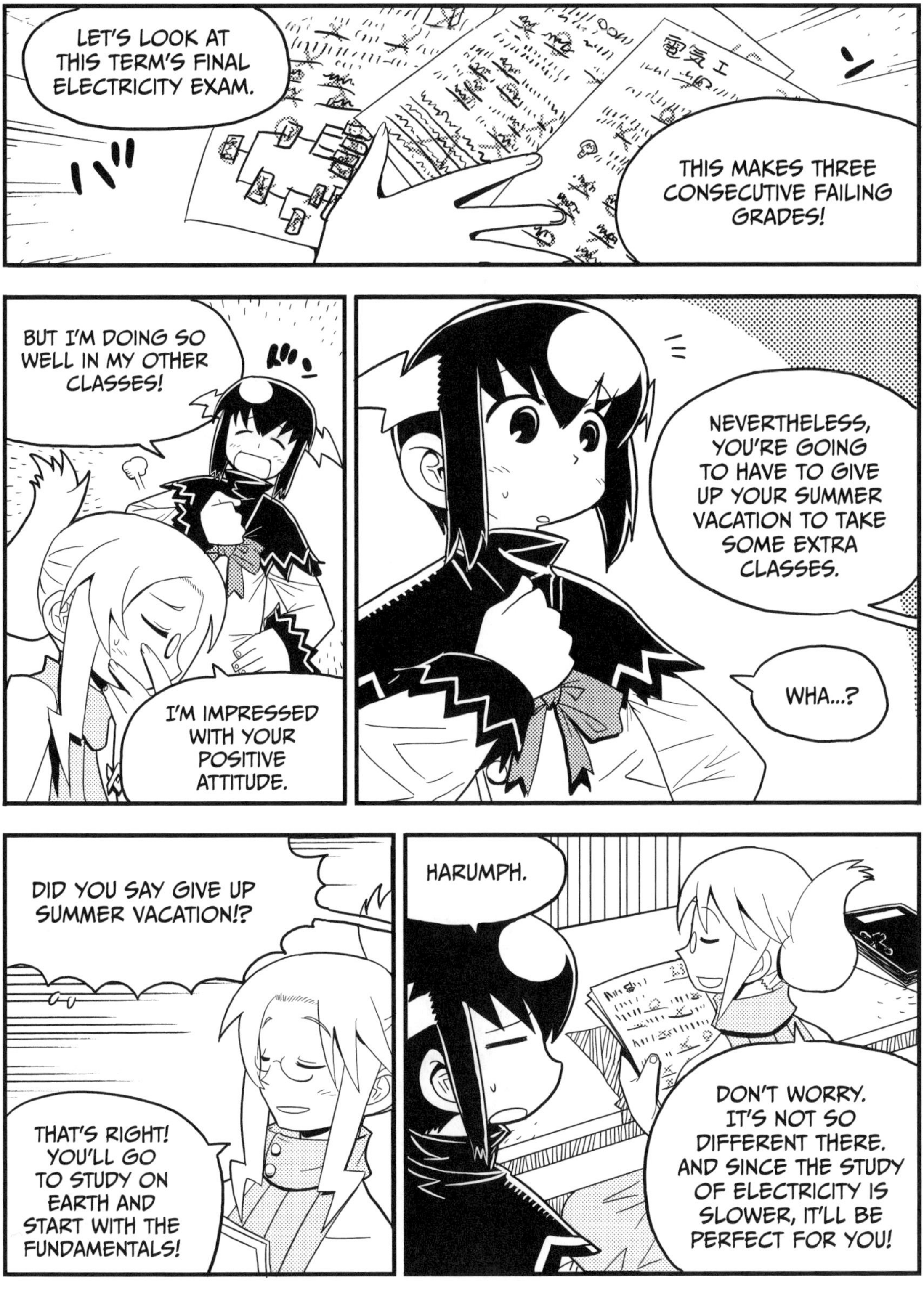
LET'S LOOK AT THIS TERM'S FINAL ELECTRICITY EXAM.
THIS MAKES THREE CONSECUTIVE FAILING GRADES!
BUT I'M DOING SO WELL IN MY OTHER CLASSES!
I'M IMPRESSED WITH YOUR POSITIVE ATTITUDE.
NEVERTHELESS, YOU'RE GOING TO HAVE TO GIVE UP YOUR SUMMER VACATION TO TAKE SOME EXTRA CLASSES.
WHA...?
DID YOU SAY GIVE UP SUMMER VACATION!?
THAT'S RIGHT! YOU'LL GO TO STUDY ON EARTH AND START WITH THE FUNDAMENTALS!
HARUMPH.
DON'T WORRY. IT'S NOT SO DIFFERENT THERE. AND SINCE THE STUDY OF ELECTRICITY IS SLOWER, IT'LL BE PERFECT FOR YOU!

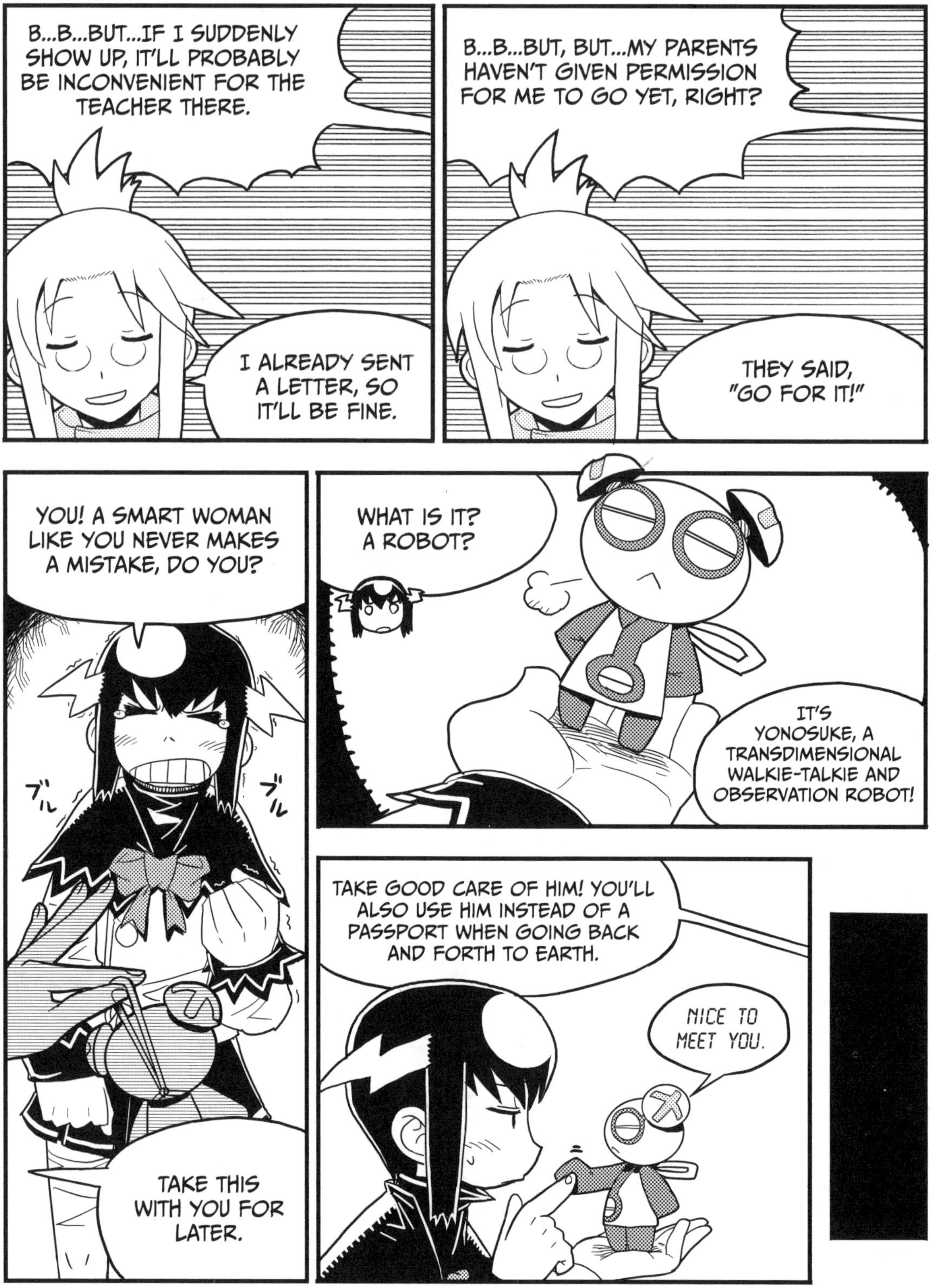
B...B...BUT...IF I SUDDENLY SHOW UP, IT'LL PROBABLY BE INCONVENIENT FOR THE TEACHER THERE.
I ALREADY SENT A LETTER, SO IT'LL BE FINE.
B...B...BUT, BUT...MY PARENTS HAVEN'T GIVEN PERMISSION FOR ME TO GO YET, RIGHT?
THEY SAID, "GO FOR IT!"
YOU! A SMART WOMAN LIKE YOU NEVER MAKES A MISTAKE, DO YOU?
ブル
ブル
TAKE THIS WITH YOU FOR LATER.
WHAT IS IT? A ROBOT?
IT'S YONOSUKE, A TRANSDIMENSIONAL WALKIE-TALKIE AND OBSERVATION ROBOT!
TAKE GOOD CARE OF HIM! YOU'LL ALSO USE HIM INSTEAD OF A PASSPORT WHEN GOING BACK AND FORTH TO EARTH.
NICE TO MEET YOU.

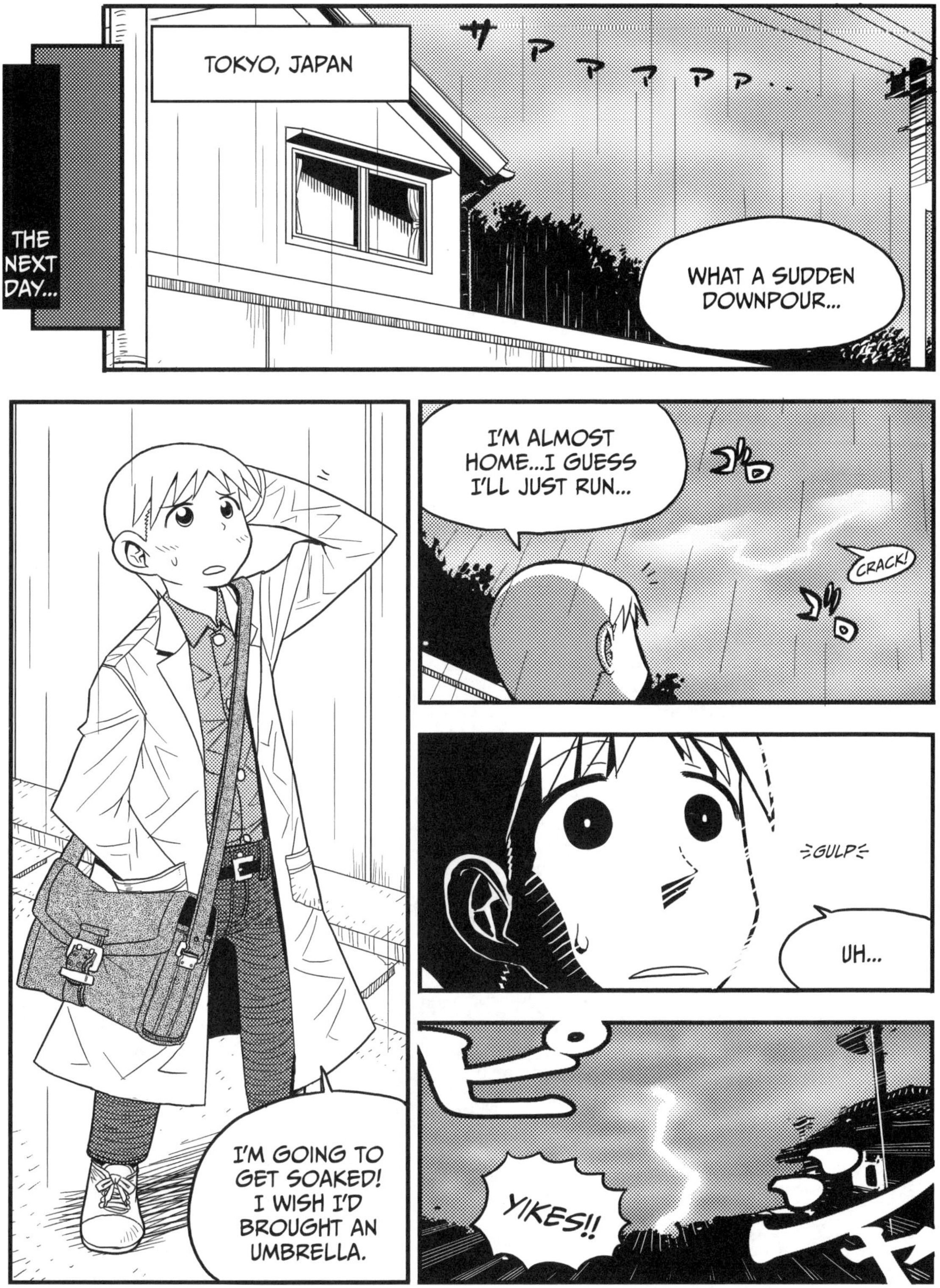
THE NEXT DAY...
TOKYO, JAPAN
WHAT A SUDDEN DOWNPOUR...
I'M GOING TO GET SOAKED! I WISH I'D BROUGHT AN UMBRELLA.
I'M ALMOST HOME...I GUESS I'LL JUST RUN...
CRACK!
GULP
UH...
YIKES!!

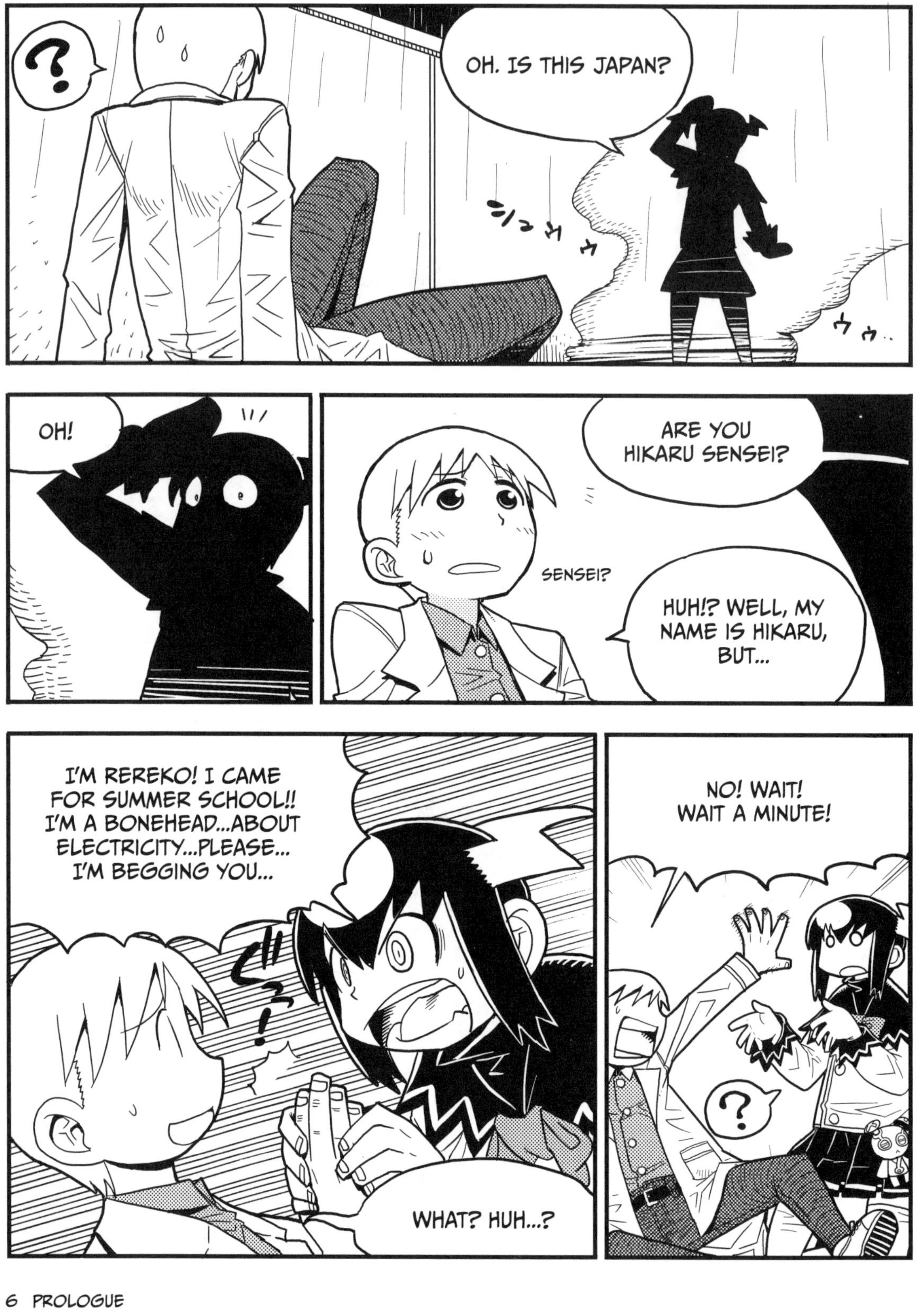
?
OH. IS THIS JAPAN?
OH!
ARE YOU HIKARU SENSEI?
SENSEI?
HUH!? WELL, MY NAME IS HIKARU, BUT...
I'M REREKO! I CAME FOR SUMMER SCHOOL!! I'M A BONEHEAD...ABOUT ELECTRICITY...PLEASE... I'M BEGGING YOU...
WHAT? HUH...?
NO! WAIT! WAIT A MINUTE!
?

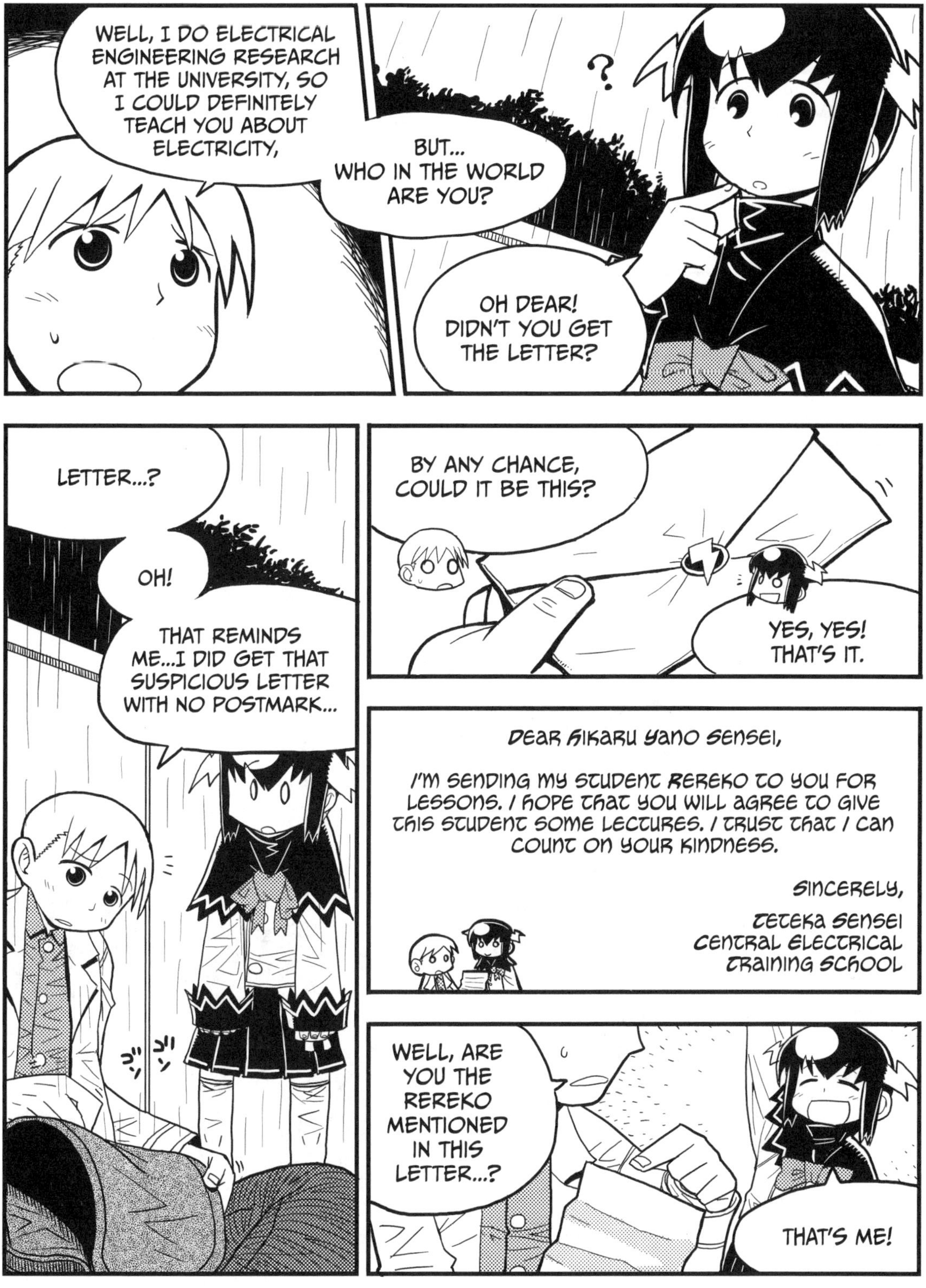
WELL, I DO ELECTRICAL ENGINEERING RESEARCH AT THE UNIVERSITY, SO I COULD DEFINITELY TEACH YOU ABOUT ELECTRICITY,
BUT... WHO IN THE WORLD ARE YOU?
?
OH DEAR! DIDN'T YOU GET THE LETTER?
LETTER...?
OH!
THAT REMINDS ME...I DID GET THAT SUSPICIOUS LETTER WITH NO POSTMARK...
BY ANY CHANCE, COULD IT BE THIS?
YES, YES! THAT'S IT.
Dear Hikaru Yano Sensei,
I'm sending my student Rereko to you for lessons. I hope that you will agree to give this student some lectures. I trust that I can count on your kindness.
Sincerely,
Teteka Sensei
Central Electrical Training School
WELL, ARE YOU THE REREKO MENTIONED IN THIS LETTER...?
THAT'S ME!

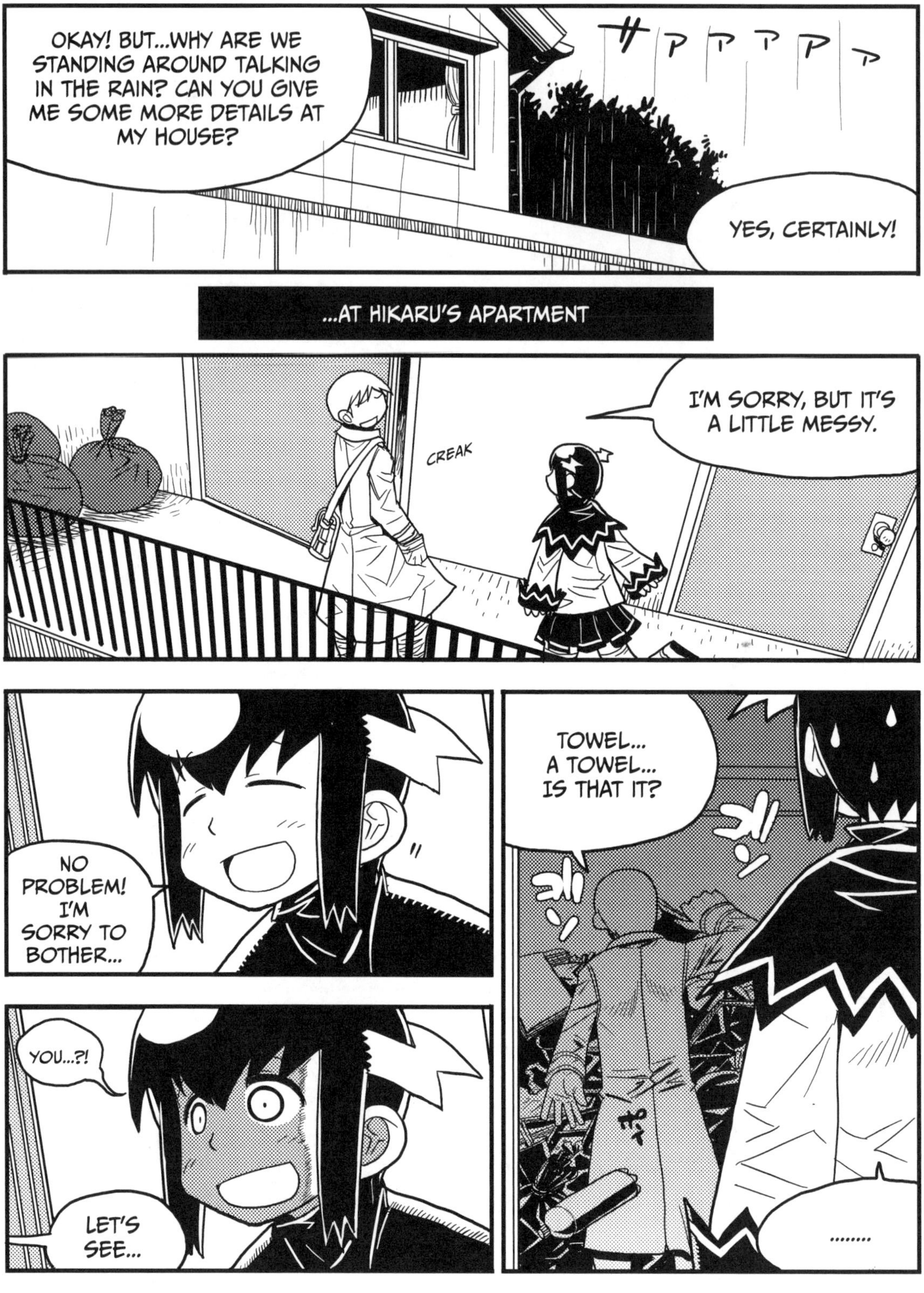
OKAY! BUT...WHY ARE WE STANDING AROUND TALKING IN THE RAIN? CAN YOU GIVE ME SOME MORE DETAILS AT MY HOUSE?
YES, CERTAINLY!
...AT HIKARU'S APARTMENT
CREAK
I'M SORRY, BUT IT'S A LITTLE MESSY.
NO PROBLEM! I'M SORRY TO BOTHER...
YOU...?!
LET'S SEE...
TOWEL... A TOWEL... IS THAT IT?
....

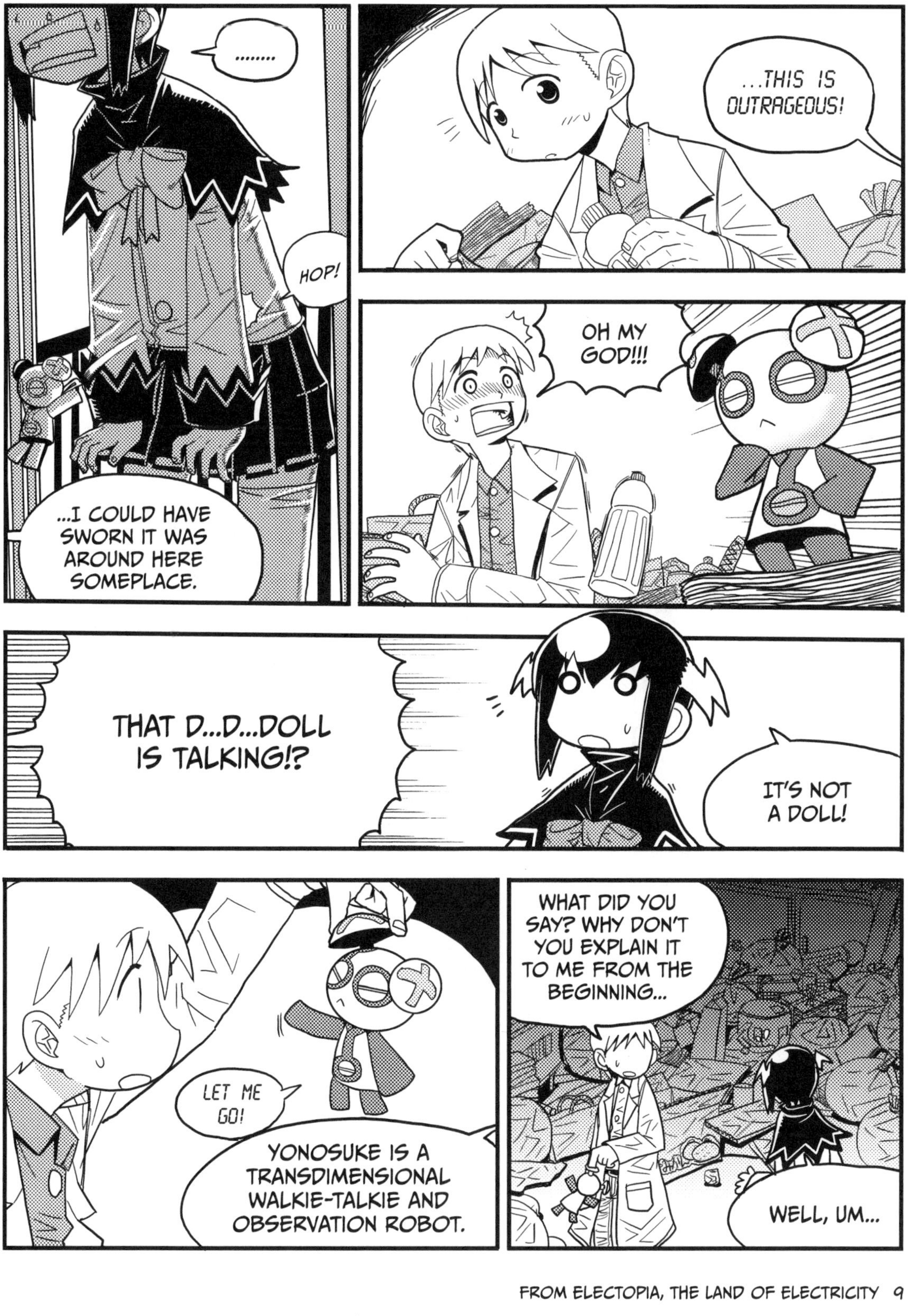

....
HOP!
...I COULD HAVE SWORN IT WAS AROUND HERE SOMEPLACE.
...THIS IS OUTRAGEOUS!
OH MY GOD!!!
THAT D...D...DOLL IS TALKING!?
IT'S NOT A DOLL!
LET ME GO!
YONOSUKE IS A TRANSDIMENSIONAL WALKIE-TALKIE AND OBSERVATION ROBOT.
WHAT DID YOU SAY? WHY DON'T YOU EXPLAIN IT TO ME FROM THE BEGINNING...
WELL, UM...

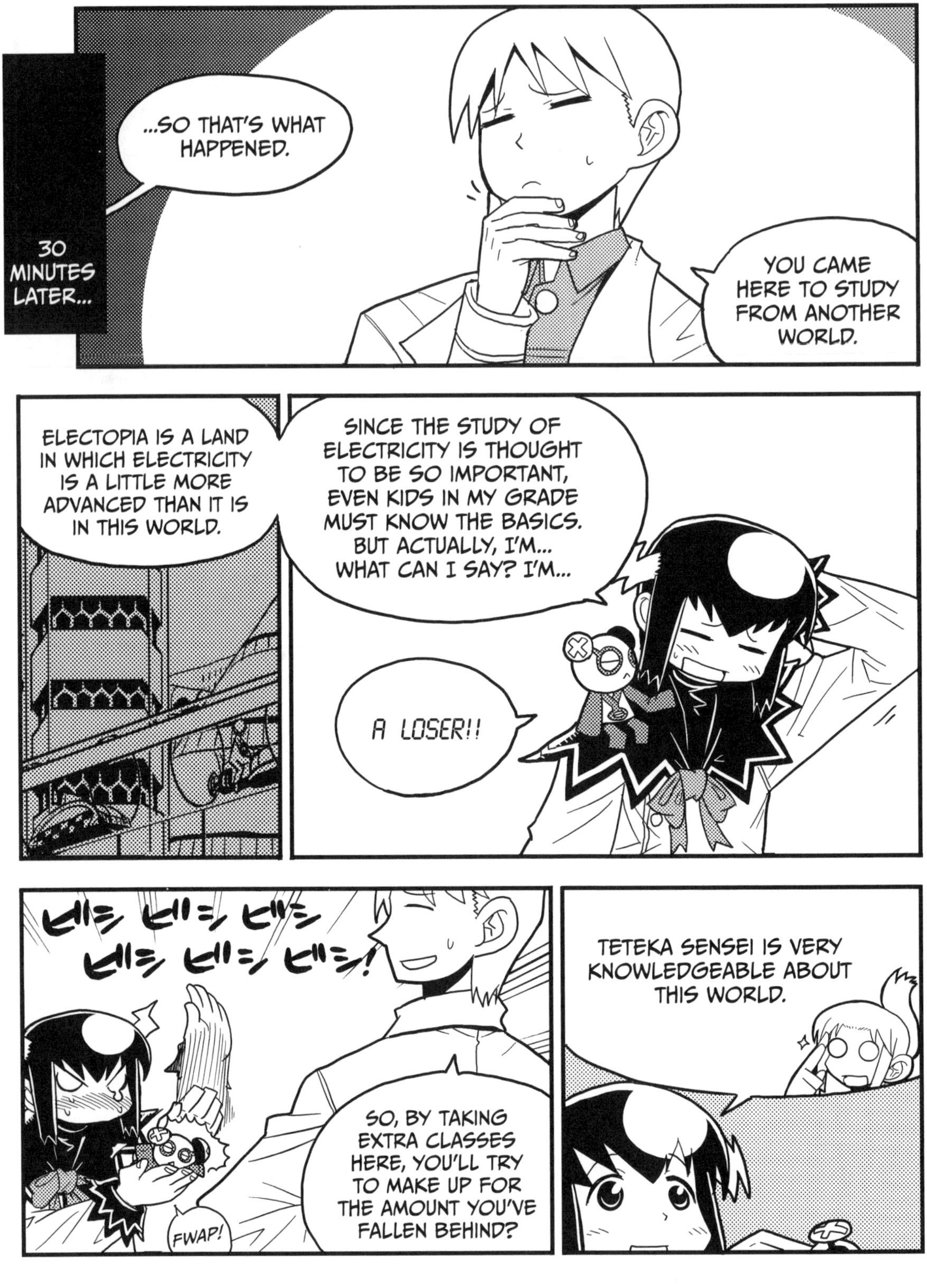
30 MINUTES LATER...
...SO THAT'S WHAT HAPPENED.
YOU CAME HERE TO STUDY FROM ANOTHER WORLD.
ELECTOPIA IS A LAND IN WHICH ELECTRICITY IS A LITTLE MORE ADVANCED THAN IT IS IN THIS WORLD.
SINCE THE STUDY OF ELECTRICITY IS THOUGHT TO BE SO IMPORTANT, EVEN KIDS IN MY GRADE MUST KNOW THE BASICS. BUT ACTUALLY, I'M... WHAT CAN I SAY? I'M...
A LOSER!!
ビシ ビシ ビシ
ビシ ビシ ビシ!
FWAP!
SO, BY TAKING EXTRA CLASSES HERE, YOU'LL TRY TO MAKE UP FOR THE AMOUNT YOU'VE FALLEN BEHIND?
TETEKA SENSEI IS VERY KNOWLEDGEABLE ABOUT THIS WORLD.

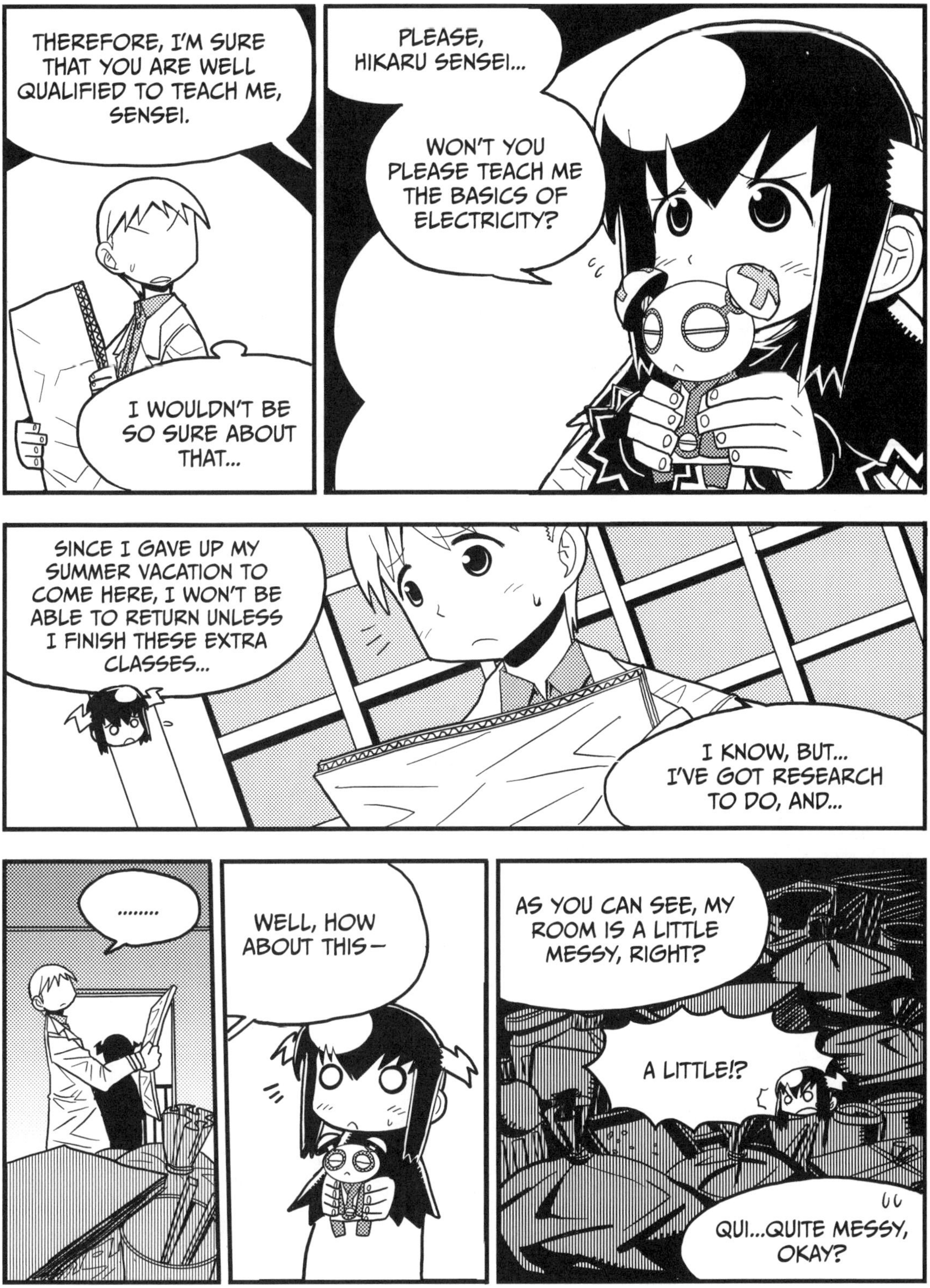
THEREFORE, I'M SURE THAT YOU ARE WELL QUALIFIED TO TEACH ME, SENSEI.
I WOULDN'T BE SO SURE ABOUT THAT...
PLEASE, HIKARU SENSEI...
WON'T YOU PLEASE TEACH ME THE BASICS OF ELECTRICITY?
SINCE I GAVE UP MY SUMMER VACATION TO COME HERE, I WON'T BE ABLE TO RETURN UNLESS I FINISH THESE EXTRA CLASSES...
I KNOW, BUT... I'VE GOT RESEARCH TO DO, AND...
....
WELL, HOW ABOUT THIS—
AS YOU CAN SEE, MY ROOM IS A LITTLE MESSY, RIGHT?
A LITTLE!?
QUI...QUITE MESSY, OKAY?

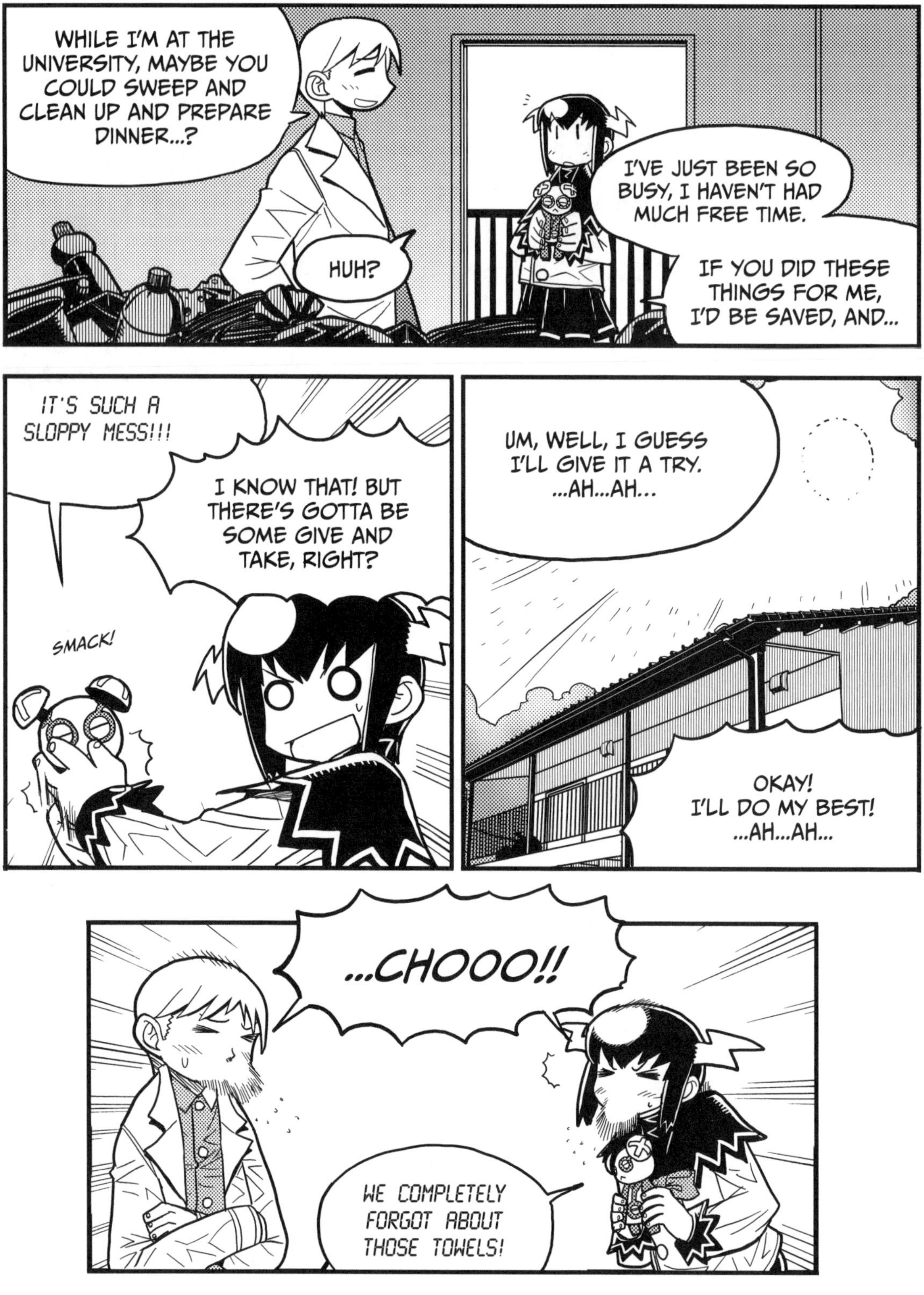
WHILE I'M AT THE UNIVERSITY, MAYBE YOU COULD SWEEP AND CLEAN UP AND PREPARE DINNER...?
HUH?
I'VE JUST BEEN SO BUSY, I HAVEN'T HAD MUCH FREE TIME.
IF YOU DID THESE THINGS FOR ME, I'D BE SAVED, AND...
IT'S SUCH A SLOPPY MESS!!!
I KNOW THAT! BUT THERE'S GOTTA BE SOME GIVE AND TAKE, RIGHT?
SMACK!
UM, WELL, I GUESS I'LL GIVE IT A TRY. ...AH...AH...
OKAY! I'LL DO MY BEST! ...AH...AH...
...CHOOO!!
WE COMPLETELY FORGOT ABOUT THOSE TOWELS!

1

WHAT IS ELECTRICITY?

ELECTRICITY AND EVERYDAY LIFE

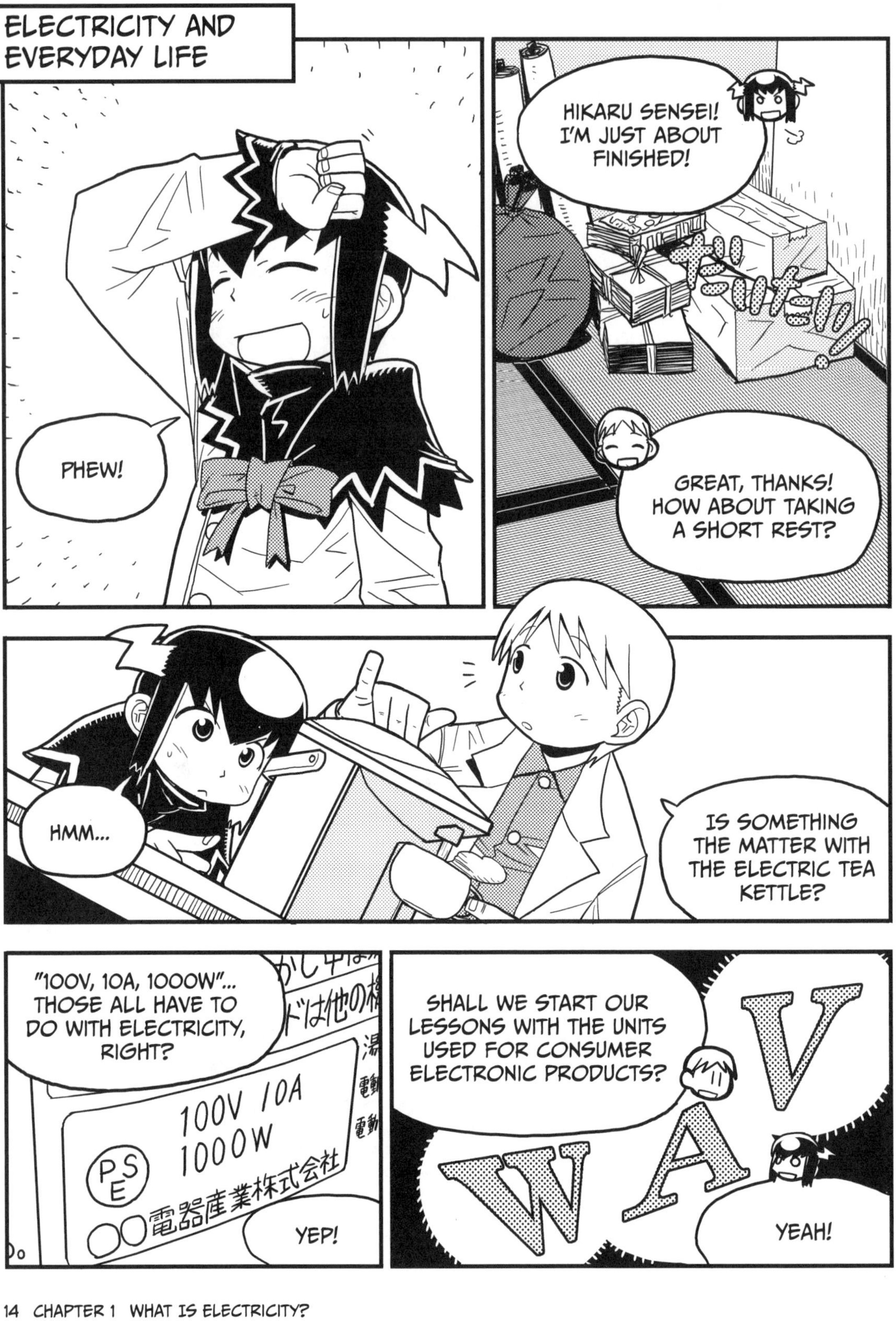

PHEW!
HIKARU SENSEI! I'M JUST ABOUT FINISHED!
GREAT, THANKS! HOW ABOUT TAKING A SHORT REST?
HMM...
IS SOMETHING THE MATTER WITH THE ELECTRIC TEA KETTLE?
"100V, 10A, 1000W"... THOSE ALL HAVE TO DO WITH ELECTRICITY, RIGHT?
100V 10A
1000W
PSE
○○電器産業株式会社
YEP!
SHALL WE START OUR LESSONS WITH THE UNITS USED FOR CONSUMER ELECTRONIC PRODUCTS?
V
A
W
YEAH!

ELECTRICAL UNITS

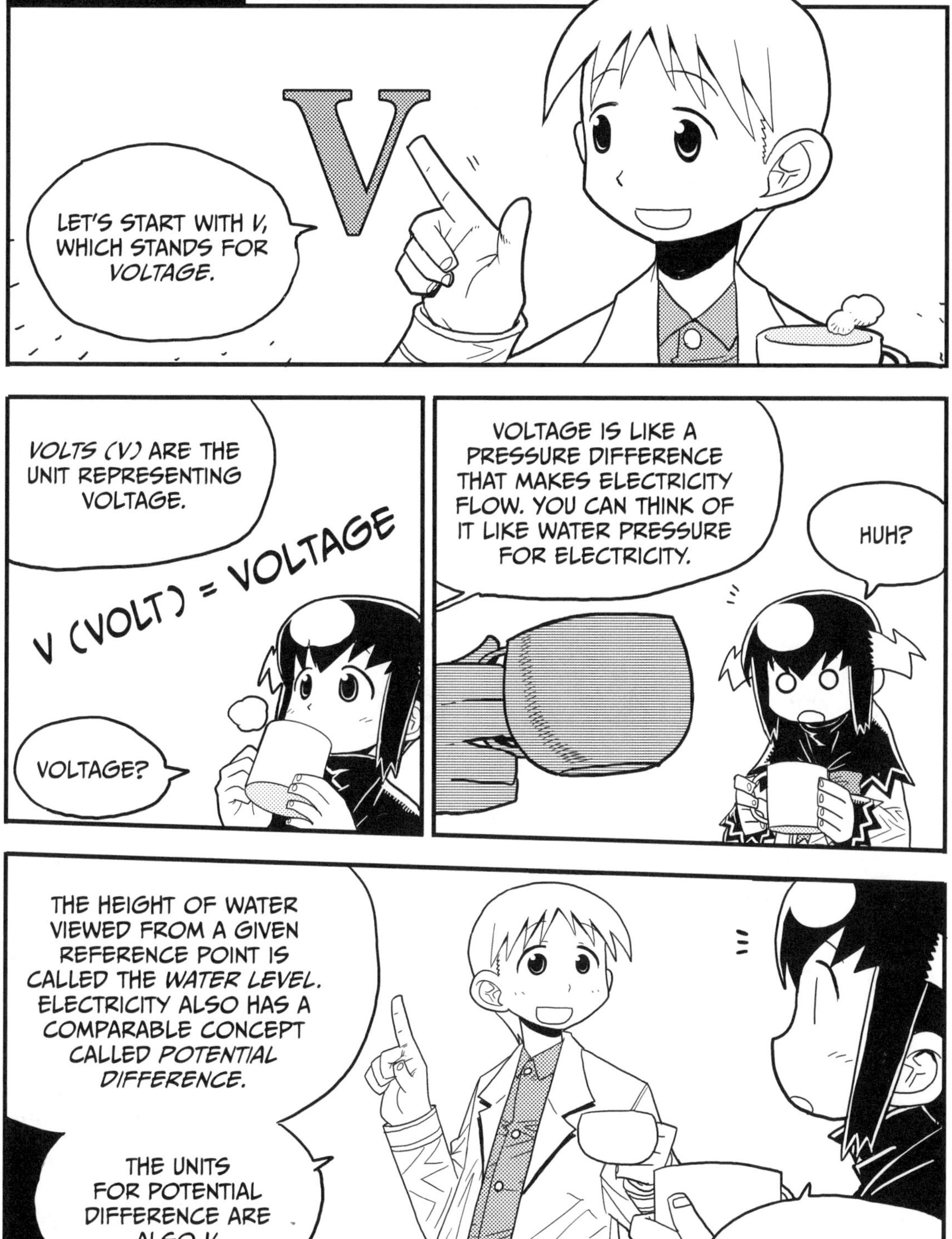

LET'S START WITH *V*, WHICH STANDS FOR *VOLTAGE*.
VOLTS (V) ARE THE UNIT REPRESENTING VOLTAGE.
V (VOLT) = VOLTAGE
VOLTAGE?
VOLTAGE IS LIKE A PRESSURE DIFFERENCE THAT MAKES ELECTRICITY FLOW. YOU CAN THINK OF IT LIKE WATER PRESSURE FOR ELECTRICITY.
HUH?
THE HEIGHT OF WATER VIEWED FROM A GIVEN REFERENCE POINT IS CALLED THE *WATER LEVEL*. ELECTRICITY ALSO HAS A COMPARABLE CONCEPT CALLED *POTENTIAL DIFFERENCE*.
THE UNITS FOR POTENTIAL DIFFERENCE ARE ALSO *V*.
IS THAT SO?

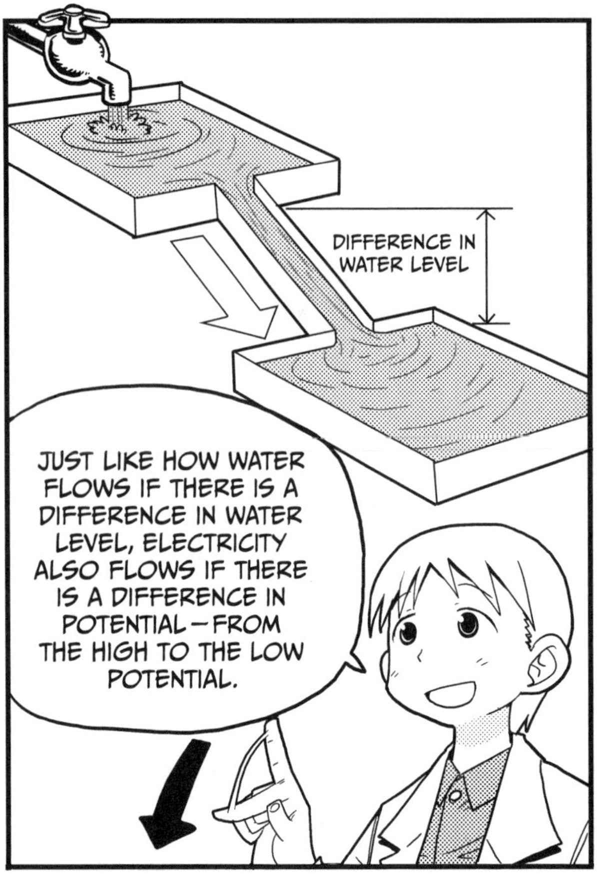
DIFFERENCE IN WATER LEVEL
JUST LIKE HOW WATER FLOWS IF THERE IS A DIFFERENCE IN WATER LEVEL, ELECTRICITY ALSO FLOWS IF THERE IS A DIFFERENCE IN POTENTIAL—FROM THE HIGH TO THE LOW POTENTIAL.

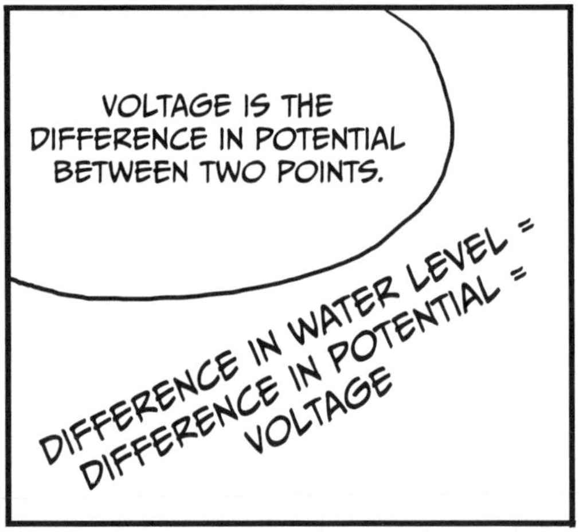
VOLTAGE IS THE DIFFERENCE IN POTENTIAL BETWEEN TWO POINTS.
DIFFERENCE IN WATER LEVEL = DIFFERENCE IN POTENTIAL = VOLTAGE

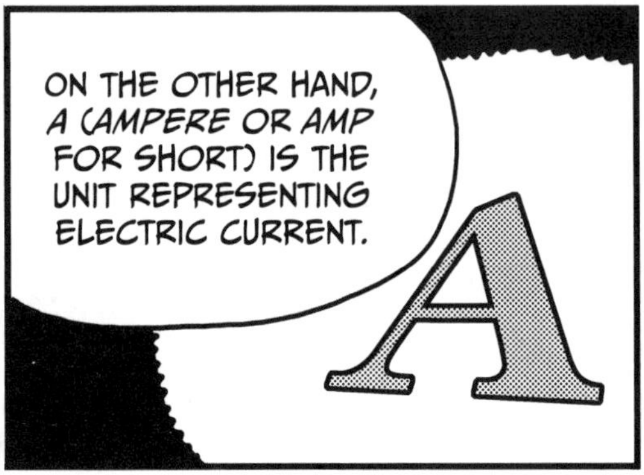
ON THE OTHER HAND, *A (AMPERE OR AMP* FOR SHORT) IS THE UNIT REPRESENTING ELECTRIC CURRENT.
A

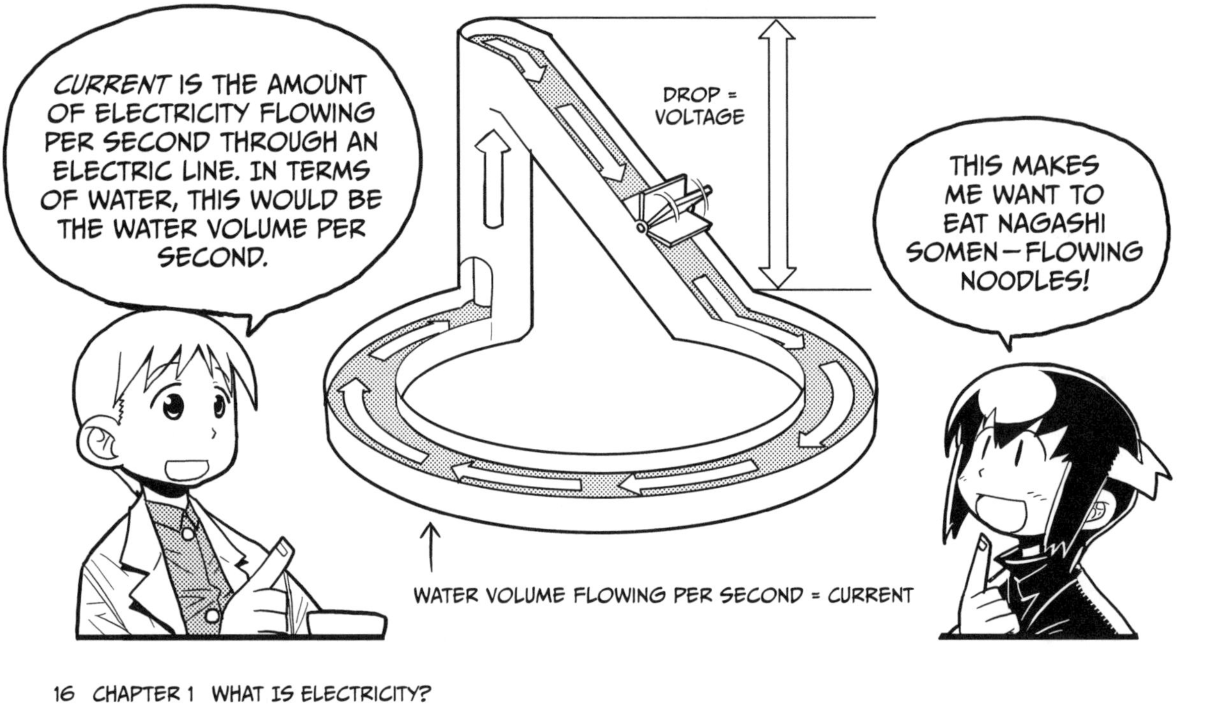
CURRENT IS THE AMOUNT OF ELECTRICITY FLOWING PER SECOND THROUGH AN ELECTRIC LINE. IN TERMS OF WATER, THIS WOULD BE THE WATER VOLUME PER SECOND.
DROP = VOLTAGE
THIS MAKES ME WANT TO EAT NAGASHI SOMEN—FLOWING NOODLES!
WATER VOLUME FLOWING PER SECOND = CURRENT

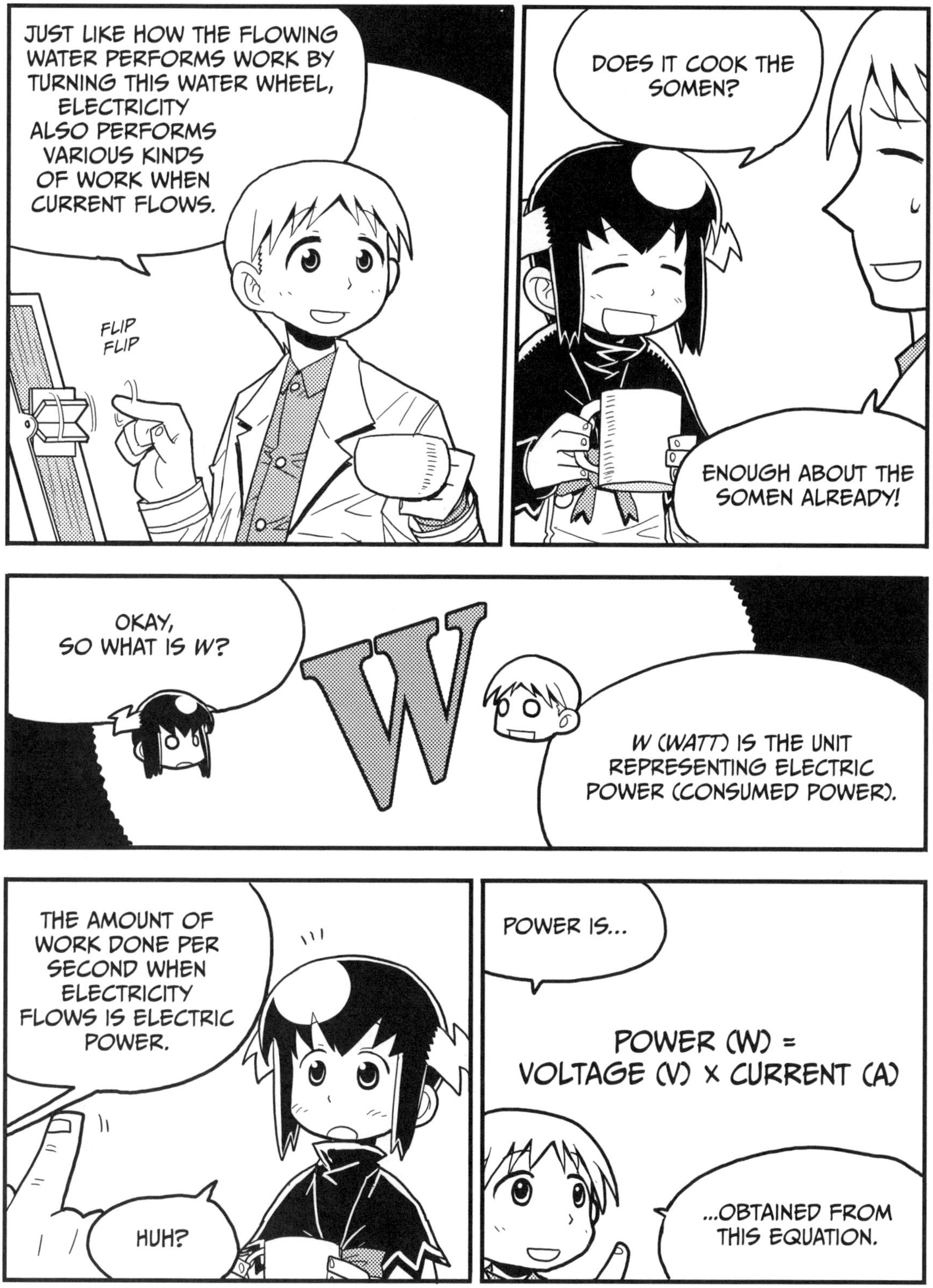
JUST LIKE HOW THE FLOWING WATER PERFORMS WORK BY TURNING THIS WATER WHEEL, ELECTRICITY ALSO PERFORMS VARIOUS KINDS OF WORK WHEN CURRENT FLOWS.
FLIP FLIP
DOES IT COOK THE SOMEN?
ENOUGH ABOUT THE SOMEN ALREADY!
OKAY, SO WHAT IS *W*?
W
W (WATT) IS THE UNIT REPRESENTING ELECTRIC POWER (CONSUMED POWER).
THE AMOUNT OF WORK DONE PER SECOND WHEN ELECTRICITY FLOWS IS ELECTRIC POWER.
HUH?
POWER IS...
POWER (W) = VOLTAGE (V) × CURRENT (A)
...OBTAINED FROM THIS EQUATION.

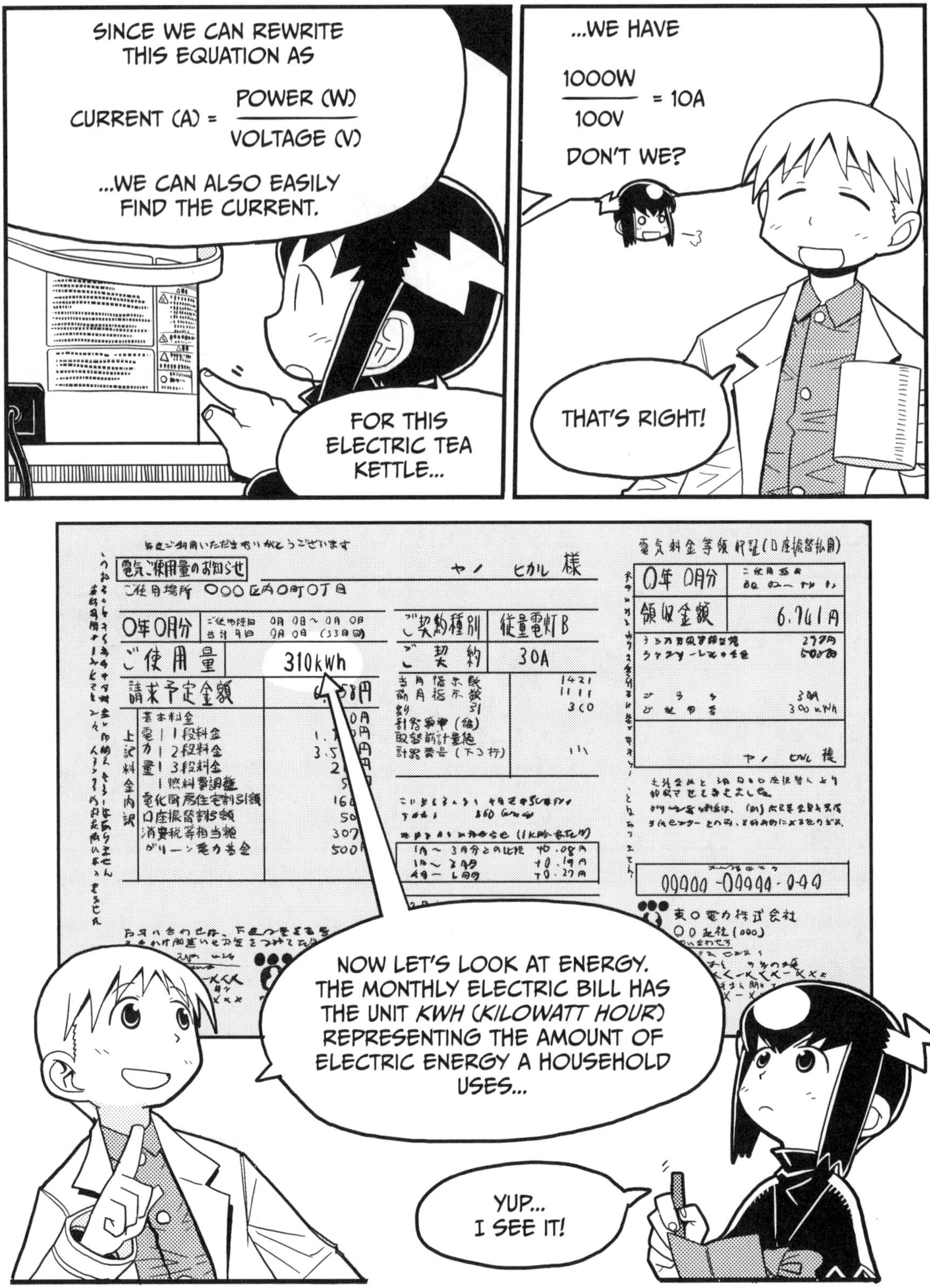
SINCE WE CAN REWRITE THIS EQUATION AS
CURRENT (A) = POWER (W) / VOLTAGE (V)
...WE CAN ALSO EASILY FIND THE CURRENT.
FOR THIS ELECTRIC TEA KETTLE...
...WE HAVE
1000W / 100V = 10A
DON'T WE?
THAT'S RIGHT!
310kWh
NOW LET'S LOOK AT ENERGY. THE MONTHLY ELECTRIC BILL HAS THE UNIT *KWH (KILOWATT HOUR)* REPRESENTING THE AMOUNT OF ELECTRIC ENERGY A HOUSEHOLD USES...
YUP... I SEE IT!

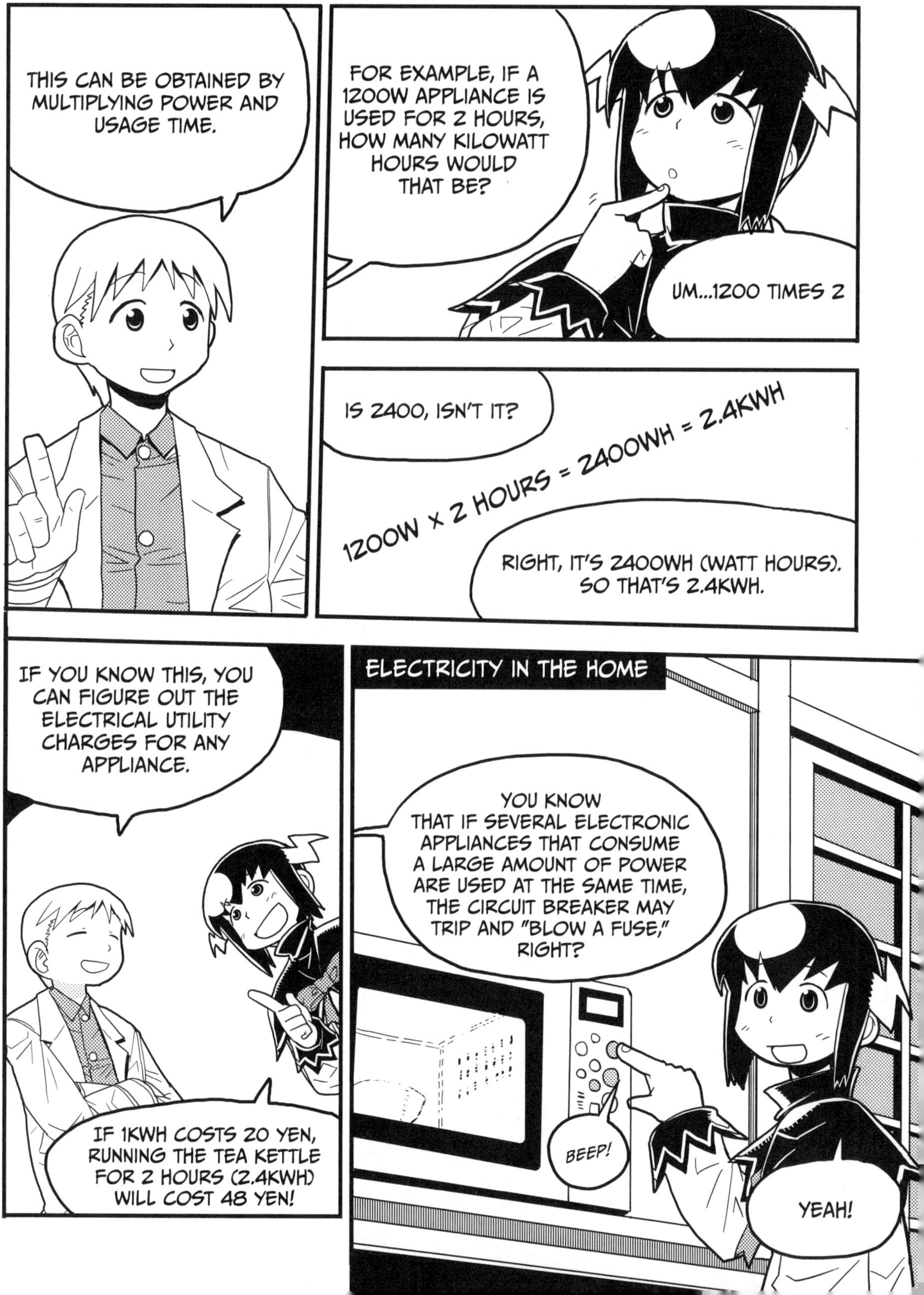
THIS CAN BE OBTAINED BY MULTIPLYING POWER AND USAGE TIME.
FOR EXAMPLE, IF A 1200W APPLIANCE IS USED FOR 2 HOURS, HOW MANY KILOWATT HOURS WOULD THAT BE?
UM...1200 TIMES 2
IS 2400, ISN'T IT?
1200W x 2 HOURS = 2400WH = 2.4KWH
RIGHT, IT'S 2400WH (WATT HOURS). SO THAT'S 2.4KWH.
IF YOU KNOW THIS, YOU CAN FIGURE OUT THE ELECTRICAL UTILITY CHARGES FOR ANY APPLIANCE.
IF 1KWH COSTS 20 YEN, RUNNING THE TEA KETTLE FOR 2 HOURS (2.4KWH) WILL COST 48 YEN!
ELECTRICITY IN THE HOME
YOU KNOW THAT IF SEVERAL ELECTRONIC APPLIANCES THAT CONSUME A LARGE AMOUNT OF POWER ARE USED AT THE SAME TIME, THE CIRCUIT BREAKER MAY TRIP AND "BLOW A FUSE," RIGHT?
BEEP!
YEAH!

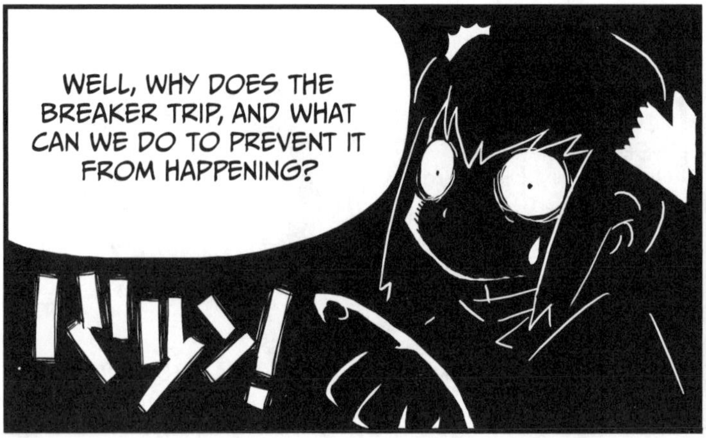
WELL, WHY DOES THE BREAKER TRIP, AND WHAT CAN WE DO TO PREVENT IT FROM HAPPENING?
バリッ!
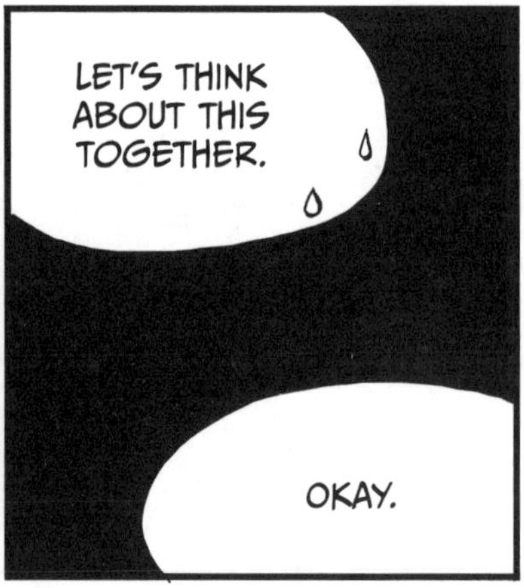
LET'S THINK ABOUT THIS TOGETHER.
OKAY.
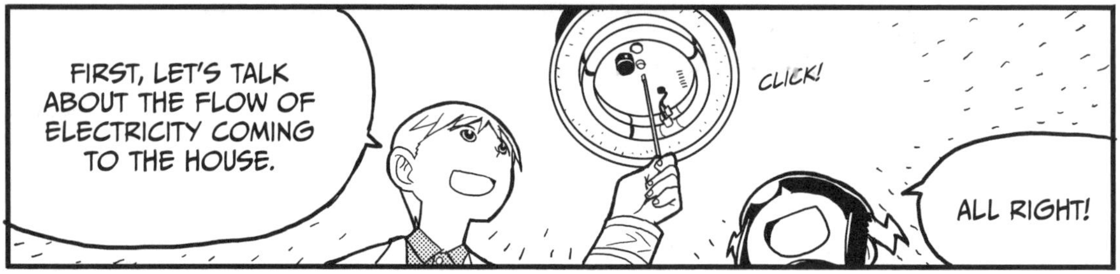
FIRST, LET'S TALK ABOUT THE FLOW OF ELECTRICITY COMING TO THE HOUSE.
CLICK!
ALL RIGHT!
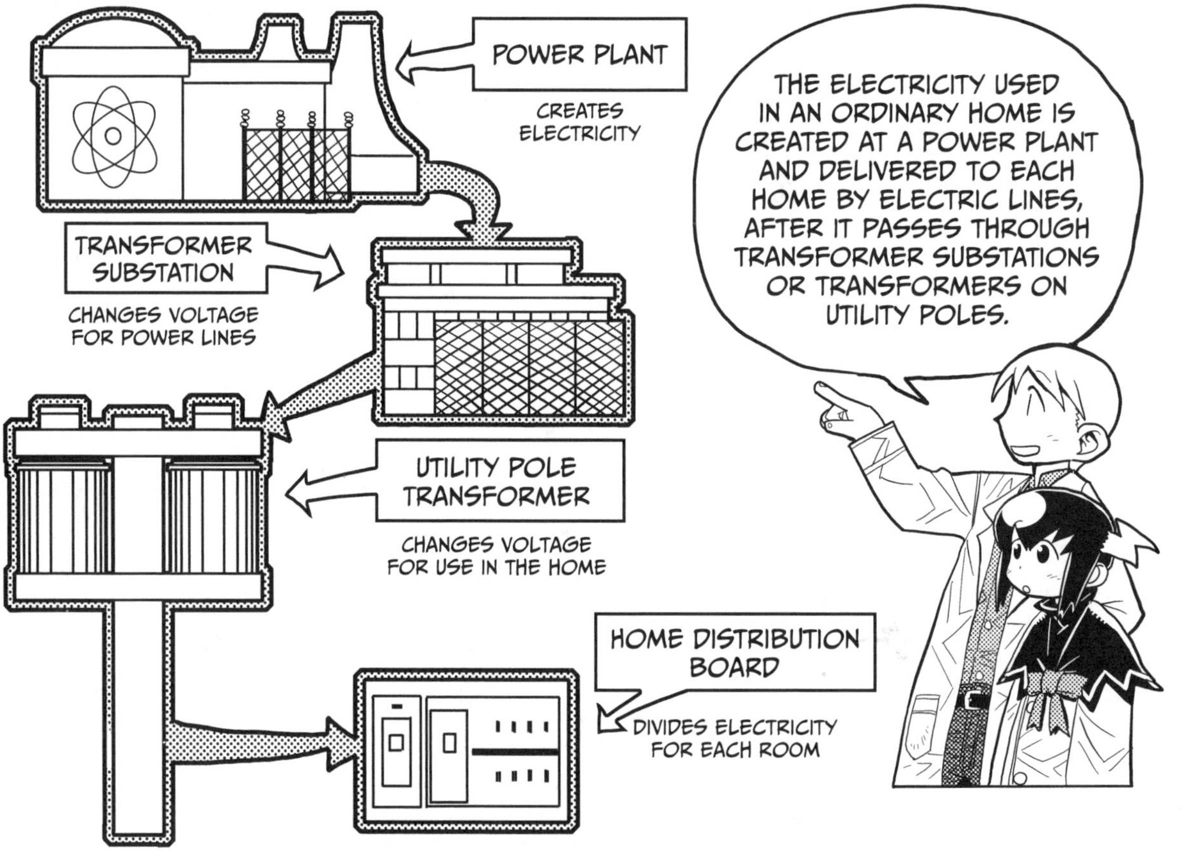
POWER PLANT
CREATES ELECTRICITY
TRANSFORMER SUBSTATION
CHANGES VOLTAGE FOR POWER LINES
UTILITY POLE TRANSFORMER
CHANGES VOLTAGE FOR USE IN THE HOME
HOME DISTRIBUTION BOARD
DIVIDES ELECTRICITY FOR EACH ROOM
THE ELECTRICITY USED IN AN ORDINARY HOME IS CREATED AT A POWER PLANT AND DELIVERED TO EACH HOME BY ELECTRIC LINES, AFTER IT PASSES THROUGH TRANSFORMER SUBSTATIONS OR TRANSFORMERS ON UTILITY POLES.

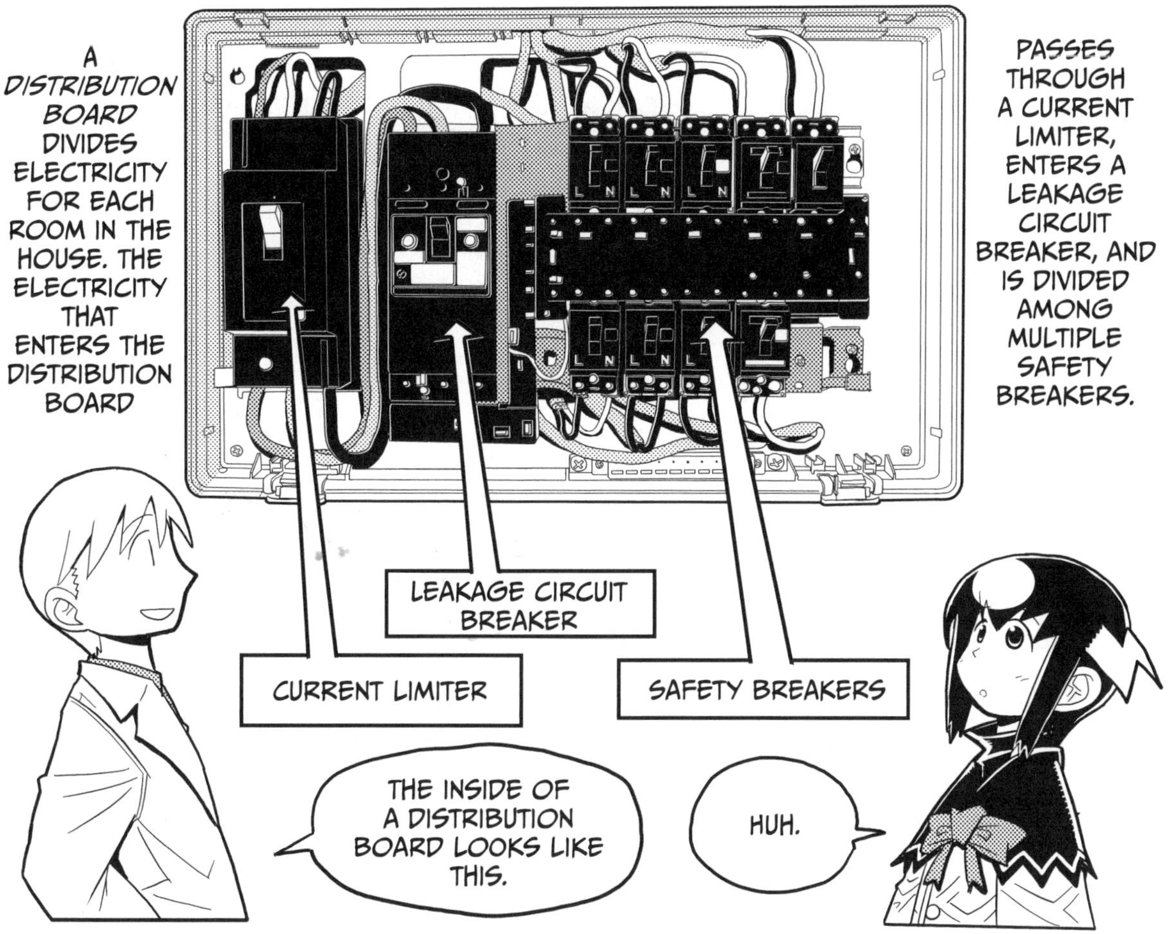
A DISTRIBUTION BOARD DIVIDES ELECTRICITY FOR EACH ROOM IN THE HOUSE. THE ELECTRICITY THAT ENTERS THE DISTRIBUTION BOARD
PASSES THROUGH A CURRENT LIMITER, ENTERS A LEAKAGE CIRCUIT BREAKER, AND IS DIVIDED AMONG MULTIPLE SAFETY BREAKERS.
LEAKAGE CIRCUIT BREAKER
CURRENT LIMITER
SAFETY BREAKERS
THE INSIDE OF A DISTRIBUTION BOARD LOOKS LIKE THIS.
HUH.

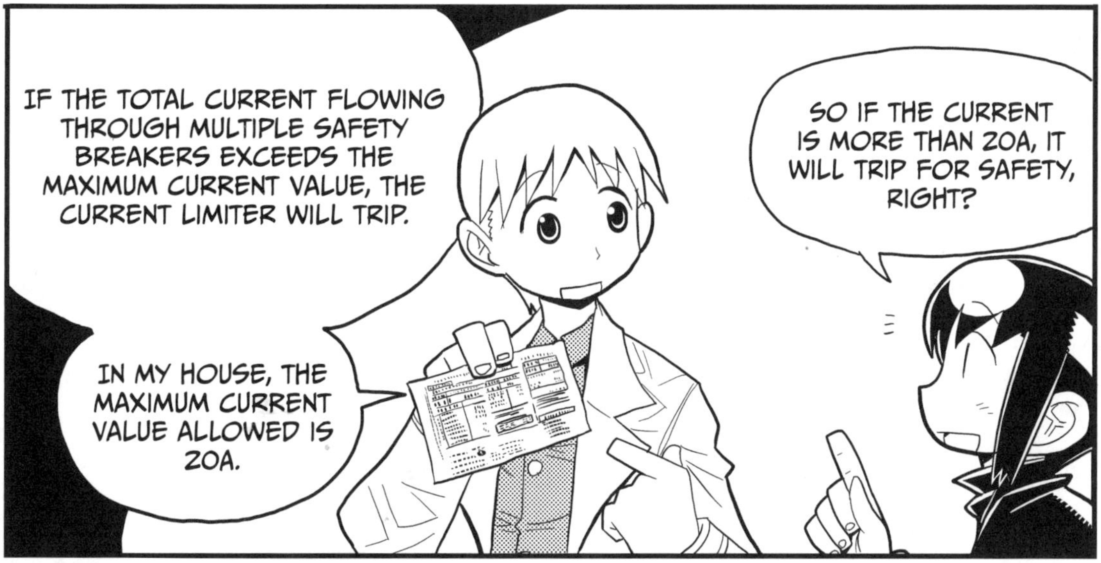
IF THE TOTAL CURRENT FLOWING THROUGH MULTIPLE SAFETY BREAKERS EXCEEDS THE MAXIMUM CURRENT VALUE, THE CURRENT LIMITER WILL TRIP.
IN MY HOUSE, THE MAXIMUM CURRENT VALUE ALLOWED IS 20A.
SO IF THE CURRENT IS MORE THAN 20A, IT WILL TRIP FOR SAFETY, RIGHT?

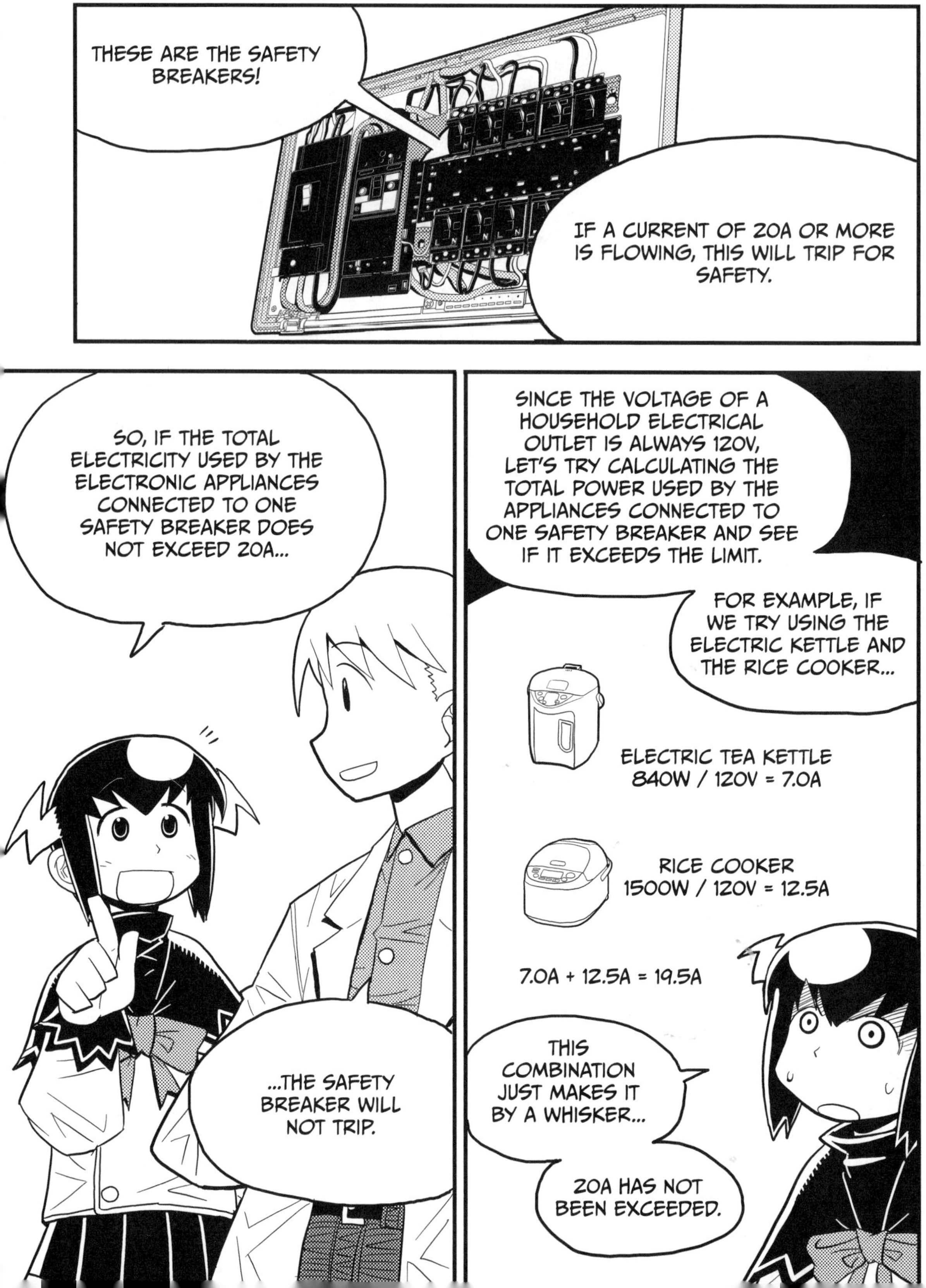
THESE ARE THE SAFETY BREAKERS!
IF A CURRENT OF 20A OR MORE IS FLOWING, THIS WILL TRIP FOR SAFETY.
SO, IF THE TOTAL ELECTRICITY USED BY THE ELECTRONIC APPLIANCES CONNECTED TO ONE SAFETY BREAKER DOES NOT EXCEED 20A...
...THE SAFETY BREAKER WILL NOT TRIP.
SINCE THE VOLTAGE OF A HOUSEHOLD ELECTRICAL OUTLET IS ALWAYS 120V, LET'S TRY CALCULATING THE TOTAL POWER USED BY THE APPLIANCES CONNECTED TO ONE SAFETY BREAKER AND SEE IF IT EXCEEDS THE LIMIT.
FOR EXAMPLE, IF WE TRY USING THE ELECTRIC KETTLE AND THE RICE COOKER...
ELECTRIC TEA KETTLE
840W / 120V = 7.0A
RICE COOKER
1500W / 120V = 12.5A
7.0A + 12.5A = 19.5A
THIS COMBINATION JUST MAKES IT BY A WHISKER...
20A HAS NOT BEEN EXCEEDED.

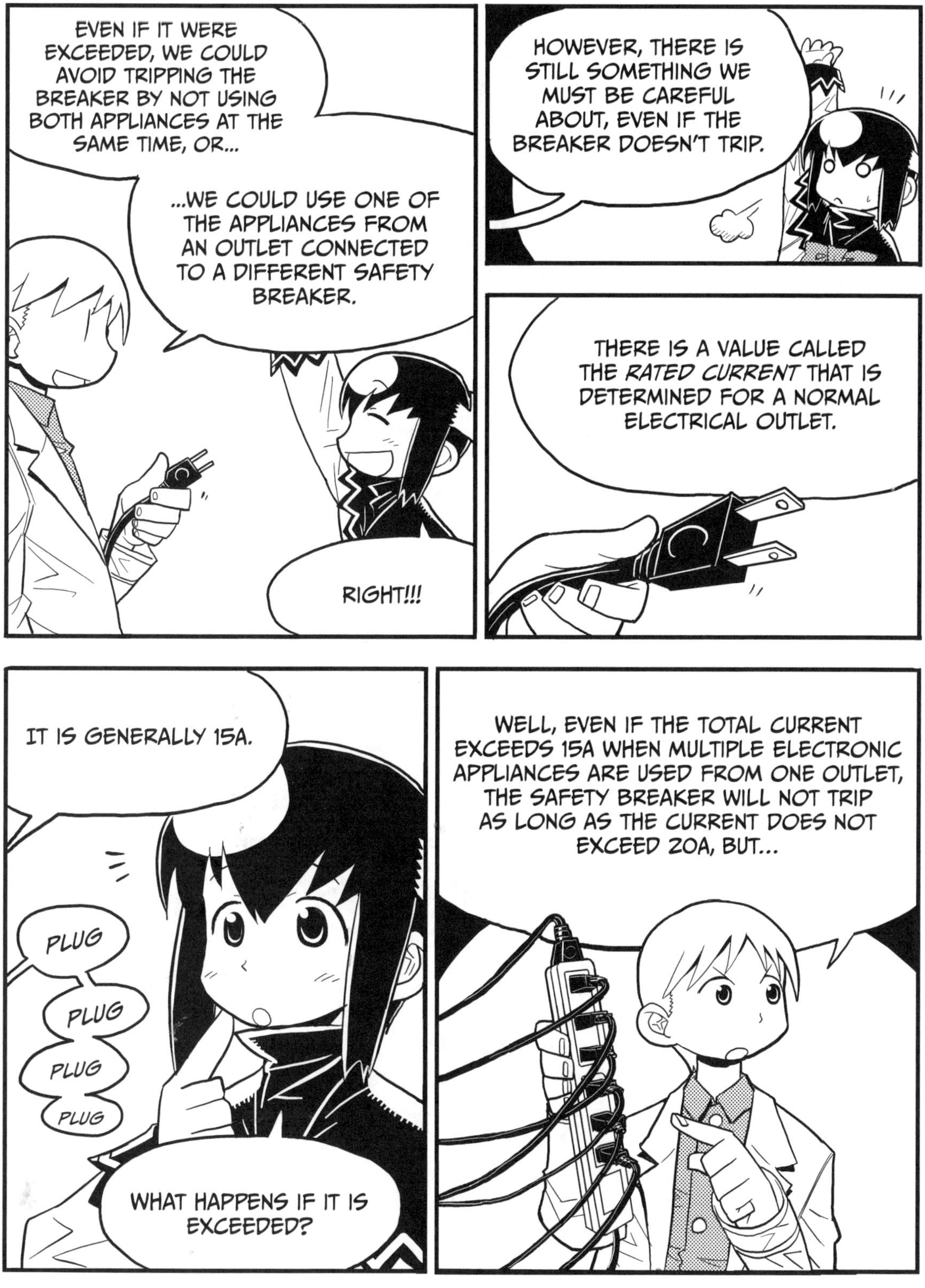
EVEN IF IT WERE EXCEEDED, WE COULD AVOID TRIPPING THE BREAKER BY NOT USING BOTH APPLIANCES AT THE SAME TIME, OR...
...WE COULD USE ONE OF THE APPLIANCES FROM AN OUTLET CONNECTED TO A DIFFERENT SAFETY BREAKER.
RIGHT!!!
HOWEVER, THERE IS STILL SOMETHING WE MUST BE CAREFUL ABOUT, EVEN IF THE BREAKER DOESN'T TRIP.
THERE IS A VALUE CALLED THE *RATED CURRENT* THAT IS DETERMINED FOR A NORMAL ELECTRICAL OUTLET.
IT IS GENERALLY 15A.
PLUG
PLUG
PLUG
PLUG
WHAT HAPPENS IF IT IS EXCEEDED?
WELL, EVEN IF THE TOTAL CURRENT EXCEEDS 15A WHEN MULTIPLE ELECTRONIC APPLIANCES ARE USED FROM ONE OUTLET, THE SAFETY BREAKER WILL NOT TRIP AS LONG AS THE CURRENT DOES NOT EXCEED 20A, BUT...

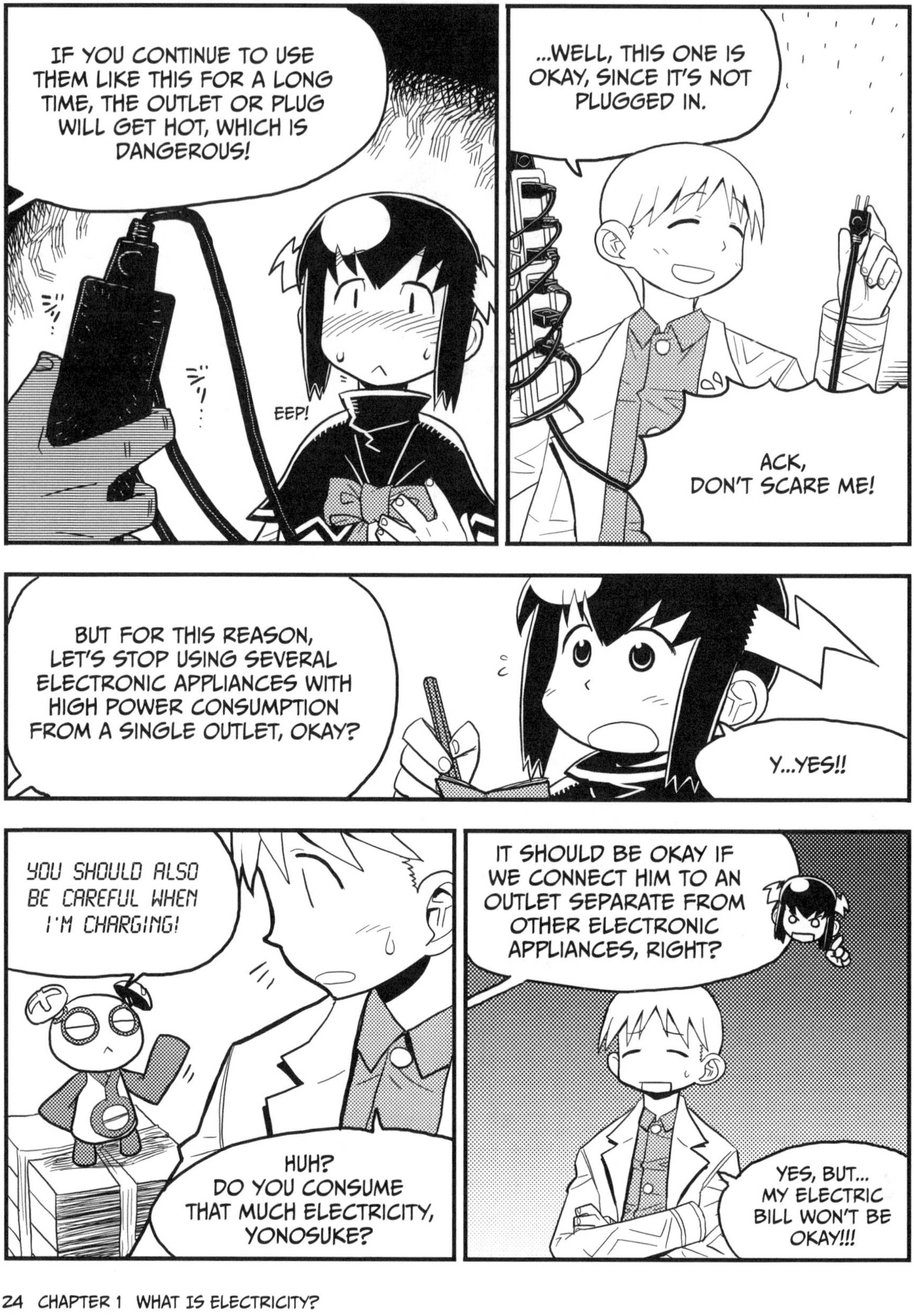
IF YOU CONTINUE TO USE THEM LIKE THIS FOR A LONG TIME, THE OUTLET OR PLUG WILL GET HOT, WHICH IS DANGEROUS!
EEP!
...WELL, THIS ONE IS OKAY, SINCE IT'S NOT PLUGGED IN.
ACK, DON'T SCARE ME!
BUT FOR THIS REASON, LET'S STOP USING SEVERAL ELECTRONIC APPLIANCES WITH HIGH POWER CONSUMPTION FROM A SINGLE OUTLET, OKAY?
Y...YES!!
YOU SHOULD ALSO BE CAREFUL WHEN I'M CHARGING!
HUH? DO YOU CONSUME THAT MUCH ELECTRICITY, YONOSUKE?
IT SHOULD BE OKAY IF WE CONNECT HIM TO AN OUTLET SEPARATE FROM OTHER ELECTRONIC APPLIANCES, RIGHT?
YES, BUT... MY ELECTRIC BILL WON'T BE OKAY!!!

HOW ELECTRICITY WORKS

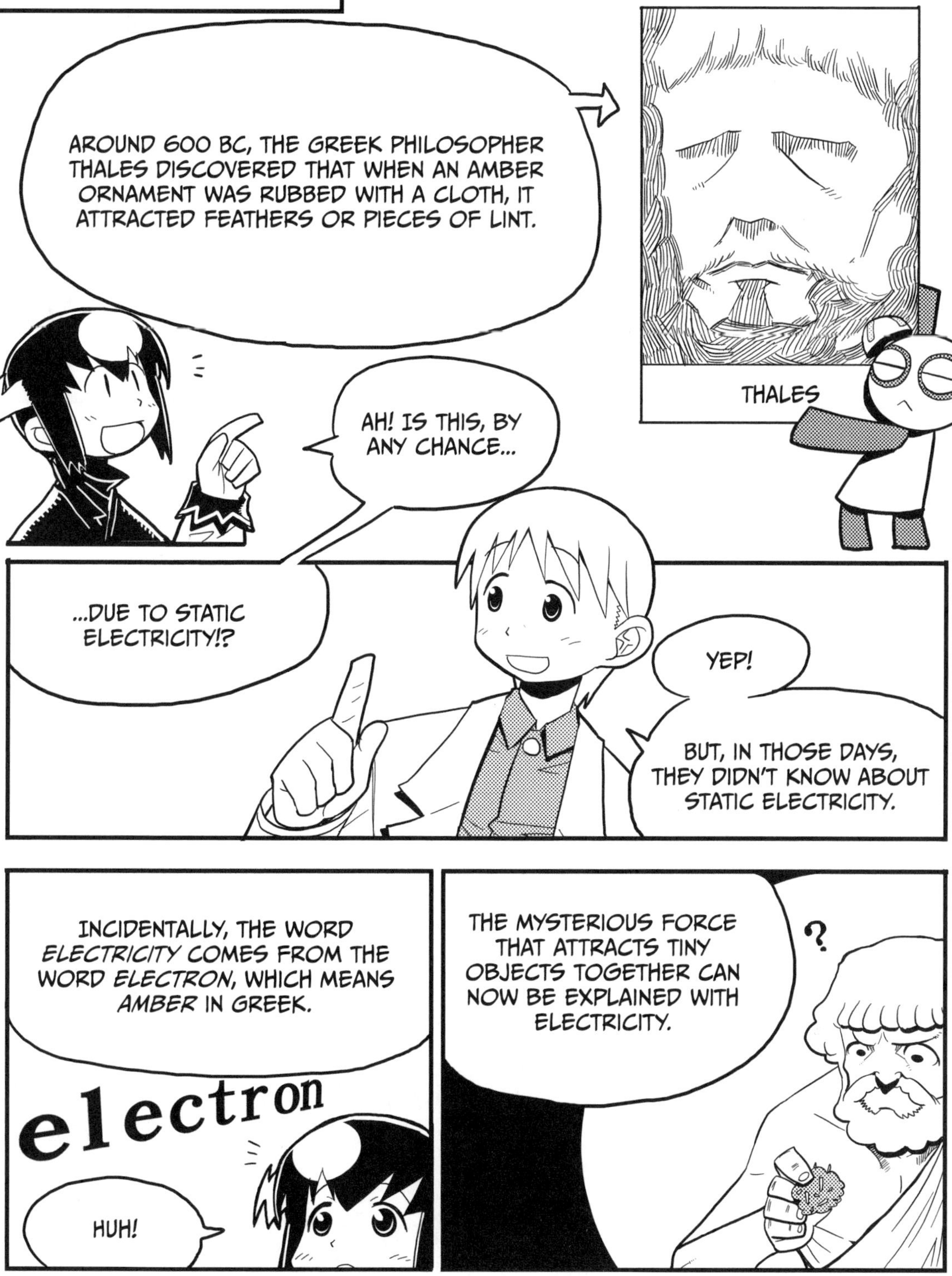
AROUND 600 BC, THE GREEK PHILOSOPHER THALES DISCOVERED THAT WHEN AN AMBER ORNAMENT WAS RUBBED WITH A CLOTH, IT ATTRACTED FEATHERS OR PIECES OF LINT.
THALES
AH! IS THIS, BY ANY CHANCE...
...DUE TO STATIC ELECTRICITY!?
YEP!
BUT, IN THOSE DAYS, THEY DIDN'T KNOW ABOUT STATIC ELECTRICITY.
INCIDENTALLY, THE WORD *ELECTRICITY* COMES FROM THE WORD *ELECTRON*, WHICH MEANS *AMBER* IN GREEK.
electron
HUH!
THE MYSTERIOUS FORCE THAT ATTRACTS TINY OBJECTS TOGETHER CAN NOW BE EXPLAINED WITH ELECTRICITY.
?

THE TRUE NATURE OF ELECTRICITY

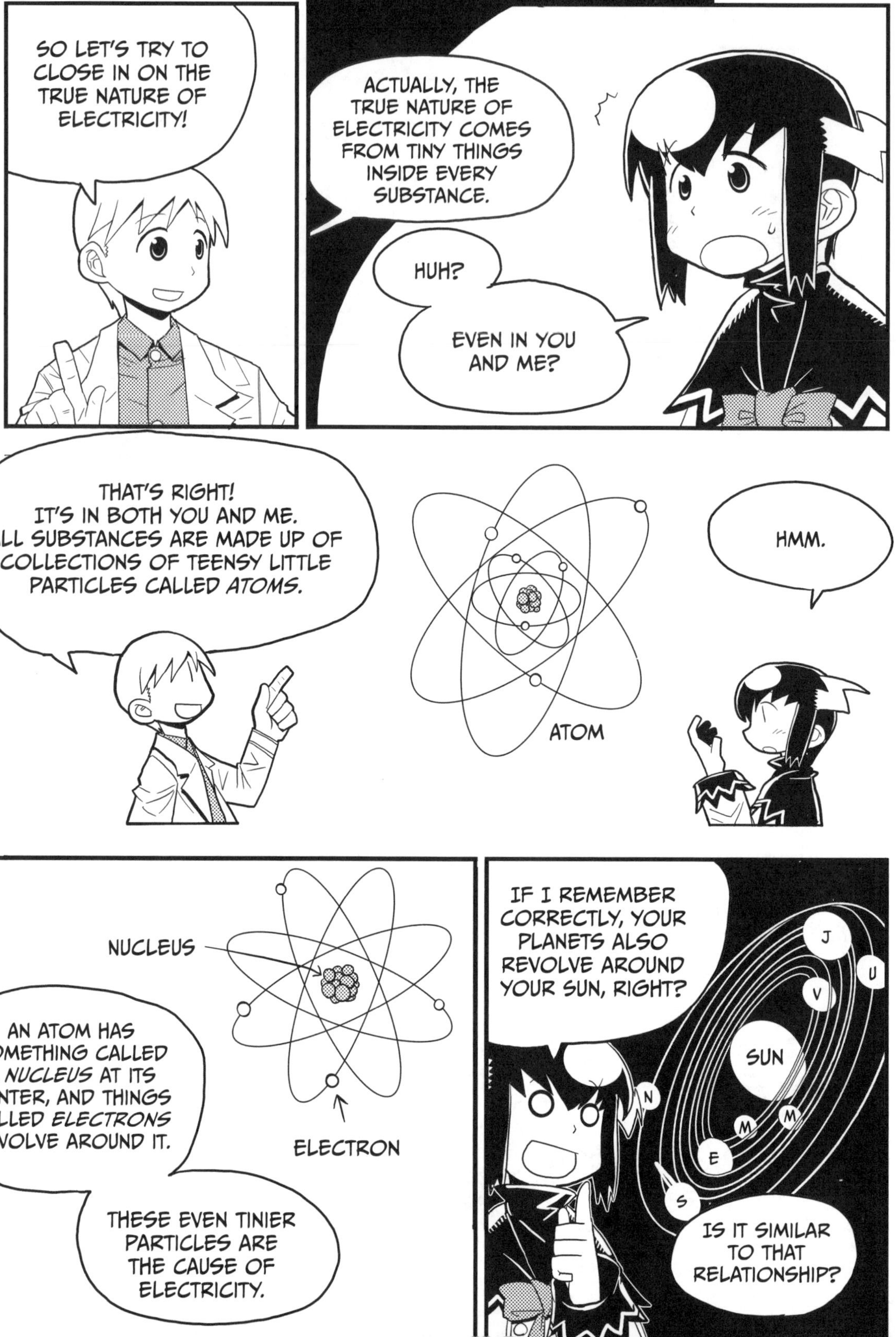

SO LET'S TRY TO CLOSE IN ON THE TRUE NATURE OF ELECTRICITY!
ACTUALLY, THE TRUE NATURE OF ELECTRICITY COMES FROM TINY THINGS INSIDE EVERY SUBSTANCE.
HUH?
EVEN IN YOU AND ME?
THAT'S RIGHT! IT'S IN BOTH YOU AND ME. ALL SUBSTANCES ARE MADE UP OF COLLECTIONS OF TEENSY LITTLE PARTICLES CALLED *ATOMS*.
ATOM
HMM.
NUCLEUS
ELECTRON
AN ATOM HAS SOMETHING CALLED A *NUCLEUS* AT ITS CENTER, AND THINGS CALLED *ELECTRONS* REVOLVE AROUND IT.
THESE EVEN TINIER PARTICLES ARE THE CAUSE OF ELECTRICITY.
IF I REMEMBER CORRECTLY, YOUR PLANETS ALSO REVOLVE AROUND YOUR SUN, RIGHT?
J
U
V
SUN
N
M
M
E
S
IS IT SIMILAR TO THAT RELATIONSHIP?

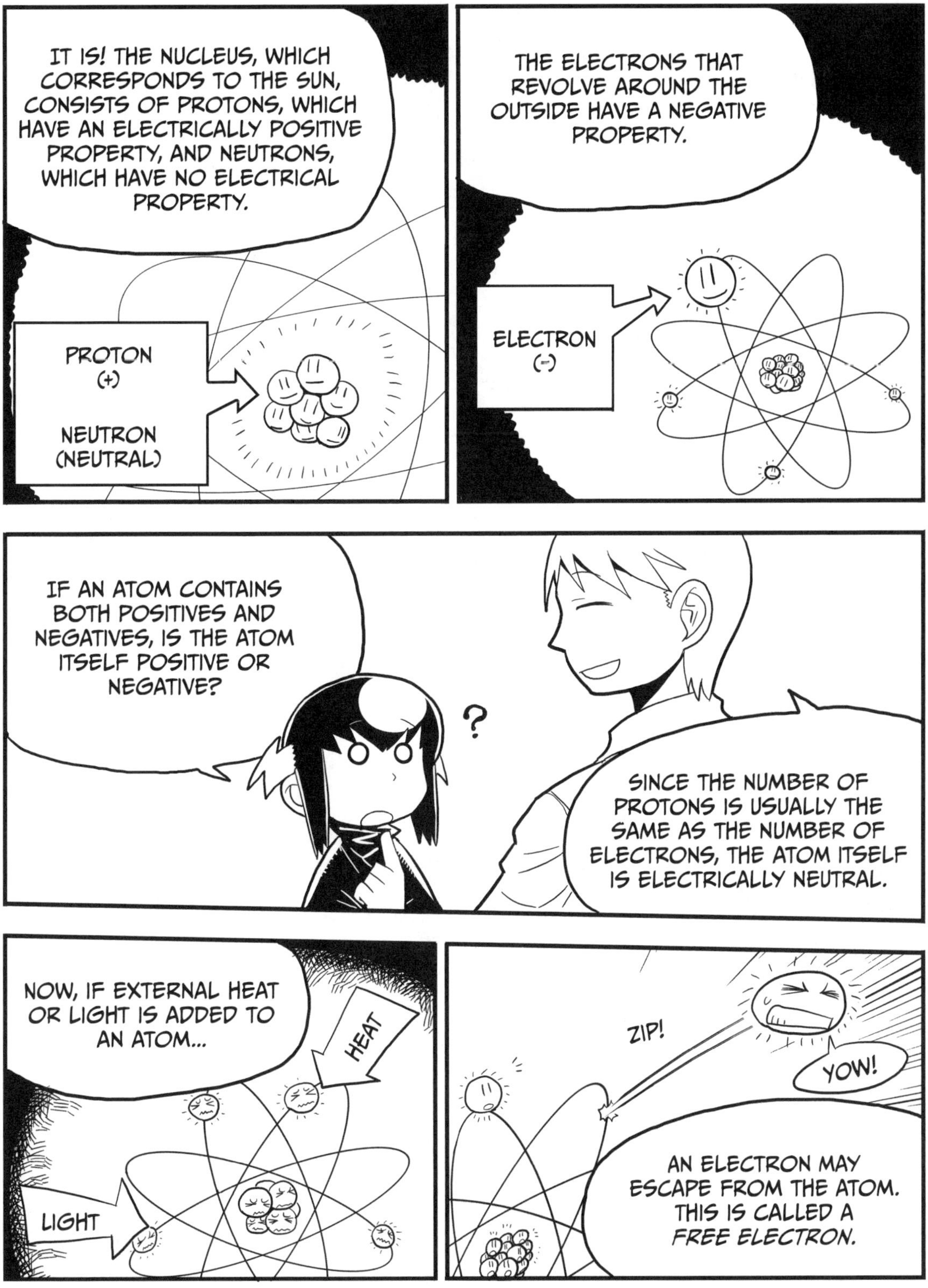
IT IS! THE NUCLEUS, WHICH CORRESPONDS TO THE SUN, CONSISTS OF PROTONS, WHICH HAVE AN ELECTRICALLY POSITIVE PROPERTY, AND NEUTRONS, WHICH HAVE NO ELECTRICAL PROPERTY.
PROTON (+)
NEUTRON (NEUTRAL)
THE ELECTRONS THAT REVOLVE AROUND THE OUTSIDE HAVE A NEGATIVE PROPERTY.
ELECTRON (−)
IF AN ATOM CONTAINS BOTH POSITIVES AND NEGATIVES, IS THE ATOM ITSELF POSITIVE OR NEGATIVE?
?
SINCE THE NUMBER OF PROTONS IS USUALLY THE SAME AS THE NUMBER OF ELECTRONS, THE ATOM ITSELF IS ELECTRICALLY NEUTRAL.
NOW, IF EXTERNAL HEAT OR LIGHT IS ADDED TO AN ATOM...
HEAT
LIGHT
ZIP!
YOW!
AN ELECTRON MAY ESCAPE FROM THE ATOM. THIS IS CALLED A *FREE ELECTRON*.

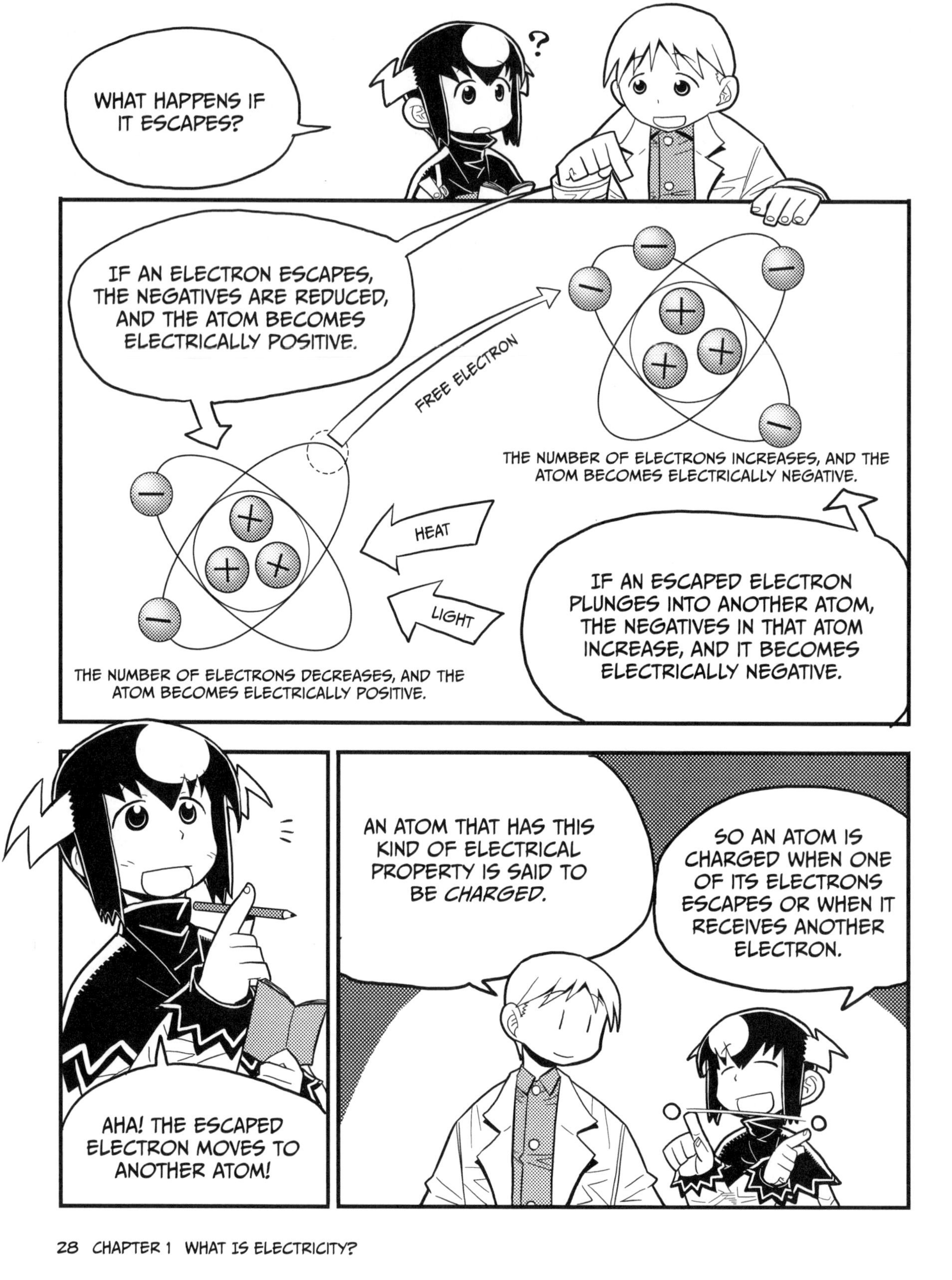
WHAT HAPPENS IF IT ESCAPES?
IF AN ELECTRON ESCAPES, THE NEGATIVES ARE REDUCED, AND THE ATOM BECOMES ELECTRICALLY POSITIVE.
FREE ELECTRON
HEAT
LIGHT
THE NUMBER OF ELECTRONS DECREASES, AND THE ATOM BECOMES ELECTRICALLY POSITIVE.
THE NUMBER OF ELECTRONS INCREASES, AND THE ATOM BECOMES ELECTRICALLY NEGATIVE.
IF AN ESCAPED ELECTRON PLUNGES INTO ANOTHER ATOM, THE NEGATIVES IN THAT ATOM INCREASE, AND IT BECOMES ELECTRICALLY NEGATIVE.
AHA! THE ESCAPED ELECTRON MOVES TO ANOTHER ATOM!
AN ATOM THAT HAS THIS KIND OF ELECTRICAL PROPERTY IS SAID TO BE *CHARGED*.
SO AN ATOM IS CHARGED WHEN ONE OF ITS ELECTRONS ESCAPES OR WHEN IT RECEIVES ANOTHER ELECTRON.

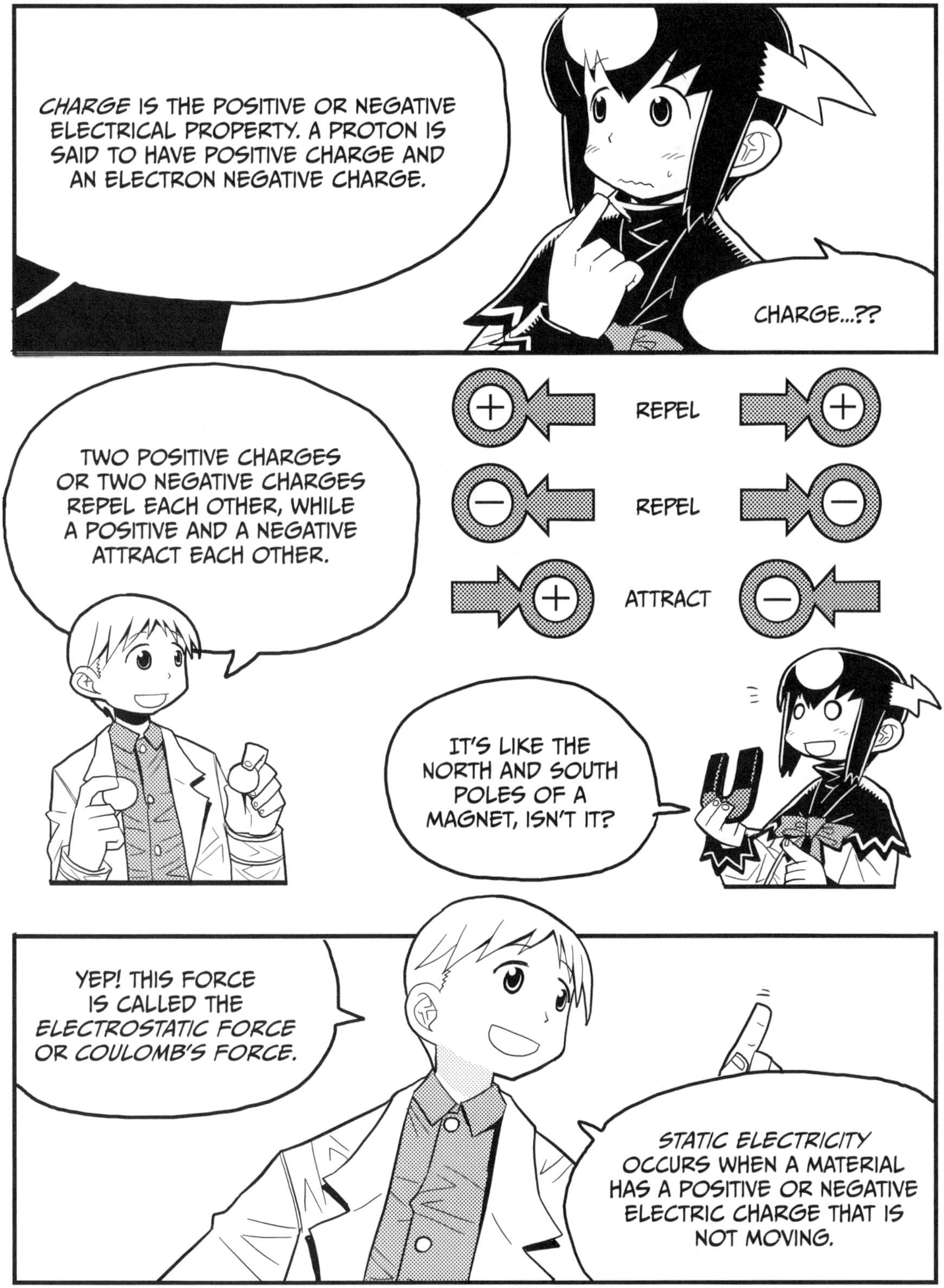
CHARGE IS THE POSITIVE OR NEGATIVE ELECTRICAL PROPERTY. A PROTON IS SAID TO HAVE POSITIVE CHARGE AND AN ELECTRON NEGATIVE CHARGE.
CHARGE...??
TWO POSITIVE CHARGES OR TWO NEGATIVE CHARGES REPEL EACH OTHER, WHILE A POSITIVE AND A NEGATIVE ATTRACT EACH OTHER.
REPEL
REPEL
ATTRACT
IT'S LIKE THE NORTH AND SOUTH POLES OF A MAGNET, ISN'T IT?
YEP! THIS FORCE IS CALLED THE *ELECTROSTATIC FORCE* OR *COULOMB'S FORCE*.
STATIC ELECTRICITY OCCURS WHEN A MATERIAL HAS A POSITIVE OR NEGATIVE ELECTRIC CHARGE THAT IS NOT MOVING.

CURRENT AND ELECTRICAL DISCHARGE

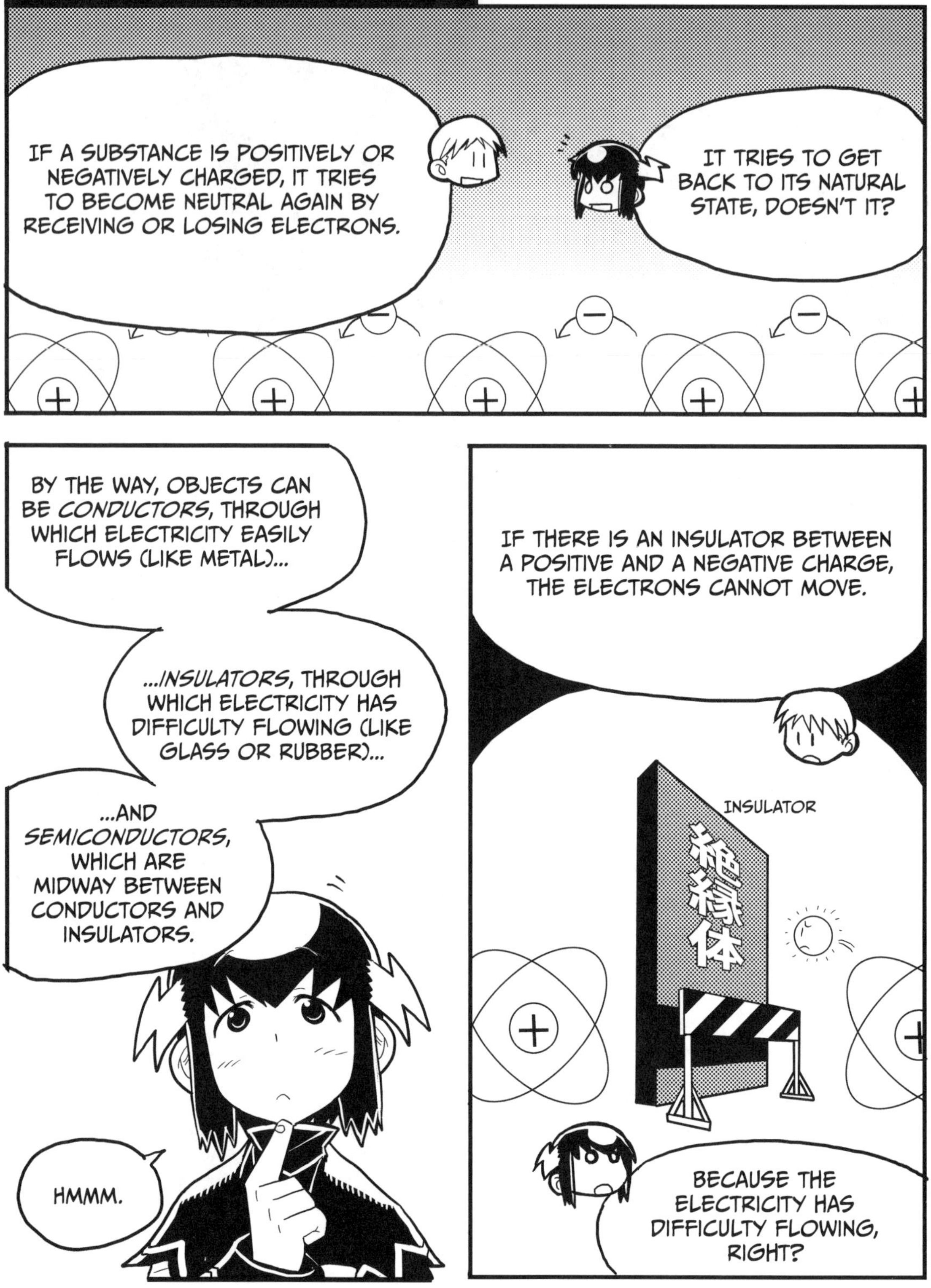
IF A SUBSTANCE IS POSITIVELY OR NEGATIVELY CHARGED, IT TRIES TO BECOME NEUTRAL AGAIN BY RECEIVING OR LOSING ELECTRONS.
IT TRIES TO GET BACK TO ITS NATURAL STATE, DOESN'T IT?
BY THE WAY, OBJECTS CAN BE *CONDUCTORS*, THROUGH WHICH ELECTRICITY EASILY FLOWS (LIKE METAL)...
...*INSULATORS*, THROUGH WHICH ELECTRICITY HAS DIFFICULTY FLOWING (LIKE GLASS OR RUBBER)...
...AND *SEMICONDUCTORS*, WHICH ARE MIDWAY BETWEEN CONDUCTORS AND INSULATORS.
HMMM.
IF THERE IS AN INSULATOR BETWEEN A POSITIVE AND A NEGATIVE CHARGE, THE ELECTRONS CANNOT MOVE.
INSULATOR
絶縁体
BECAUSE THE ELECTRICITY HAS DIFFICULTY FLOWING, RIGHT?

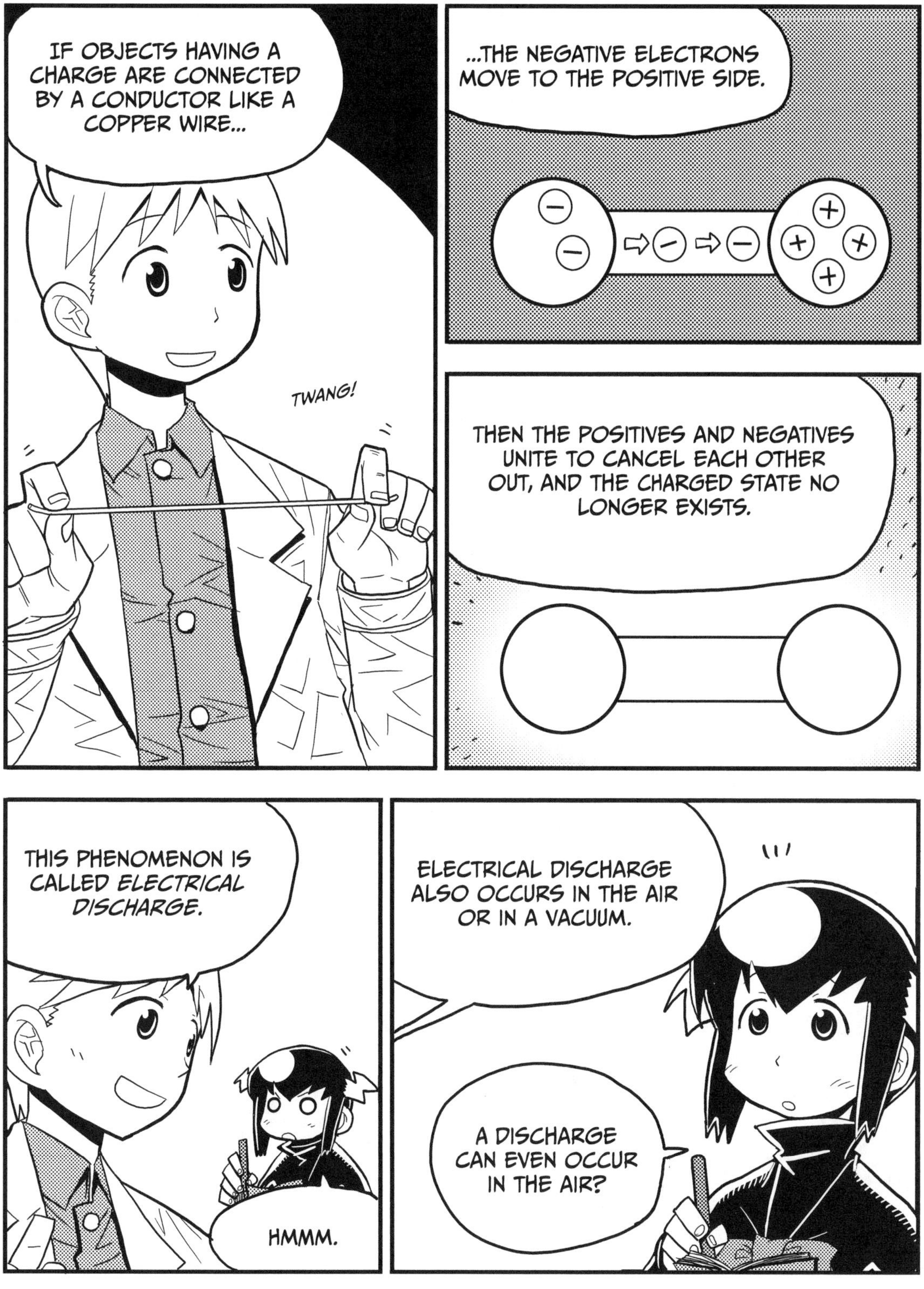
IF OBJECTS HAVING A CHARGE ARE CONNECTED BY A CONDUCTOR LIKE A COPPER WIRE...
TWANG!
...THE NEGATIVE ELECTRONS MOVE TO THE POSITIVE SIDE.
THEN THE POSITIVES AND NEGATIVES UNITE TO CANCEL EACH OTHER OUT, AND THE CHARGED STATE NO LONGER EXISTS.
THIS PHENOMENON IS CALLED *ELECTRICAL DISCHARGE*.
HMMM.
ELECTRICAL DISCHARGE ALSO OCCURS IN THE AIR OR IN A VACUUM.
A DISCHARGE CAN EVEN OCCUR IN THE AIR?

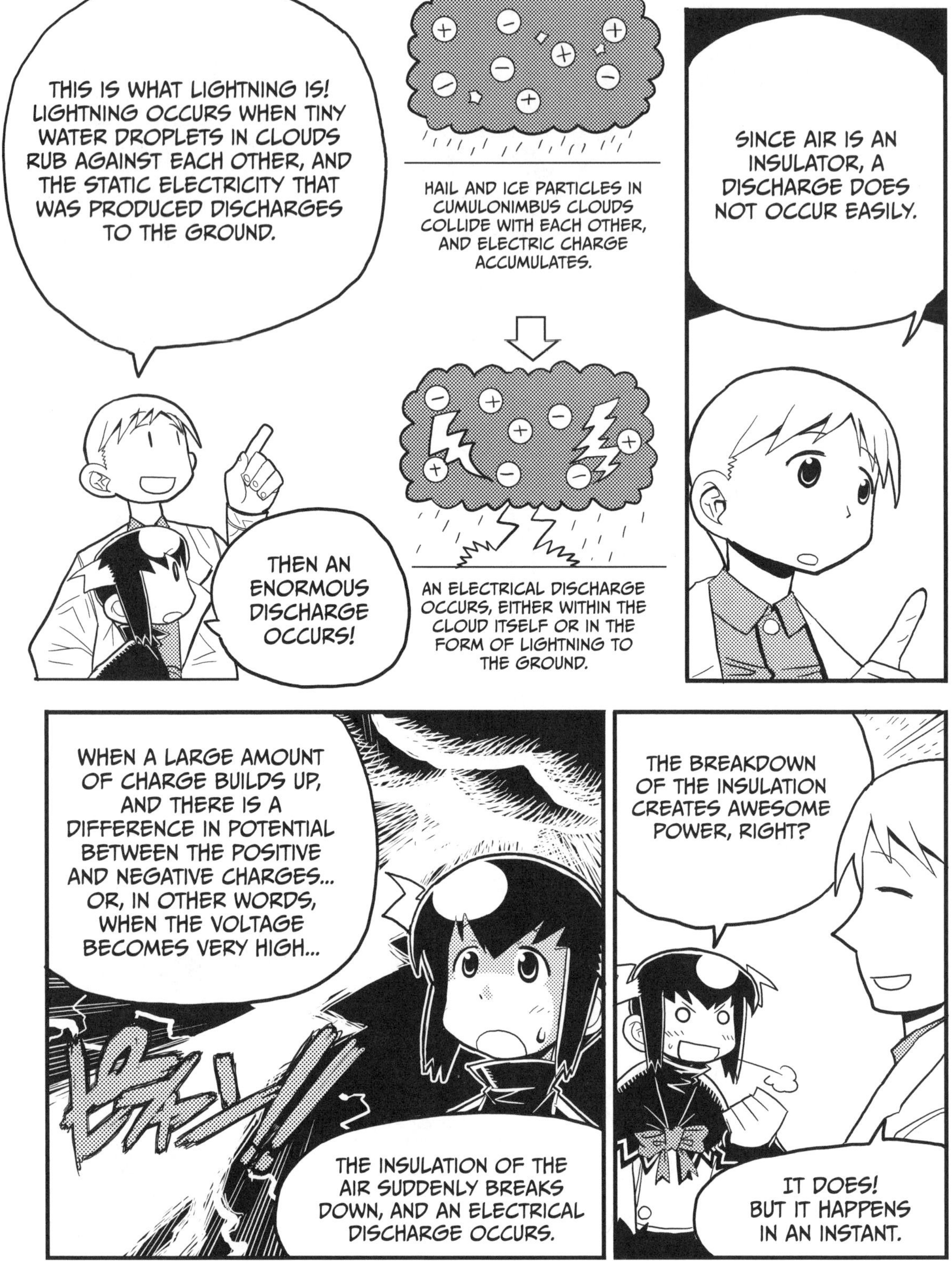
THIS IS WHAT LIGHTNING IS! LIGHTNING OCCURS WHEN TINY WATER DROPLETS IN CLOUDS RUB AGAINST EACH OTHER, AND THE STATIC ELECTRICITY THAT WAS PRODUCED DISCHARGES TO THE GROUND.
THEN AN ENORMOUS DISCHARGE OCCURS!
HAIL AND ICE PARTICLES IN CUMULONIMBUS CLOUDS COLLIDE WITH EACH OTHER, AND ELECTRIC CHARGE ACCUMULATES.
AN ELECTRICAL DISCHARGE OCCURS, EITHER WITHIN THE CLOUD ITSELF OR IN THE FORM OF LIGHTNING TO THE GROUND.
SINCE AIR IS AN INSULATOR, A DISCHARGE DOES NOT OCCUR EASILY.
WHEN A LARGE AMOUNT OF CHARGE BUILDS UP, AND THERE IS A DIFFERENCE IN POTENTIAL BETWEEN THE POSITIVE AND NEGATIVE CHARGES... OR, IN OTHER WORDS, WHEN THE VOLTAGE BECOMES VERY HIGH...
THE INSULATION OF THE AIR SUDDENLY BREAKS DOWN, AND AN ELECTRICAL DISCHARGE OCCURS.
THE BREAKDOWN OF THE INSULATION CREATES AWESOME POWER, RIGHT?
IT DOES! BUT IT HAPPENS IN AN INSTANT.

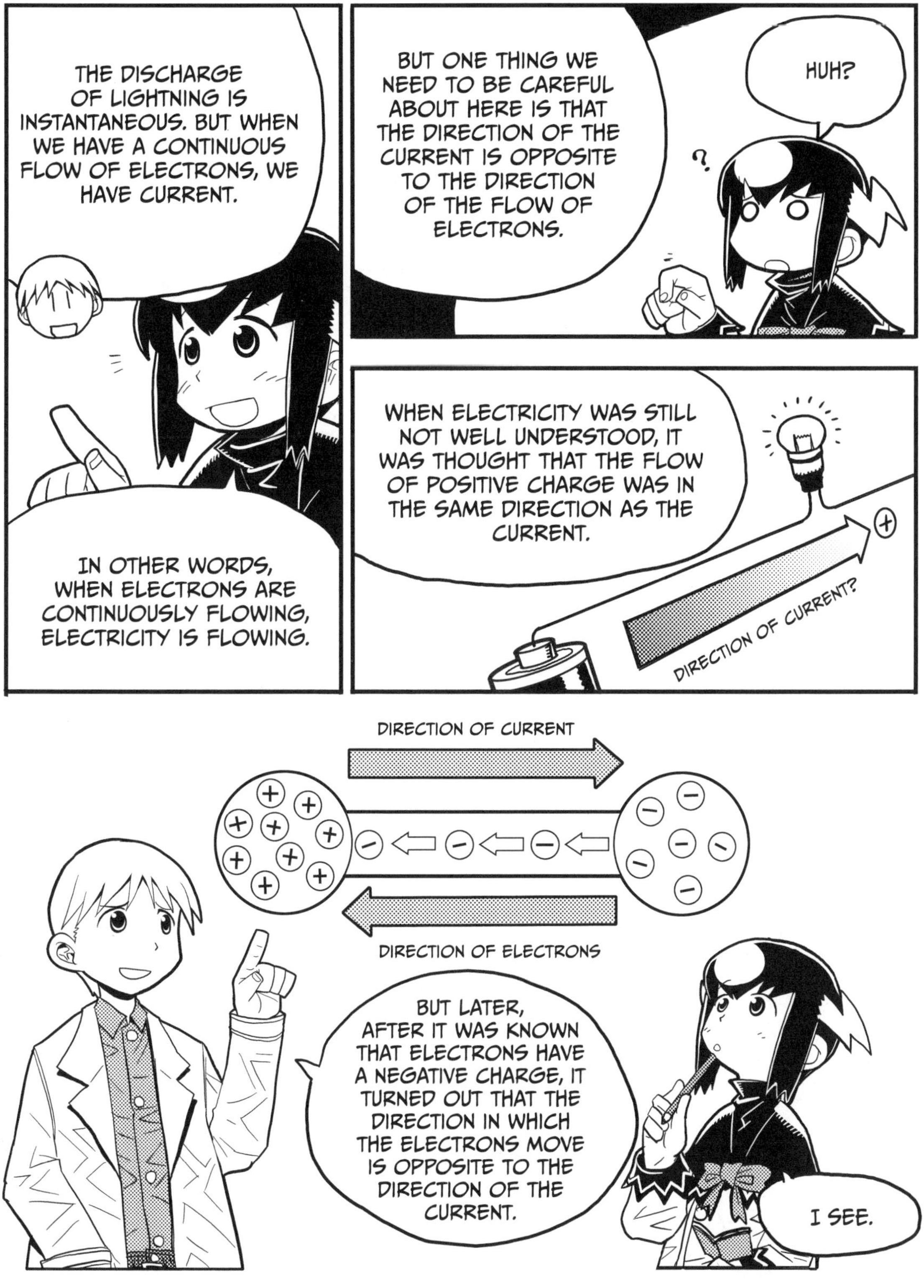
THE DISCHARGE OF LIGHTNING IS INSTANTANEOUS. BUT WHEN WE HAVE A CONTINUOUS FLOW OF ELECTRONS, WE HAVE CURRENT.
IN OTHER WORDS, WHEN ELECTRONS ARE CONTINUOUSLY FLOWING, ELECTRICITY IS FLOWING.
BUT ONE THING WE NEED TO BE CAREFUL ABOUT HERE IS THAT THE DIRECTION OF THE CURRENT IS OPPOSITE TO THE DIRECTION OF THE FLOW OF ELECTRONS.
HUH?
?
WHEN ELECTRICITY WAS STILL NOT WELL UNDERSTOOD, IT WAS THOUGHT THAT THE FLOW OF POSITIVE CHARGE WAS IN THE SAME DIRECTION AS THE CURRENT.
+
DIRECTION OF CURRENT?
DIRECTION OF CURRENT
DIRECTION OF ELECTRONS
BUT LATER, AFTER IT WAS KNOWN THAT ELECTRONS HAVE A NEGATIVE CHARGE, IT TURNED OUT THAT THE DIRECTION IN WHICH THE ELECTRONS MOVE IS OPPOSITE TO THE DIRECTION OF THE CURRENT.
I SEE.

ATOMIC STRUCTURE AND CONDUCTIVITY

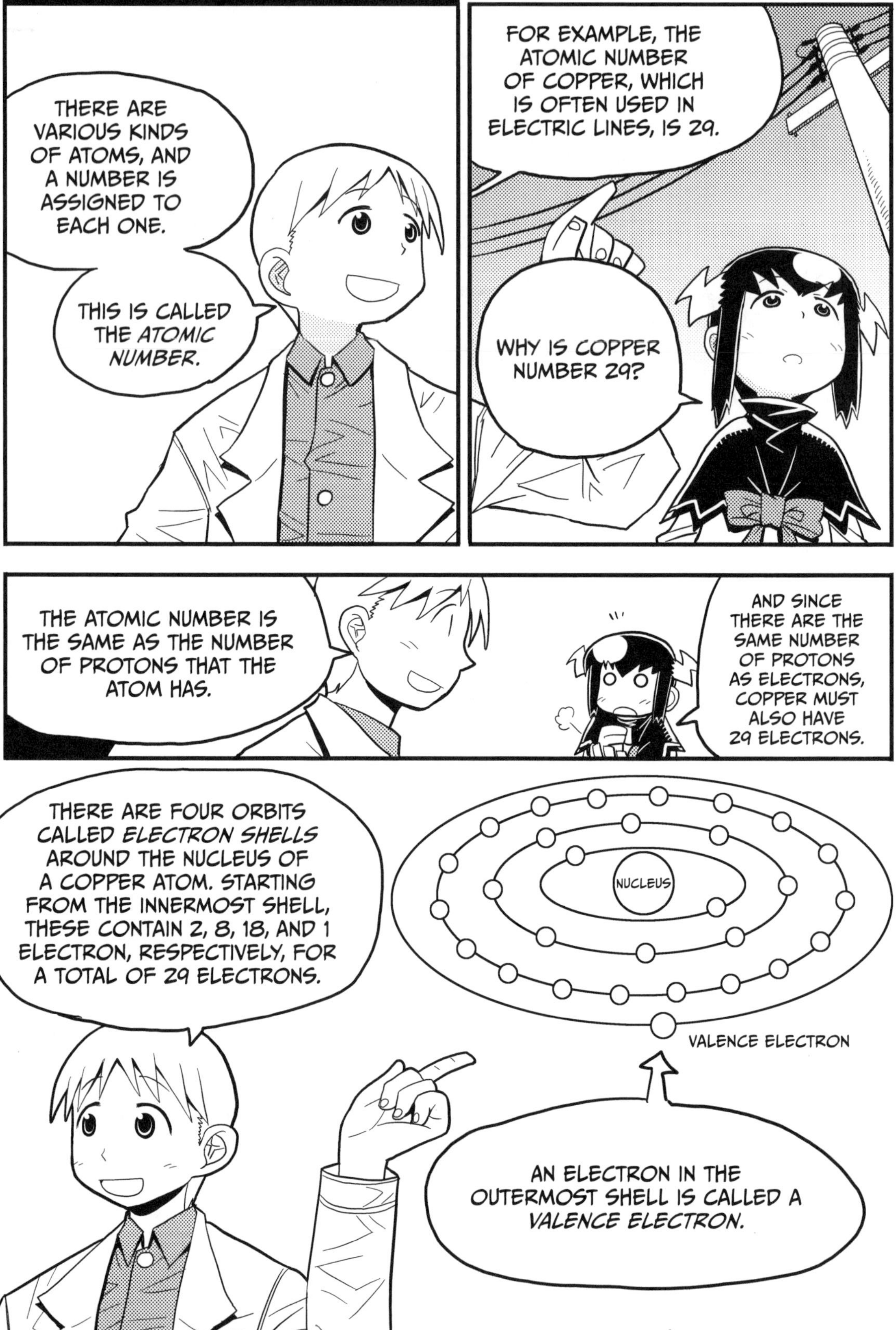

THERE ARE VARIOUS KINDS OF ATOMS, AND A NUMBER IS ASSIGNED TO EACH ONE.
THIS IS CALLED THE *ATOMIC NUMBER*.
FOR EXAMPLE, THE ATOMIC NUMBER OF COPPER, WHICH IS OFTEN USED IN ELECTRIC LINES, IS 29.
WHY IS COPPER NUMBER 29?
THE ATOMIC NUMBER IS THE SAME AS THE NUMBER OF PROTONS THAT THE ATOM HAS.
AND SINCE THERE ARE THE SAME NUMBER OF PROTONS AS ELECTRONS, COPPER MUST ALSO HAVE 29 ELECTRONS.
THERE ARE FOUR ORBITS CALLED *ELECTRON SHELLS* AROUND THE NUCLEUS OF A COPPER ATOM. STARTING FROM THE INNERMOST SHELL, THESE CONTAIN 2, 8, 18, AND 1 ELECTRON, RESPECTIVELY, FOR A TOTAL OF 29 ELECTRONS.
NUCLEUS
VALENCE ELECTRON
AN ELECTRON IN THE OUTERMOST SHELL IS CALLED A *VALENCE ELECTRON*.

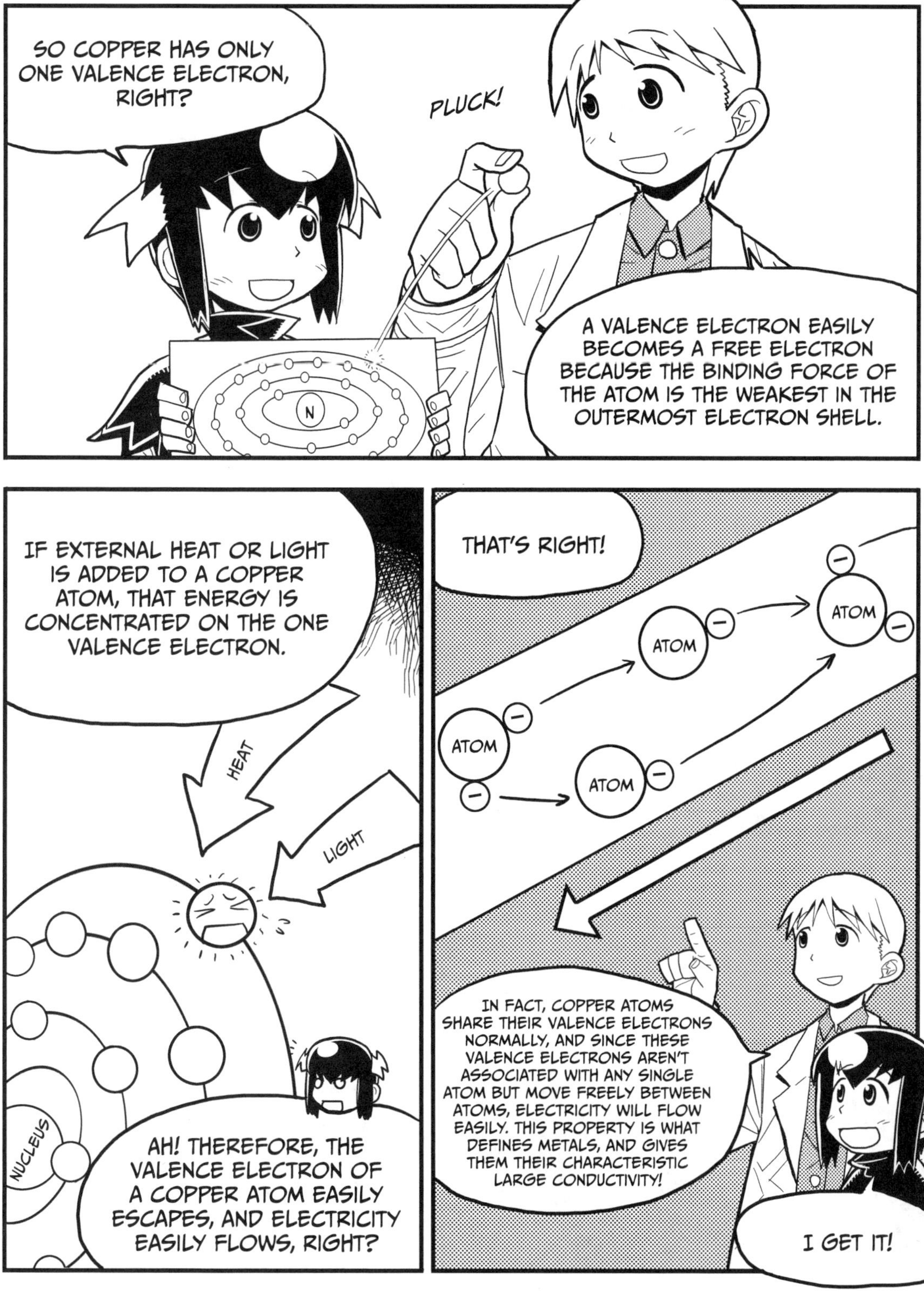
SO COPPER HAS ONLY ONE VALENCE ELECTRON, RIGHT?
PLUCK!
N
A VALENCE ELECTRON EASILY BECOMES A FREE ELECTRON BECAUSE THE BINDING FORCE OF THE ATOM IS THE WEAKEST IN THE OUTERMOST ELECTRON SHELL.
IF EXTERNAL HEAT OR LIGHT IS ADDED TO A COPPER ATOM, THAT ENERGY IS CONCENTRATED ON THE ONE VALENCE ELECTRON.
HEAT
LIGHT
NUCLEUS
AH! THEREFORE, THE VALENCE ELECTRON OF A COPPER ATOM EASILY ESCAPES, AND ELECTRICITY EASILY FLOWS, RIGHT?
THAT'S RIGHT!
ATOM
ATOM
ATOM
ATOM
IN FACT, COPPER ATOMS SHARE THEIR VALENCE ELECTRONS NORMALLY, AND SINCE THESE VALENCE ELECTRONS AREN'T ASSOCIATED WITH ANY SINGLE ATOM BUT MOVE FREELY BETWEEN ATOMS, ELECTRICITY WILL FLOW EASILY. THIS PROPERTY IS WHAT DEFINES METALS, AND GIVES THEM THEIR CHARACTERISTIC LARGE CONDUCTIVITY!
I GET IT!

STATIC ELECTRICITY

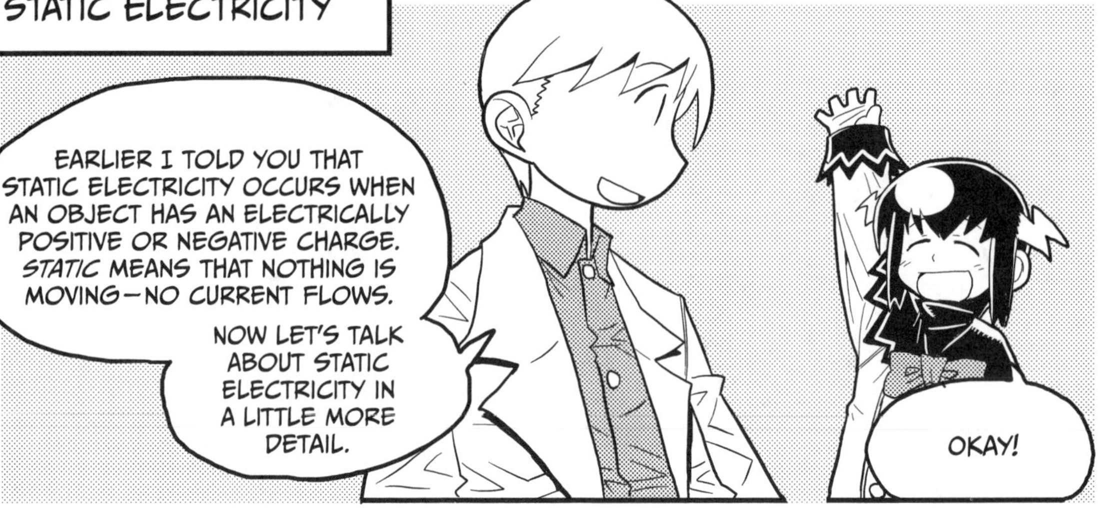
EARLIER I TOLD YOU THAT STATIC ELECTRICITY OCCURS WHEN AN OBJECT HAS AN ELECTRICALLY POSITIVE OR NEGATIVE CHARGE. *STATIC* MEANS THAT NOTHING IS MOVING—NO CURRENT FLOWS.
NOW LET'S TALK ABOUT STATIC ELECTRICITY IN A LITTLE MORE DETAIL.
OKAY!

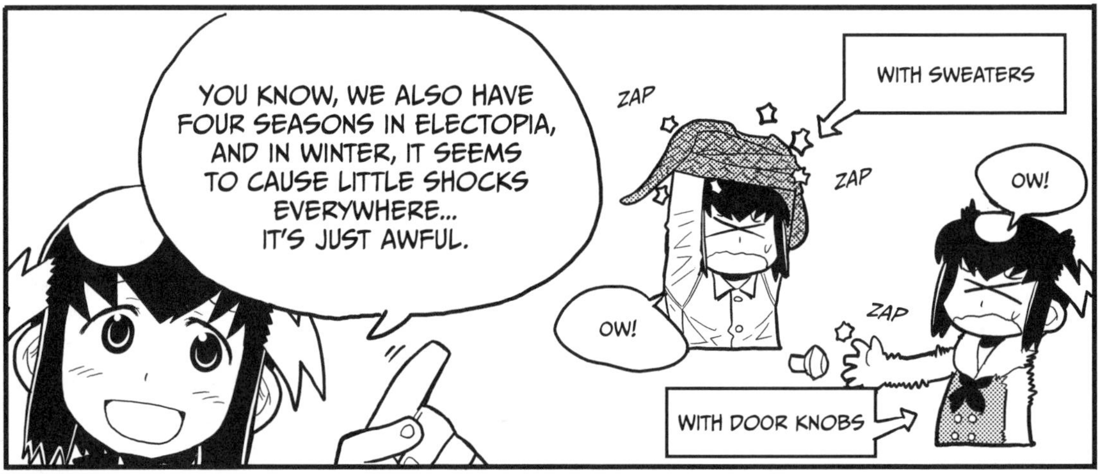
YOU KNOW, WE ALSO HAVE FOUR SEASONS IN ELECTOPIA, AND IN WINTER, IT SEEMS TO CAUSE LITTLE SHOCKS EVERYWHERE... IT'S JUST AWFUL.
ZAP
WITH SWEATERS
ZAP
OW!
OW!
ZAP
WITH DOOR KNOBS

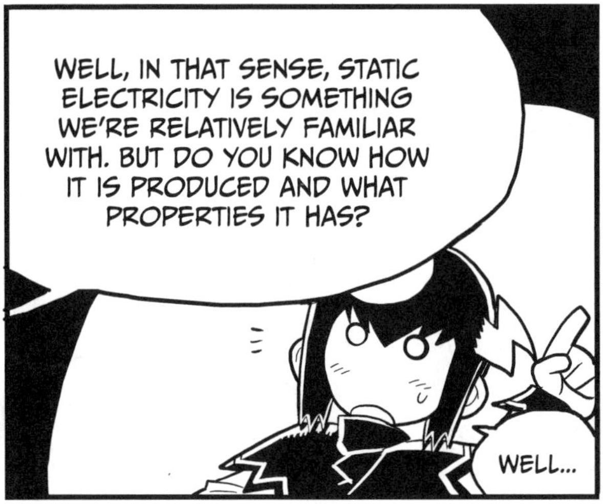
WELL, IN THAT SENSE, STATIC ELECTRICITY IS SOMETHING WE'RE RELATIVELY FAMILIAR WITH. BUT DO YOU KNOW HOW IT IS PRODUCED AND WHAT PROPERTIES IT HAS?
WELL...

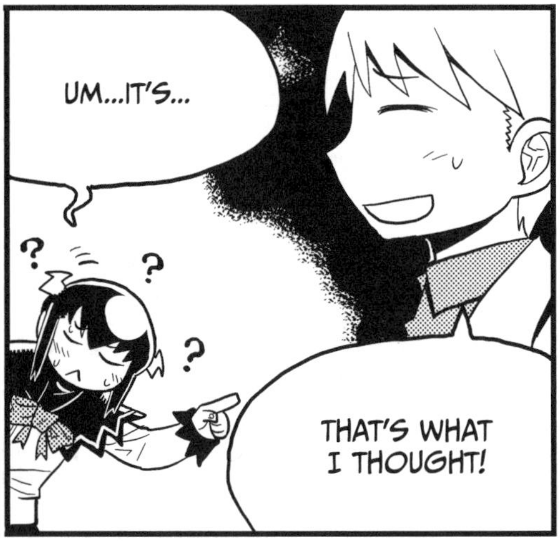
UM...IT'S...
THAT'S WHAT I THOUGHT!

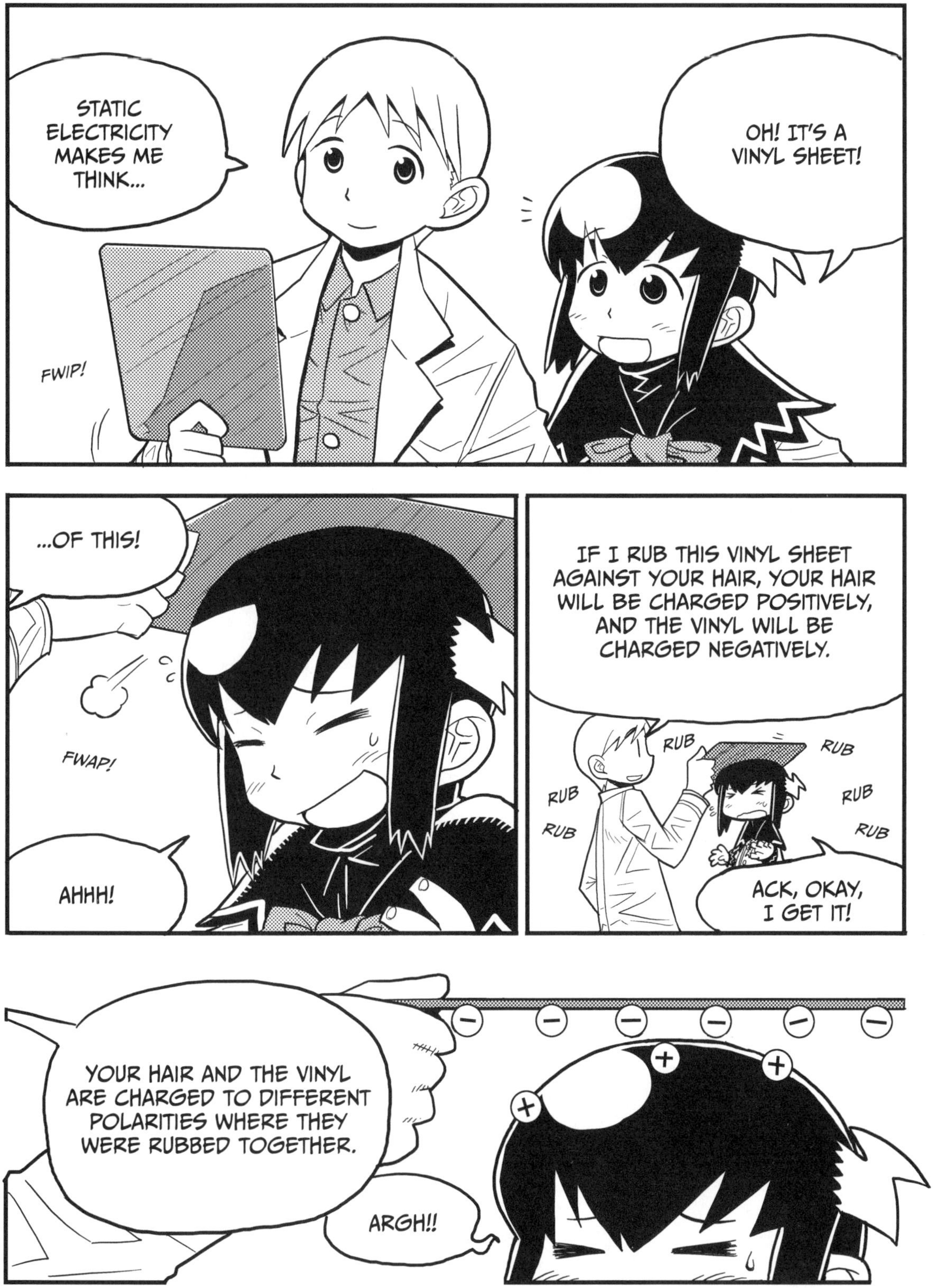
STATIC ELECTRICITY MAKES ME THINK...
FWIP!
OH! IT'S A VINYL SHEET!
...OF THIS!
FWAP!
AHHH!
IF I RUB THIS VINYL SHEET AGAINST YOUR HAIR, YOUR HAIR WILL BE CHARGED POSITIVELY, AND THE VINYL WILL BE CHARGED NEGATIVELY.
RUB RUB RUB RUB RUB RUB
ACK, OKAY, I GET IT!
YOUR HAIR AND THE VINYL ARE CHARGED TO DIFFERENT POLARITIES WHERE THEY WERE RUBBED TOGETHER.
ARGH!!

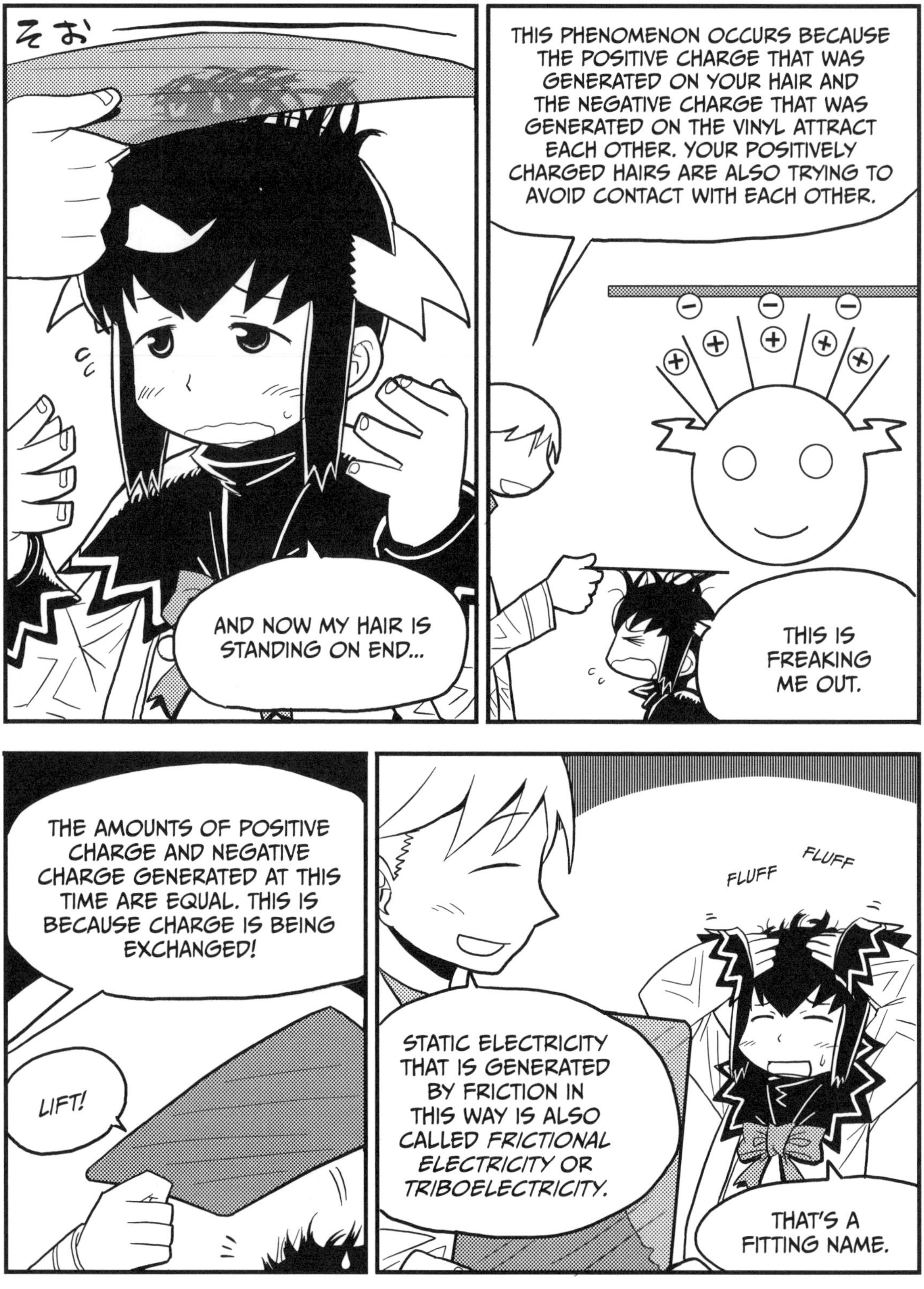
そお
AND NOW MY HAIR IS STANDING ON END...
THIS PHENOMENON OCCURS BECAUSE THE POSITIVE CHARGE THAT WAS GENERATED ON YOUR HAIR AND THE NEGATIVE CHARGE THAT WAS GENERATED ON THE VINYL ATTRACT EACH OTHER. YOUR POSITIVELY CHARGED HAIRS ARE ALSO TRYING TO AVOID CONTACT WITH EACH OTHER.
THIS IS FREAKING ME OUT.
THE AMOUNTS OF POSITIVE CHARGE AND NEGATIVE CHARGE GENERATED AT THIS TIME ARE EQUAL. THIS IS BECAUSE CHARGE IS BEING EXCHANGED!
LIFT!
FLUFF
FLUFF
STATIC ELECTRICITY THAT IS GENERATED BY FRICTION IN THIS WAY IS ALSO CALLED *FRICTIONAL ELECTRICITY* OR *TRIBOELECTRICITY*.
THAT'S A FITTING NAME.

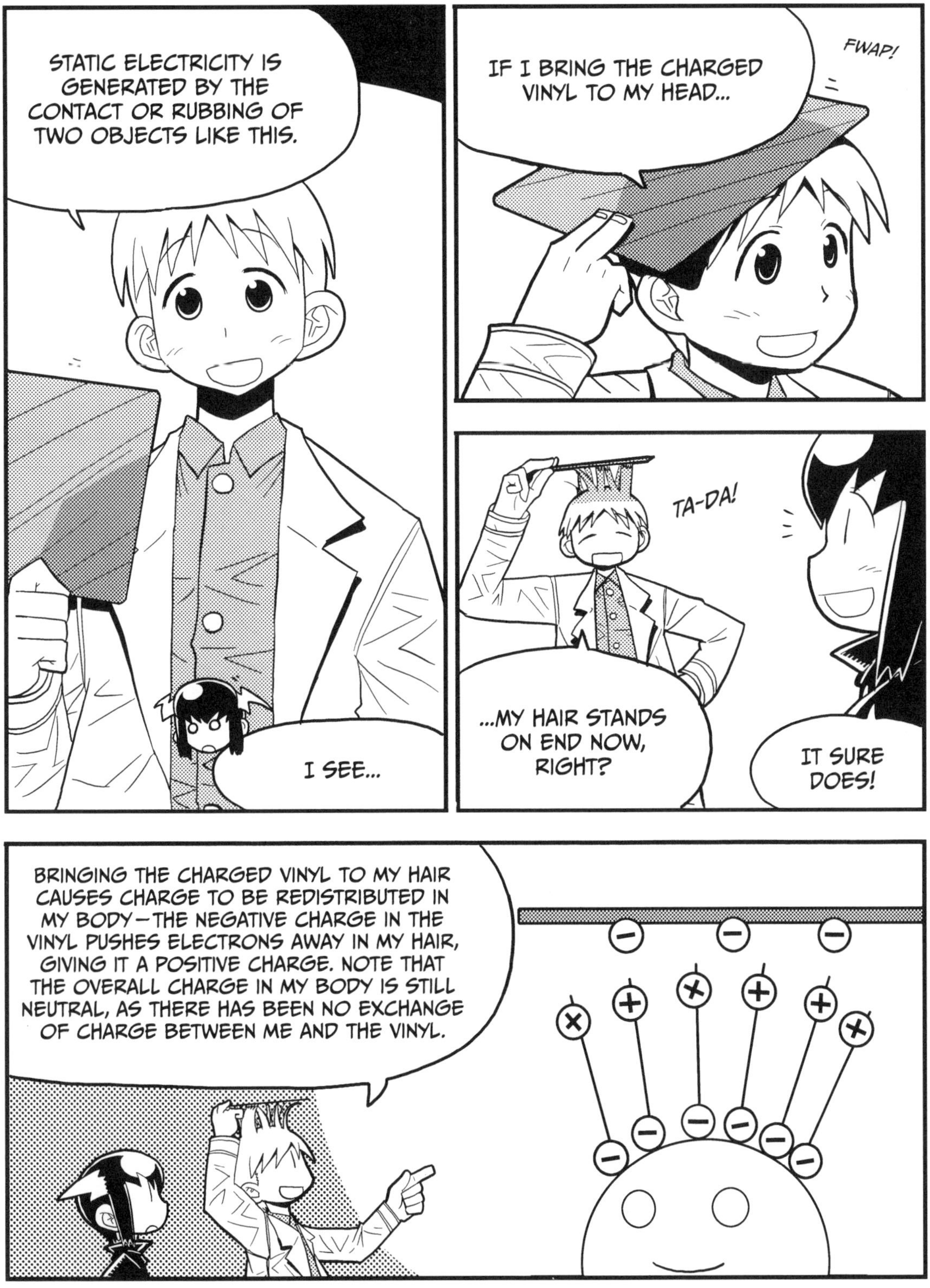
STATIC ELECTRICITY IS GENERATED BY THE CONTACT OR RUBBING OF TWO OBJECTS LIKE THIS.
I SEE...
IF I BRING THE CHARGED VINYL TO MY HEAD...
FWAP!
TA-DA!
...MY HAIR STANDS ON END NOW, RIGHT?
IT SURE DOES!
BRINGING THE CHARGED VINYL TO MY HAIR CAUSES CHARGE TO BE REDISTRIBUTED IN MY BODY—THE NEGATIVE CHARGE IN THE VINYL PUSHES ELECTRONS AWAY IN MY HAIR, GIVING IT A POSITIVE CHARGE. NOTE THAT THE OVERALL CHARGE IN MY BODY IS STILL NEUTRAL, AS THERE HAS BEEN NO EXCHANGE OF CHARGE BETWEEN ME AND THE VINYL.

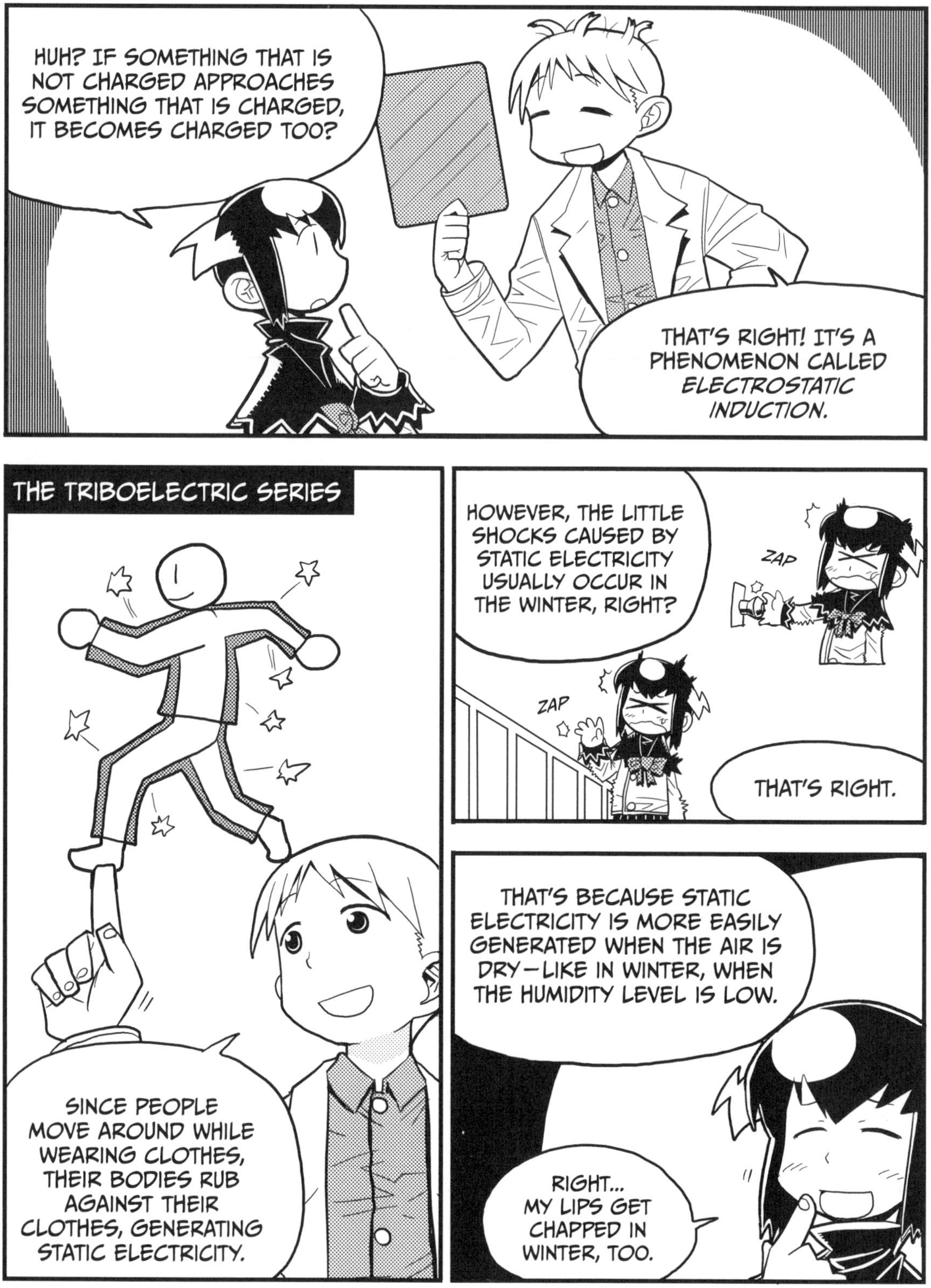
HUH? IF SOMETHING THAT IS NOT CHARGED APPROACHES SOMETHING THAT IS CHARGED, IT BECOMES CHARGED TOO?
THAT'S RIGHT! IT'S A PHENOMENON CALLED *ELECTROSTATIC INDUCTION*.
THE TRIBOELECTRIC SERIES
SINCE PEOPLE MOVE AROUND WHILE WEARING CLOTHES, THEIR BODIES RUB AGAINST THEIR CLOTHES, GENERATING STATIC ELECTRICITY.
HOWEVER, THE LITTLE SHOCKS CAUSED BY STATIC ELECTRICITY USUALLY OCCUR IN THE WINTER, RIGHT?
ZAP
ZAP
THAT'S RIGHT.
THAT'S BECAUSE STATIC ELECTRICITY IS MORE EASILY GENERATED WHEN THE AIR IS DRY—LIKE IN WINTER, WHEN THE HUMIDITY LEVEL IS LOW.
RIGHT... MY LIPS GET CHAPPED IN WINTER, TOO.

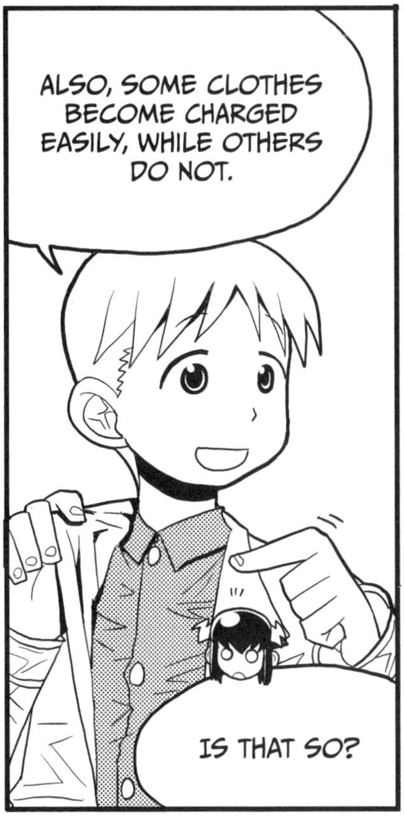
ALSO, SOME CLOTHES BECOME CHARGED EASILY, WHILE OTHERS DO NOT.
IS THAT SO?

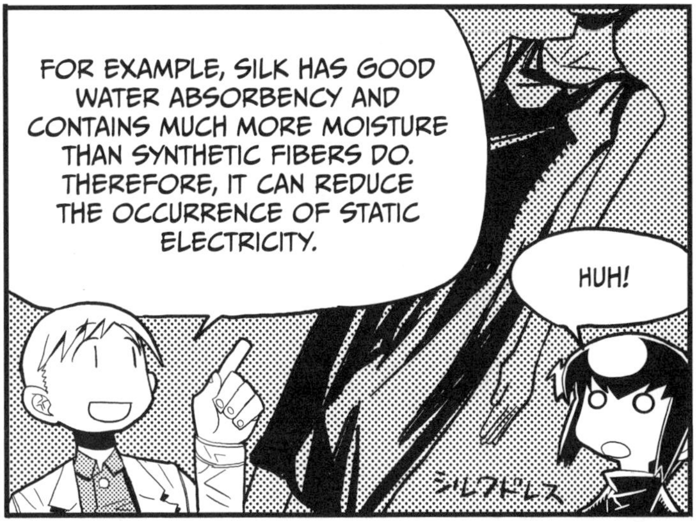
FOR EXAMPLE, SILK HAS GOOD WATER ABSORBENCY AND CONTAINS MUCH MORE MOISTURE THAN SYNTHETIC FIBERS DO. THEREFORE, IT CAN REDUCE THE OCCURRENCE OF STATIC ELECTRICITY.
HUH!

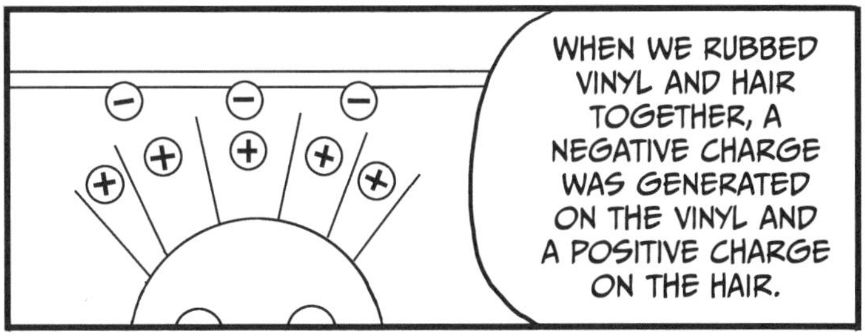
WHEN WE RUBBED VINYL AND HAIR TOGETHER, A NEGATIVE CHARGE WAS GENERATED ON THE VINYL AND A POSITIVE CHARGE ON THE HAIR.

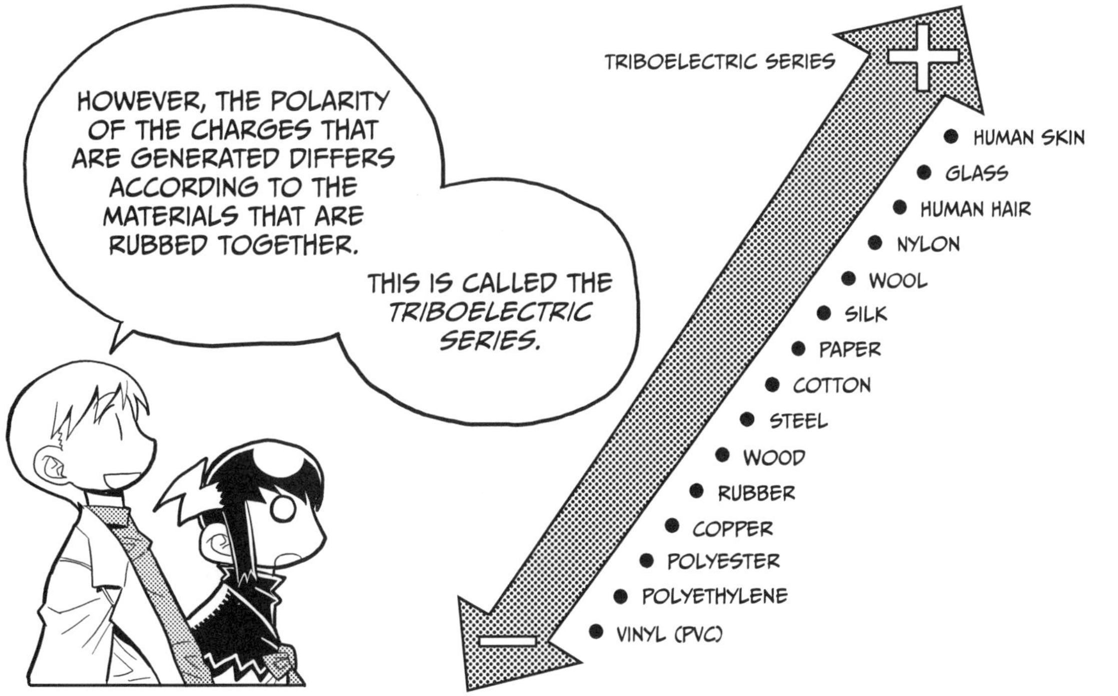
HOWEVER, THE POLARITY OF THE CHARGES THAT ARE GENERATED DIFFERS ACCORDING TO THE MATERIALS THAT ARE RUBBED TOGETHER.
THIS IS CALLED THE *TRIBOELECTRIC SERIES*.
TRIBOELECTRIC SERIES
+
HUMAN SKIN
GLASS
HUMAN HAIR
NYLON
WOOL
SILK
PAPER
COTTON
STEEL
WOOD
RUBBER
COPPER
POLYESTER
POLYETHYLENE
VINYL (PVC)
−

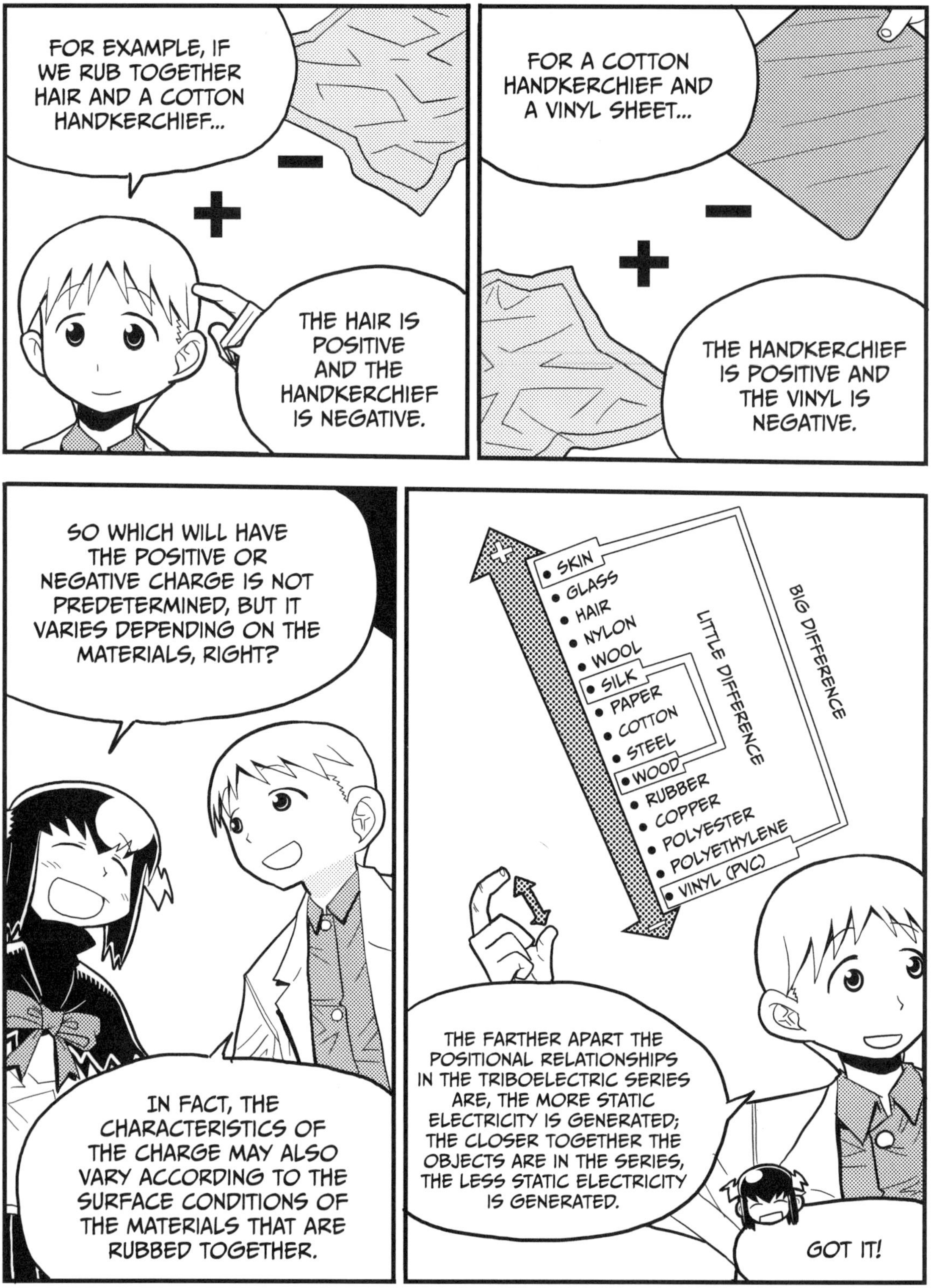
FOR EXAMPLE, IF WE RUB TOGETHER HAIR AND A COTTON HANDKERCHIEF...
THE HAIR IS POSITIVE AND THE HANDKERCHIEF IS NEGATIVE.
FOR A COTTON HANDKERCHIEF AND A VINYL SHEET...
THE HANDKERCHIEF IS POSITIVE AND THE VINYL IS NEGATIVE.
SO WHICH WILL HAVE THE POSITIVE OR NEGATIVE CHARGE IS NOT PREDETERMINED, BUT IT VARIES DEPENDING ON THE MATERIALS, RIGHT?
IN FACT, THE CHARACTERISTICS OF THE CHARGE MAY ALSO VARY ACCORDING TO THE SURFACE CONDITIONS OF THE MATERIALS THAT ARE RUBBED TOGETHER.
SKIN
GLASS
HAIR
NYLON
WOOL
SILK
PAPER
COTTON
STEEL
WOOD
RUBBER
COPPER
POLYESTER
POLYETHYLENE
VINYL (PVC)
LITTLE DIFFERENCE
BIG DIFFERENCE
THE FARTHER APART THE POSITIONAL RELATIONSHIPS IN THE TRIBOELECTRIC SERIES ARE, THE MORE STATIC ELECTRICITY IS GENERATED; THE CLOSER TOGETHER THE OBJECTS ARE IN THE SERIES, THE LESS STATIC ELECTRICITY IS GENERATED.
GOT IT!

USES OF STATIC ELECTRICITY

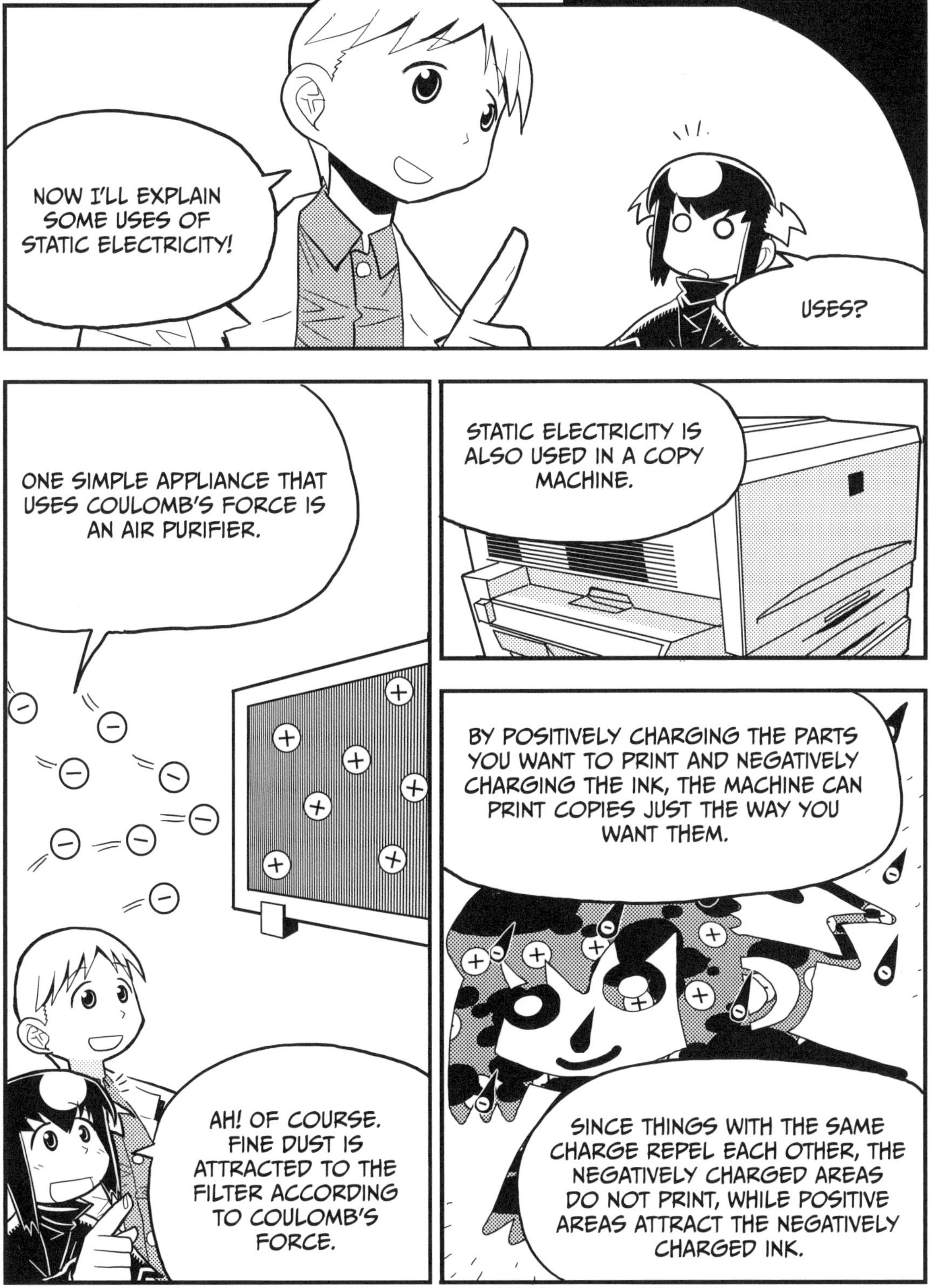
NOW I'LL EXPLAIN SOME USES OF STATIC ELECTRICITY!
USES?
ONE SIMPLE APPLIANCE THAT USES COULOMB'S FORCE IS AN AIR PURIFIER.
AH! OF COURSE. FINE DUST IS ATTRACTED TO THE FILTER ACCORDING TO COULOMB'S FORCE.
STATIC ELECTRICITY IS ALSO USED IN A COPY MACHINE.
BY POSITIVELY CHARGING THE PARTS YOU WANT TO PRINT AND NEGATIVELY CHARGING THE INK, THE MACHINE CAN PRINT COPIES JUST THE WAY YOU WANT THEM.
SINCE THINGS WITH THE SAME CHARGE REPEL EACH OTHER, THE NEGATIVELY CHARGED AREAS DO NOT PRINT, WHILE POSITIVE AREAS ATTRACT THE NEGATIVELY CHARGED INK.

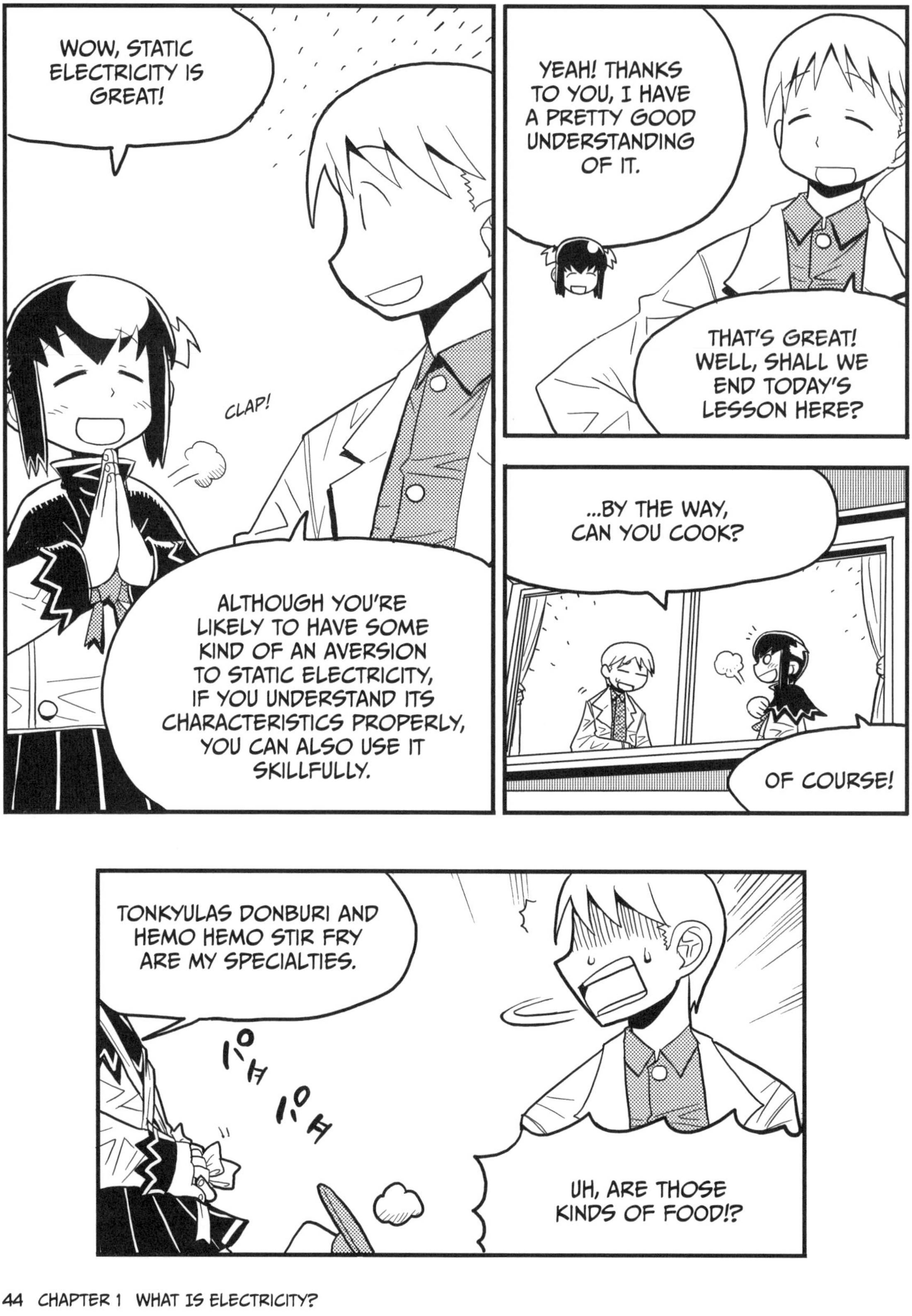
WOW, STATIC ELECTRICITY IS GREAT!
CLAP!
ALTHOUGH YOU'RE LIKELY TO HAVE SOME KIND OF AN AVERSION TO STATIC ELECTRICITY, IF YOU UNDERSTAND ITS CHARACTERISTICS PROPERLY, YOU CAN ALSO USE IT SKILLFULLY.
YEAH! THANKS TO YOU, I HAVE A PRETTY GOOD UNDERSTANDING OF IT.
THAT'S GREAT! WELL, SHALL WE END TODAY'S LESSON HERE?
...BY THE WAY, CAN YOU COOK?
OF COURSE!
TONKYULAS DONBURI AND HEMO HEMO STIR FRY ARE MY SPECIALTIES.
UH, ARE THOSE KINDS OF FOOD!?

TAGS ON CONSUMER ELECTRIC PRODUCTS

Consumer electric products have tags related to electricity with information such as voltage, current, and power—for example, 120V, 1440W, and 12A.

ELECTRIC HEATER
120V
1440W
12A

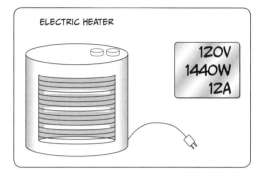

A tag on an electric heater

Voltage, the potential difference or "pressure" that makes electricity flow, is represented by the symbol *V*. The unit used to measure voltage is the *volt (V)*, which is named for the Italian physicist Alessandro Volta, who invented the battery. The voltage used in an ordinary household appliance is 120V in the United States, 240V in Europe, and 100V in Japan.

Current is the quantity of electricity flowing per second through an electric line, and it is represented using the symbol *I*, which comes from the initial letter of ***I**ntensity of electricity*. Current is measured in *amperes (A)*, or *amps* for short, which are named for the French physicist André Marie Ampère. One amp is equal to one coulomb per second.

Power, which is the electric energy consumed in one second when current flows, is represented using the symbol *P*. Power is measured in *watts (W)*, which are named for the British mechanical engineer James Watt, who invented the steam engine. One watt is equal to one joule per second.

You can determine the power a device draws by multiplying its voltage and current. The power of a 120V device in which 12A of current flows is $P = V \times I = 120V \times 12A = 1440W$.

A typical American household contains many 120V devices. If you divide the power value that is displayed on each of these devices by 120V, you can find the value of the current that flows in each device. For devices with the same power, a 240V electronic device runs using half the current of an 120V electronic device.

Since $P = V \times I$, we can rearrange this equation to look like this using simple algebra.

$$I = \frac{P}{V}$$

For a 120V electric device... $I = \frac{1440W}{120V} = 12A$ ...12A of current flows.

For a 240V electric device... $I = \frac{1440W}{240V} = 6A$ ...6A of current flows.

SI PREFIXES

1000W may also be represented by 1kW. This is because *k* stands for *kilo* and represents 1,000 or 10^3. But we can use other prefixes, too: 3,600,000 joules (J) are equal to 3.6 megajoules (MJ). These prefixes for different powers of 10 are called *SI prefixes*, and they come from internationally determined rules for units called the *International System of Units (SI units)*. The most common ones are shown in the table below.

SI PREFIXES OFTEN USED IN ELECTRICAL RELATIONSHIPS

Prefix Symbol	Name	Quantity	
T	tera	10^{12}	= 1,000,000,000,000
G	giga	10^{9}	= 1,000,000,000
M	mega	10^{6}	= 1,000,000
k	kilo	10^{3}	= 1,000
m	milli	10^{-3}	= 0.001
μ	micro	10^{-6}	= 0.000 001
n	nano	10^{-9}	= 0.000 000 001
p	pico	10^{-12}	= 0.000 000 000 001

You can find the amount of *energy*, which is the total amount of *work* done by an electrical device, by multiplying the power it draws and the time the device operates. Power is often measured by power companies in kWh (kilowatt hours). For example, if an electric heater with 1kW is used for 1 hour, the amount of energy it uses is 1kW × 1 hour = 1kWh.

However, when time is represented in seconds, Ws (watt second) can be used for the energy's unit. A watt-second is equivalent to a *joule (J)*. For example, when a 1kW electric heater is used for 1 hour, since 1 hour = 60 minutes × 60 seconds = 3600 seconds, the amount of energy used is 1kW × 3600 seconds = 3600kWs or 3,600,000 joules.

You can calculate how much it will cost to use an ordinary household appliance by multiplying the amount of energy used (in kWh) by the utility company's price per kWh (you will also need to add in any flat-rate charges, if your utility company has them). Since the average electrical utility charge for 1kWh in the United States is approximately 12 cents for 1kWh, if a device with 1kW of power is used for 1 hour, the amount of energy used is 1kWh, and the electrical utility charge will be approximately 12 cents.

VOLTAGE AND POTENTIAL

Electricity flows from a high potential to a low potential. The potential difference between two points is called *voltage*. For example, for a AA battery, if we let the negative pole be the reference point, then the potential of the negative pole is 0V and the potential of the positive pole is 1.5V. The potential difference between the positive and negative poles is the *supply voltage* of this battery.

POSITIVE POLE
1.5V
AA BATTERY
VOLTAGE =
1.5V (POTENTIAL OF POSITIVE POLE)
- 0V (POTENTIAL OF NEGATIVE POLE)
= 1.5V
0V
NEGATIVE POLE
REFERENCE POINT

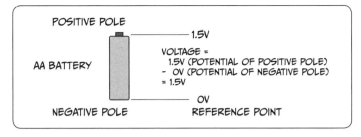

Supply voltage of a AA battery

If we stack two batteries and let the reference point be point B, the potential of point A is 1.5V, the potential of point B is 0V, and the potential of point C is -1.5V. The voltage between points A and C can be obtained by subtracting the potential of point C from the potential of point A; the voltage, in this case, is 3V. If we let point C be the reference point, the potential of point C is 0V, the potential of point B is 1.5V, and the potential of point A is 3V. The voltage is still 3V.

The larger the difference in electrical potential, the larger the voltage.

A
1.5V
B
0V
REFERENCE POINT
C
-1.5V
VOLTAGE BETWEEN A AND C =
1.5V - (-1.5V) = 3V
A
3V
B
1.5V
C
0V
REFERENCE POINT
VOLTAGE BETWEEN A AND C =
3V - 0V = 3V

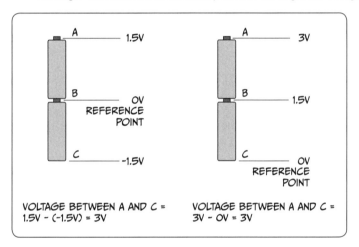

The voltage when two batteries are stacked

ATOMS AND ELECTRONS

All substances are made of atoms. An *atom* consists of a nucleus, which is made of protons and neutrons, and electrons. Since protons have a positive charge and neutrons are electrically neutral, the nucleus itself is electrically positive. Electrons, on the other hand, have a negative charge. But since protons and electrons are equally and oppositely charged, an atom is typically electrically neutral.

Electrons move around the nucleus in a series of orbits called *electron shells*. Since the attraction from the nucleus is weaker for electrons in the outermost shells than ones in the innermost shells, electrons in those outermost shells may escape from orbit if external energy such as heat or light is applied. An electron that has escaped from orbit can move around freely and is called a *free electron*. In substances like copper and other metals,

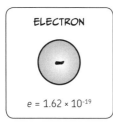

ELECTRON
-
e = 1.62 × 10-19

The smallest quantity of electricity that exists in the natural world is a single electron.

LIGHT OR HEAT
FREE ELECTRON

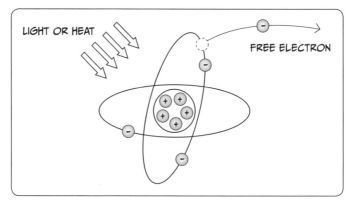

The escape of an electron

through which electricity easily flows, there are many free electrons, and if a voltage is applied to this substance, the free electrons all flow in one direction. This is how electricity flows through an electric line. The outermost electron shell of an atom is called the *valence shell*, and the electrons that are in it are called *valence electrons*.

The total number of electrons in an atom is the same as that atom's atomic number. Although there are many atoms with high atomic numbers and a lot of electrons, those substances are not necessarily ones through which electricity easily flows—the flow of electricity depends on the number of valence electrons.

MATERIAL THROUGH WHICH ELECTRICITY EASILY FLOWS HAS LOTS OF FREE ELECTRONS.
NUCLEUS
FREE ELECTRONS
NUCLEUS
FREE ELECTRON
ELECTRICITY HARDLY FLOWS AT ALL THROUGH MATERIAL WITH FEW FREE ELECTRONS.

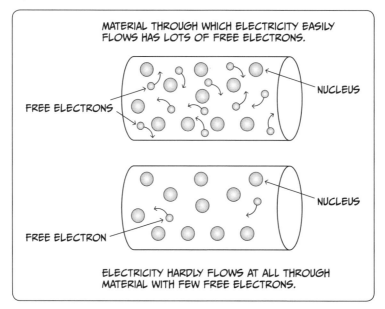

Different materials have different amounts of free electrons.

STATIC ELECTRICITY

When two different substances are rubbed together, atoms collide, and electrons that are easily separated from the atoms of one substance may escape and move to the atoms of the other substance. At this time, the substance that lost electrons becomes positively charged, and the substance that gained electrons becomes negatively charged. A substance that carries electricity in this way is said to be *charged*, and since this electricity is stationary (that is, it's not flowing), it is called *static electricity*. The quantity of the positive charge that is generated by this process is always the same as the quantity of the negative charge. Since static electricity is generated by friction, it is also called *frictional electricity*.

THE NUMBER OF ELECTRONS IS REDUCED BY ONE, SO THIS ATOM HAS A POSITIVE CHARGE.
LIGHT OR HEAT
THE NUMBER OF ELECTRONS IS INCREASED BY ONE, SO THIS ATOM HAS A NEGATIVE CHARGE.

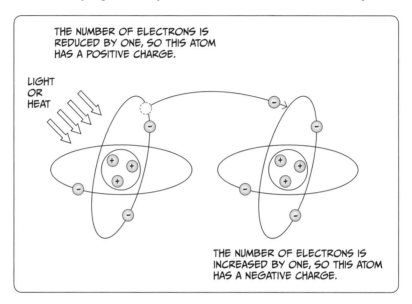

Electron movement and electric charge

ELECTROSTATIC FORCE

Charge is measured in *coulombs* and is represented by Q, the quantity of charge. The name of the unit comes from Charles Augustine Coulomb, a French physicist who studied electricity.

A force called *electrostatic force* (also known as *Coulomb's force*) operates between two charges. This force causes the same types of charge to repel each other and different types of charge to attract each other. The size of the electrostatic force F of attraction or repulsion (measured in a unit called a newton) operating between charge Q_1 and Q_2 is directly proportional to the product of Q_1 and Q_2 and inversely proportional to the square of the distance (r meters) between the charges. The stronger the charges, and the smaller the distance, the larger the resulting electrostatic force. This is called *Coulomb's law* with respect to static electricity.

SAME TYPES OF CHARGE REPEL.
Q1
Q2
CHARGE
CHARGE
r
$F = 9 \times 10^9 \times \frac{Q_1 \times Q_2}{r^2}$
DIFFERENT TYPES OF CHARGE ATTRACT.
ELECTROSTATIC FORCE
F

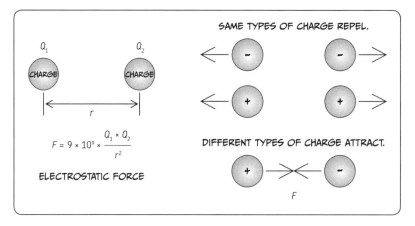

Electrostatic force operating between charges and Coulomb's law

If static electricity is generated by rubbing a vinyl sheet on a person's hair, the hair has a positive charge, the vinyl has a negative charge, and the hair clings to the vinyl due to the electrostatic force.

Also, if the negatively charged vinyl sheet is brought close to hair that has not been charged, the hair will become positively charged and will cling to the vinyl. This phenomenon, in which something that is not charged becomes charged when it is in close proximity to something else that is charged, is called *electrostatic induction*.

RUBBING A VINYL SHEET ON YOUR HAIR GENERATES STATIC ELECTRICITY.
BRINGING THE SHEET CLOSE TO SOMEONE ELSE'S HAIR THAT HAS NOT BEEN CHARGED...
...CHARGES THAT PERSON'S HAIR BY ELECTROSTATIC INDUCTION SO THAT IT CLINGS TO THE VINYL.

Electrostatic induction

THE TRIBOELECTRIC SERIES

Static electricity is more easily generated as the air gets drier—humidity prevents static electricity from gathering on a surface. Also, some clothes easily become charged, while others do not, depending on the material they are made from. Since silk has good water absorbency and contains much more moisture than synthetic fibers, it can reduce the occurrence of static electricity.

The polarities of the charges that are generated by friction differ according to the materials that are rubbed together. These differences are represented by the *triboelectric series*. For example, if hair and cotton are rubbed together, the hair will become positively charged and the cotton will become negatively charged, but for cotton and vinyl, the cotton will become positively charged and the vinyl will become negatively charged.

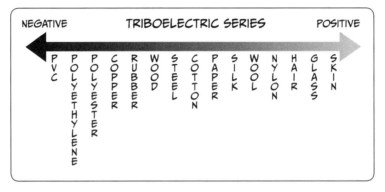

NEGATIVE
TRIBOELECTRIC SERIES
POSITIVE
PVC
POLYETHYLENE
POLYESTER
COPPER
RUBBER
WOOD
STEEL
COTTON
PAPER
SILK
WOOL
NYLON
HAIR
GLASS
SKIN

The triboelectric series

The farther apart the materials are in the triboelectric series, the more static electricity is generated between them, and the closer together the objects are in the triboelectric series, the less static electricity is generated. In other words, you can reduce the occurrence of static electricity by wearing clothes that are made of materials that are close together in the triboelectric series.

MOVEMENT OF CHARGE AND DIRECTION OF CURRENT

Lightning is also a result of static electricity. Lightning occurs when the static electricity that is produced by the friction between hail and ice particles in a cloud discharges between the cloud and ground. In the case of lightning, air (which is an insulator through which electricity has difficulty flowing) exists between the positive and negative charges, so a discharge does not easily occur.

When a large amount of charge builds up and the potential difference between the positive and negative charges is extremely large, the insulation of the air suddenly breaks down and an electrical discharge occurs. *Electrical discharge* is the phenomenon in which charge flows continuously. This continuous flow of electricity is called *current*.

Electric current flows from positive to negative. Scientists have discovered that the movement of electrons, however, is from negative to positive. Therefore, the direction in which the electrons move is actually opposite to the direction in which current flows.

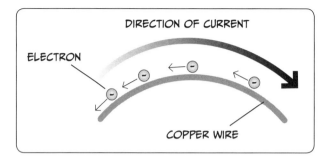

DIRECTION OF CURRENT
ELECTRON
COPPER WIRE

Direction of current and direction of electron movement

The amount of current is represented by the quantity of electricity passing through a wire in a second.

THE SIZE OF THE CURRENT IS THE QUANTITY OF ELECTRICITY PASSING THROUGH A WIRE IN ONE SECOND.

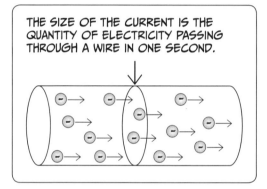

Size of the current

For example, when a charge of 1C passes through a given point, the current I can be obtained by dividing the charge (Q) in coulombs by the time (t) in seconds as follows.

$$I = \frac{Q}{t} = \frac{1C}{1s} = 1A$$

Also, the number of electrons flowing at 1A can be obtained by dividing 1C by the quantity of charge in 1 electron, as follows:

$$\frac{1C}{1.602 \times 10^{-19} C/\text{electron}} = 6.24 \times 10^{18} \text{ electrons}$$

In other words, when a current of 1A is flowing, there are 6.24×10^{18} electrons flowing per second.

The speed at which the electrons move is very slow—less than 1 cm per second. However, the speed at which electrical motion is transmitted to neighboring electrons is the same as the speed of light: 300,000 km per second. Therefore, the current also flows at 300,000 km per second (the speed of light).

ELECTRONS MOVING AT LESS THAN 1 CM PER SECOND
NUCLEUS
CURRENT FLOWING AT 300,000 KM PER SECOND

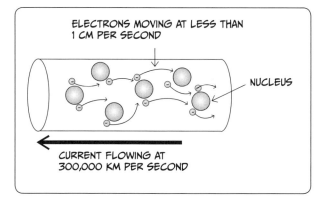

Speed of electrons and speed of current

Although electricity itself cannot be seen with the naked eye, heat or light is often produced when current flows. Therefore, we know that electricity exists by observing the phenomena caused by current.

2
WHAT ARE ELECTRIC CIRCUITS?

ELECTRIC CIRCUITS IN EVERYDAY DEVICES

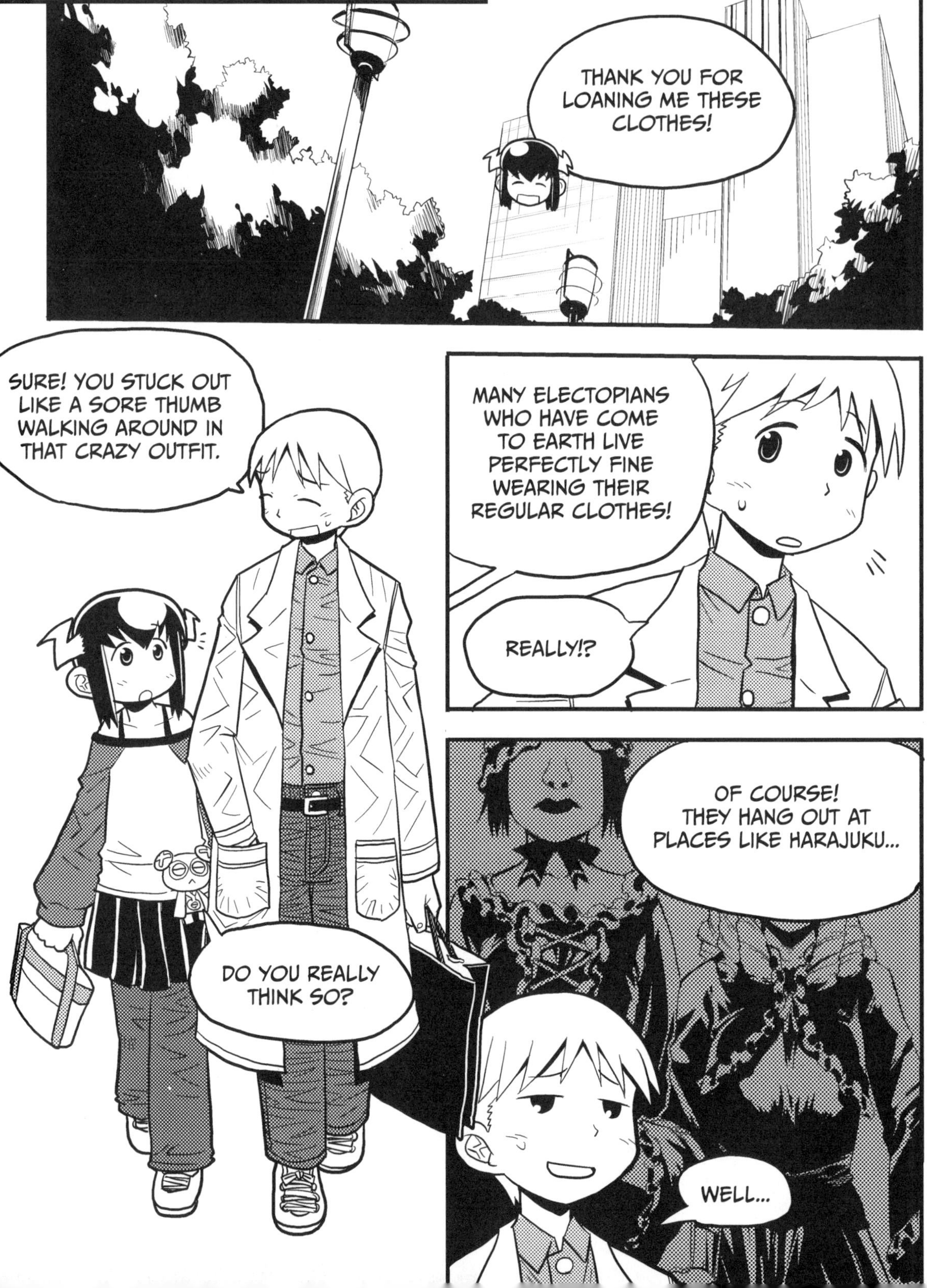
THANK YOU FOR LOANING ME THESE CLOTHES!
SURE! YOU STUCK OUT LIKE A SORE THUMB WALKING AROUND IN THAT CRAZY OUTFIT.
DO YOU REALLY THINK SO?
MANY ELECTOPIANS WHO HAVE COME TO EARTH LIVE PERFECTLY FINE WEARING THEIR REGULAR CLOTHES!
REALLY!?
OF COURSE! THEY HANG OUT AT PLACES LIKE HARAJUKU...
WELL...

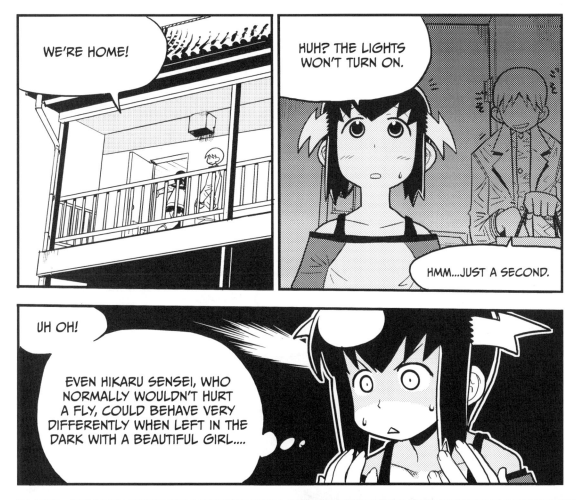

WE'RE HOME!

HUH? THE LIGHTS WON'T TURN ON.

HMM...JUST A SECOND.

UH OH!

EVEN HIKARU SENSEI, WHO NORMALLY WOULDN'T HURT A FLY, COULD BEHAVE VERY DIFFERENTLY WHEN LEFT IN THE DARK WITH A BEAUTIFUL GIRL....

HEY, REREKOOOOO...

AIEEEEE!

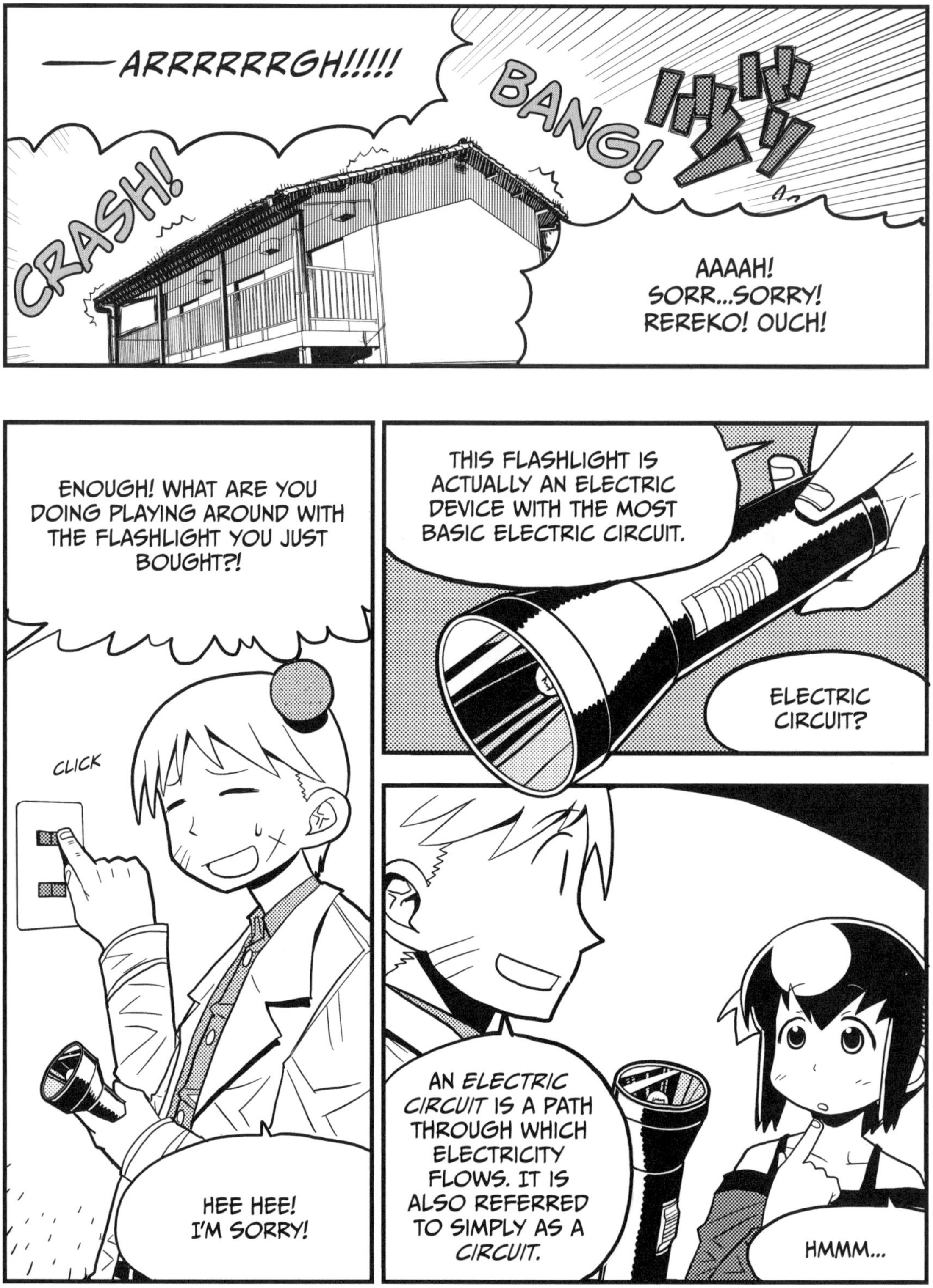
——ARRRRRRGH!!!!!
BANG!
CRASH!
AAAAH! SORR...SORRY! REREKO! OUCH!
ENOUGH! WHAT ARE YOU DOING PLAYING AROUND WITH THE FLASHLIGHT YOU JUST BOUGHT?!
CLICK
HEE HEE! I'M SORRY!
THIS FLASHLIGHT IS ACTUALLY AN ELECTRIC DEVICE WITH THE MOST BASIC ELECTRIC CIRCUIT.
ELECTRIC CIRCUIT?
AN *ELECTRIC CIRCUIT* IS A PATH THROUGH WHICH ELECTRICITY FLOWS. IT IS ALSO REFERRED TO SIMPLY AS A *CIRCUIT*.
HMMM...

A FLASHLIGHT'S CIRCUIT

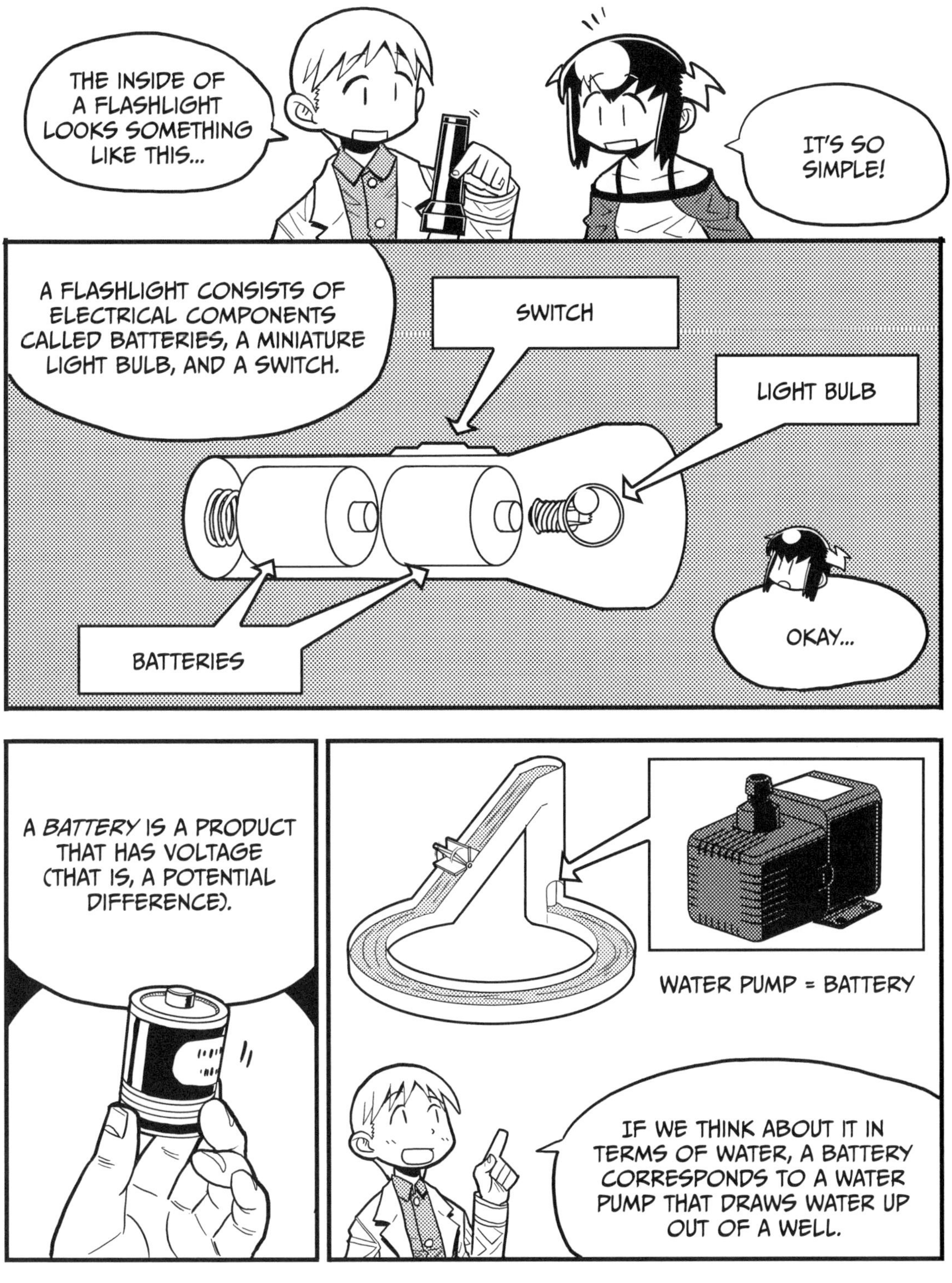
THE INSIDE OF A FLASHLIGHT LOOKS SOMETHING LIKE THIS...
IT'S SO SIMPLE!
A FLASHLIGHT CONSISTS OF ELECTRICAL COMPONENTS CALLED BATTERIES, A MINIATURE LIGHT BULB, AND A SWITCH.
SWITCH
LIGHT BULB
BATTERIES
OKAY...
A *BATTERY* IS A PRODUCT THAT HAS VOLTAGE (THAT IS, A POTENTIAL DIFFERENCE).
WATER PUMP = BATTERY
IF WE THINK ABOUT IT IN TERMS OF WATER, A BATTERY CORRESPONDS TO A WATER PUMP THAT DRAWS WATER UP OUT OF A WELL.

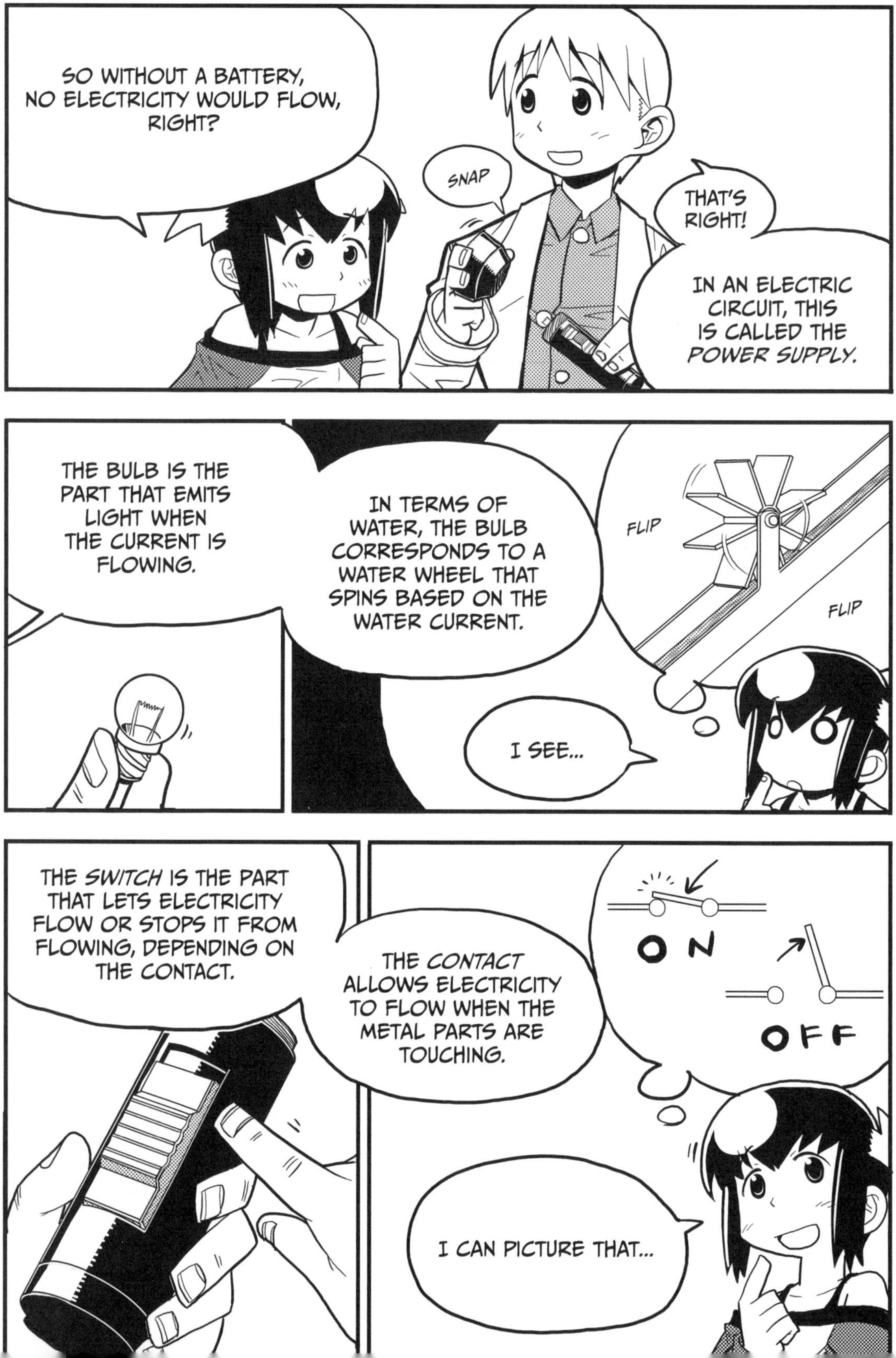
SO WITHOUT A BATTERY, NO ELECTRICITY WOULD FLOW, RIGHT?
SNAP
THAT'S RIGHT!
IN AN ELECTRIC CIRCUIT, THIS IS CALLED THE *POWER SUPPLY.*
THE BULB IS THE PART THAT EMITS LIGHT WHEN THE CURRENT IS FLOWING.
IN TERMS OF WATER, THE BULB CORRESPONDS TO A WATER WHEEL THAT SPINS BASED ON THE WATER CURRENT.
FLIP
FLIP
I SEE...
THE *SWITCH* IS THE PART THAT LETS ELECTRICITY FLOW OR STOPS IT FROM FLOWING, DEPENDING ON THE CONTACT.
THE *CONTACT* ALLOWS ELECTRICITY TO FLOW WHEN THE METAL PARTS ARE TOUCHING.
ON
OFF
I CAN PICTURE THAT...

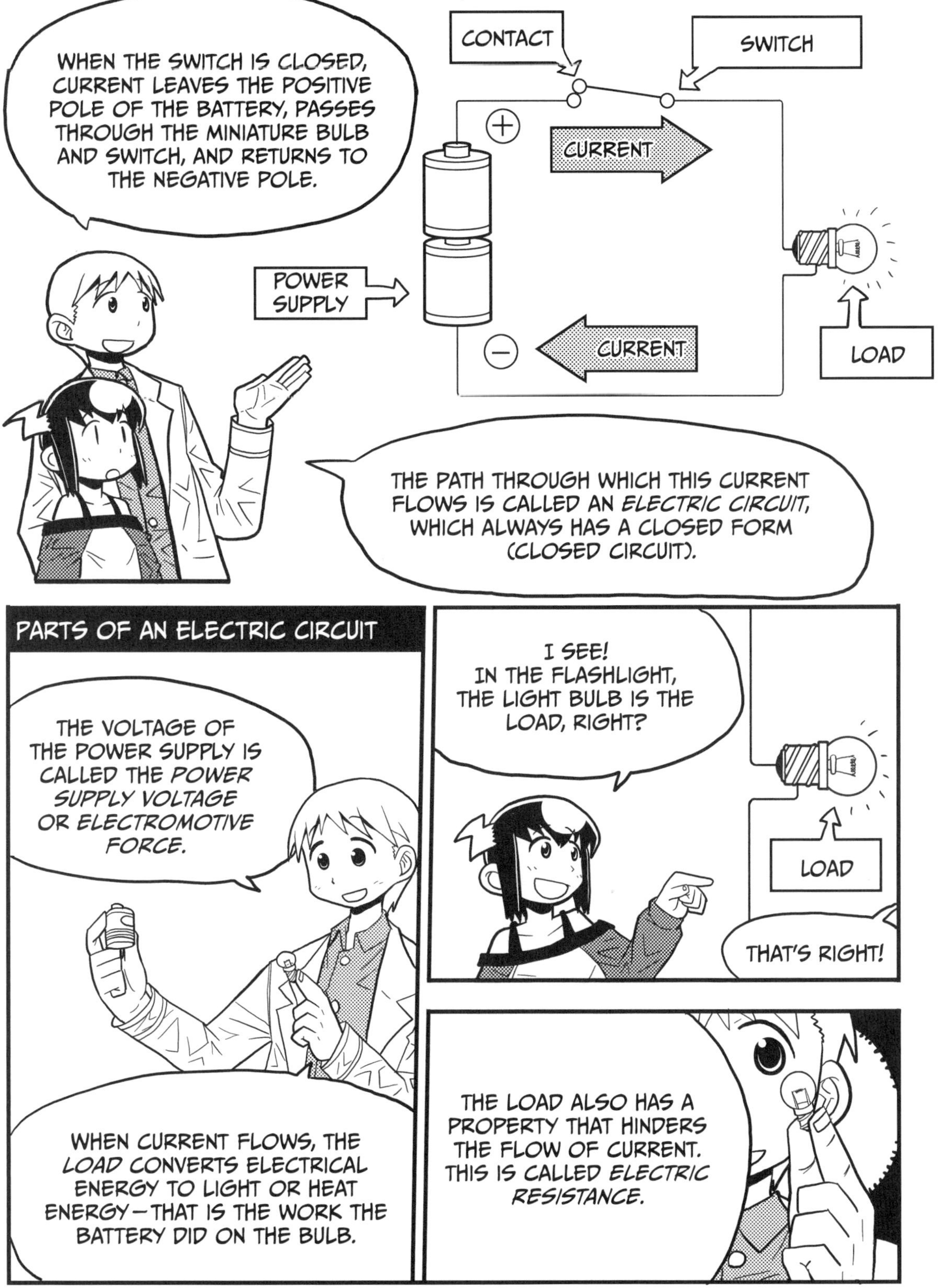
WHEN THE SWITCH IS CLOSED, CURRENT LEAVES THE POSITIVE POLE OF THE BATTERY, PASSES THROUGH THE MINIATURE BULB AND SWITCH, AND RETURNS TO THE NEGATIVE POLE.
CONTACT
SWITCH
CURRENT
POWER SUPPLY
CURRENT
LOAD
THE PATH THROUGH WHICH THIS CURRENT FLOWS IS CALLED AN *ELECTRIC CIRCUIT*, WHICH ALWAYS HAS A CLOSED FORM (CLOSED CIRCUIT).
PARTS OF AN ELECTRIC CIRCUIT
THE VOLTAGE OF THE POWER SUPPLY IS CALLED THE *POWER SUPPLY VOLTAGE* OR *ELECTROMOTIVE FORCE*.
WHEN CURRENT FLOWS, THE *LOAD* CONVERTS ELECTRICAL ENERGY TO LIGHT OR HEAT ENERGY—THAT IS THE WORK THE BATTERY DID ON THE BULB.
I SEE! IN THE FLASHLIGHT, THE LIGHT BULB IS THE LOAD, RIGHT?
LOAD
THAT'S RIGHT!
THE LOAD ALSO HAS A PROPERTY THAT HINDERS THE FLOW OF CURRENT. THIS IS CALLED *ELECTRIC RESISTANCE*.

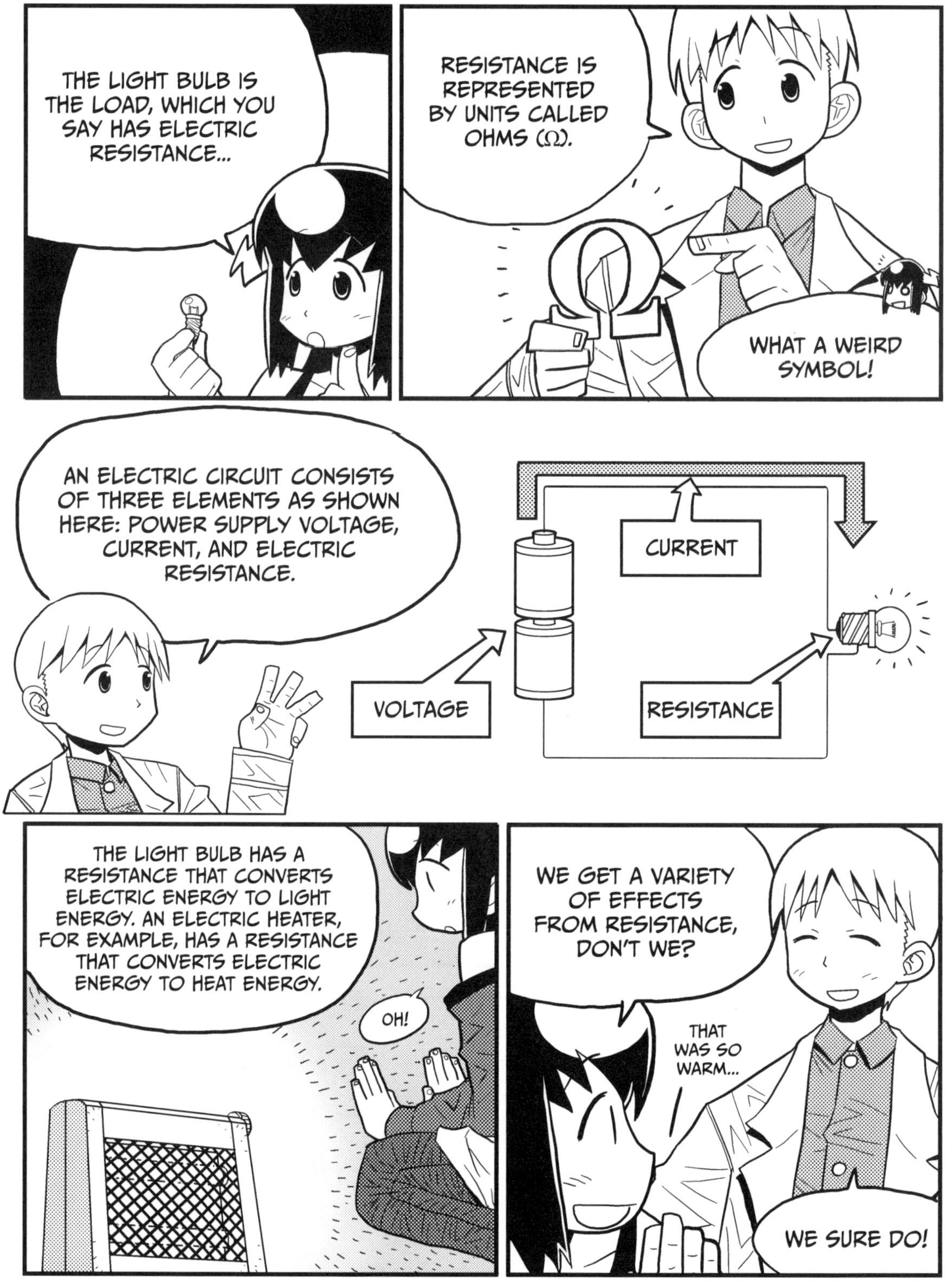
THE LIGHT BULB IS THE LOAD, WHICH YOU SAY HAS ELECTRIC RESISTANCE...
RESISTANCE IS REPRESENTED BY UNITS CALLED OHMS (Ω).
WHAT A WEIRD SYMBOL!
AN ELECTRIC CIRCUIT CONSISTS OF THREE ELEMENTS AS SHOWN HERE: POWER SUPPLY VOLTAGE, CURRENT, AND ELECTRIC RESISTANCE.
CURRENT
VOLTAGE
RESISTANCE
THE LIGHT BULB HAS A RESISTANCE THAT CONVERTS ELECTRIC ENERGY TO LIGHT ENERGY. AN ELECTRIC HEATER, FOR EXAMPLE, HAS A RESISTANCE THAT CONVERTS ELECTRIC ENERGY TO HEAT ENERGY.
OH!
WE GET A VARIETY OF EFFECTS FROM RESISTANCE, DON'T WE?
THAT WAS SO WARM...
WE SURE DO!

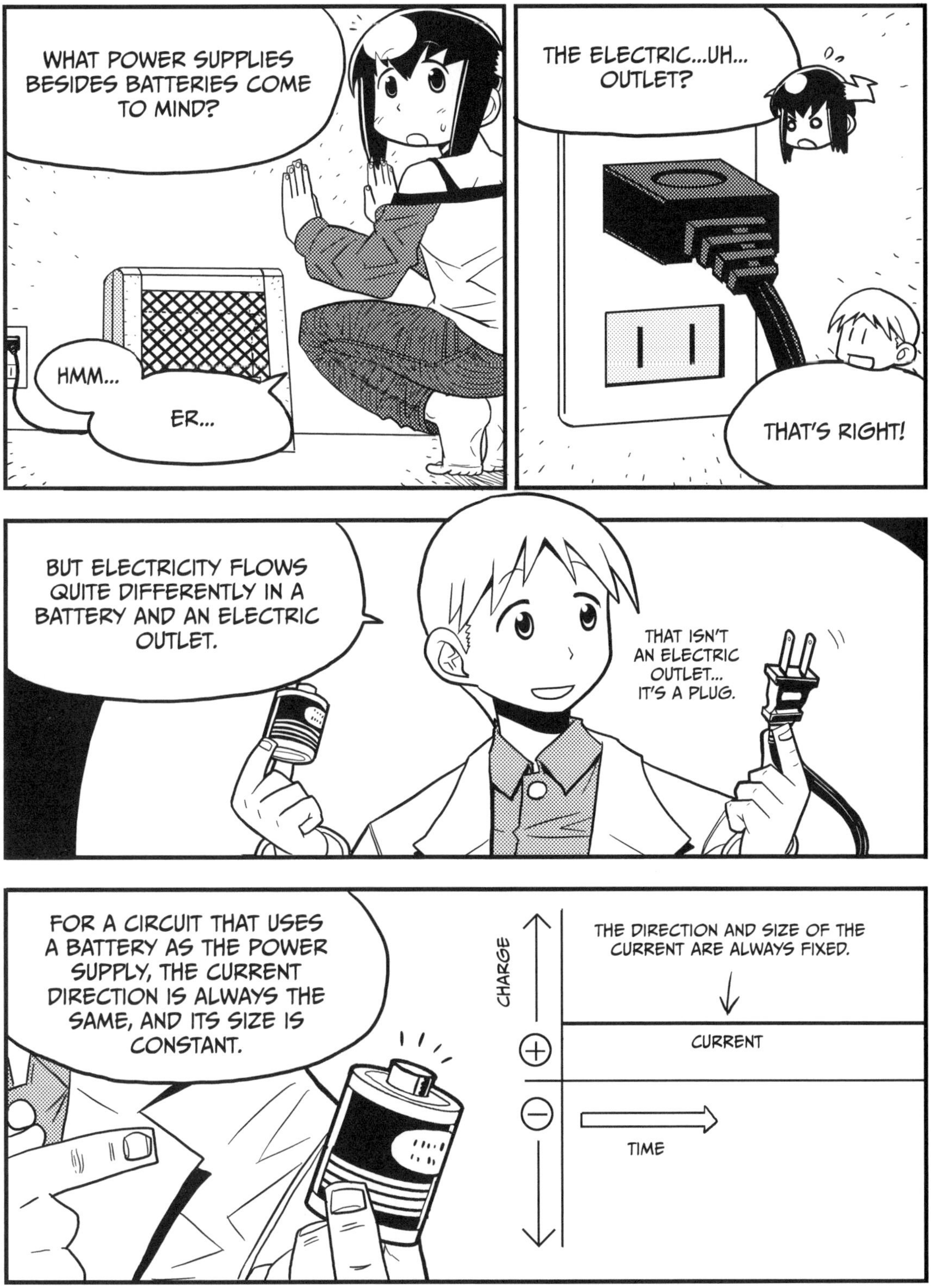

WHAT POWER SUPPLIES BESIDES BATTERIES COME TO MIND?
HMM...
ER...
THE ELECTRIC...UH... OUTLET?
THAT'S RIGHT!
BUT ELECTRICITY FLOWS QUITE DIFFERENTLY IN A BATTERY AND AN ELECTRIC OUTLET.
THAT ISN'T AN ELECTRIC OUTLET... IT'S A PLUG.
FOR A CIRCUIT THAT USES A BATTERY AS THE POWER SUPPLY, THE CURRENT DIRECTION IS ALWAYS THE SAME, AND ITS SIZE IS CONSTANT.
CHARGE
THE DIRECTION AND SIZE OF THE CURRENT ARE ALWAYS FIXED.
CURRENT
TIME

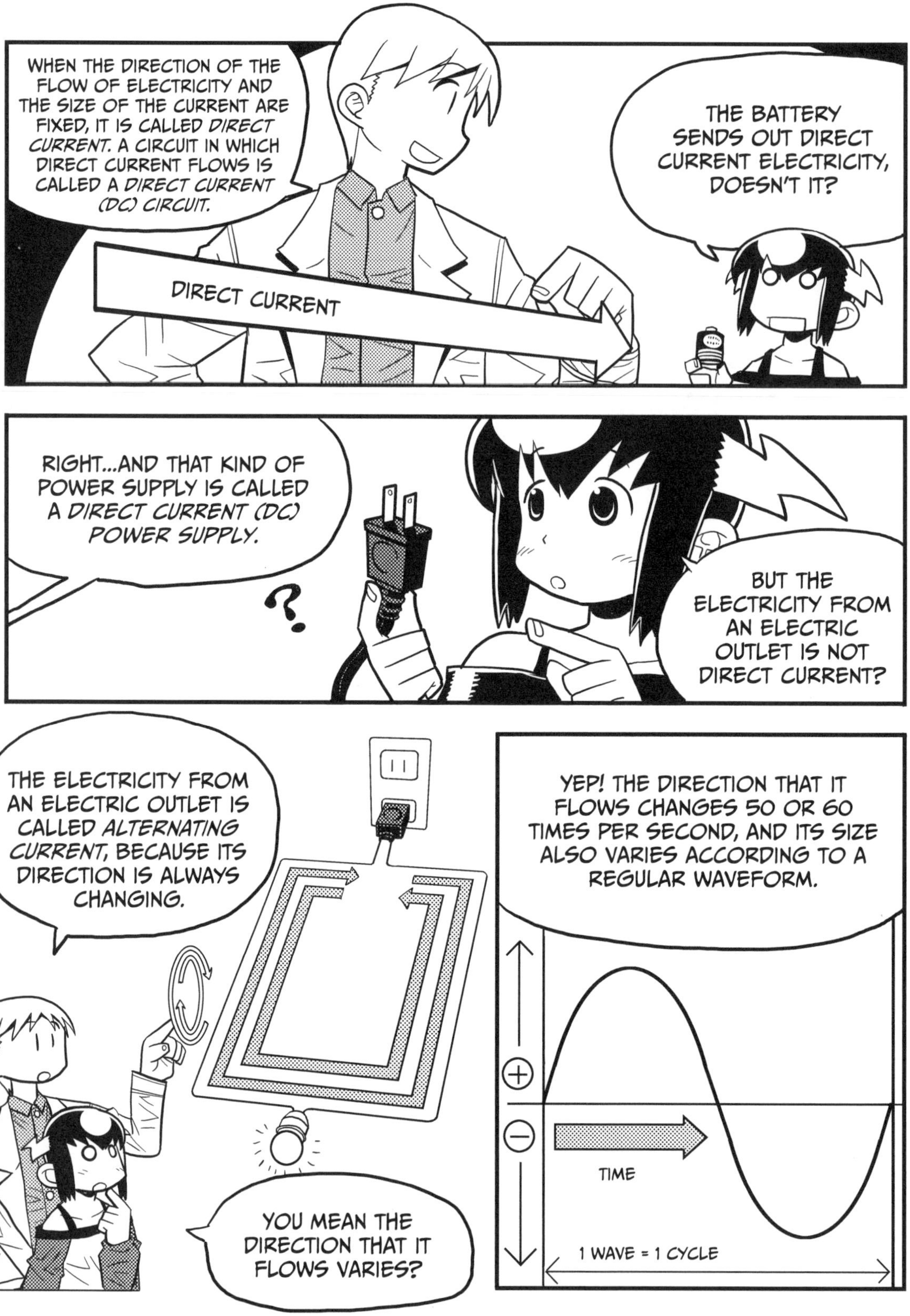
WHEN THE DIRECTION OF THE FLOW OF ELECTRICITY AND THE SIZE OF THE CURRENT ARE FIXED, IT IS CALLED DIRECT CURRENT. A CIRCUIT IN WHICH DIRECT CURRENT FLOWS IS CALLED A DIRECT CURRENT (DC) CIRCUIT.
DIRECT CURRENT
THE BATTERY SENDS OUT DIRECT CURRENT ELECTRICITY, DOESN'T IT?
RIGHT...AND THAT KIND OF POWER SUPPLY IS CALLED A DIRECT CURRENT (DC) POWER SUPPLY.
?
BUT THE ELECTRICITY FROM AN ELECTRIC OUTLET IS NOT DIRECT CURRENT?
THE ELECTRICITY FROM AN ELECTRIC OUTLET IS CALLED ALTERNATING CURRENT, BECAUSE ITS DIRECTION IS ALWAYS CHANGING.
YOU MEAN THE DIRECTION THAT IT FLOWS VARIES?
YEP! THE DIRECTION THAT IT FLOWS CHANGES 50 OR 60 TIMES PER SECOND, AND ITS SIZE ALSO VARIES ACCORDING TO A REGULAR WAVEFORM.
+
−
TIME
1 WAVE = 1 CYCLE

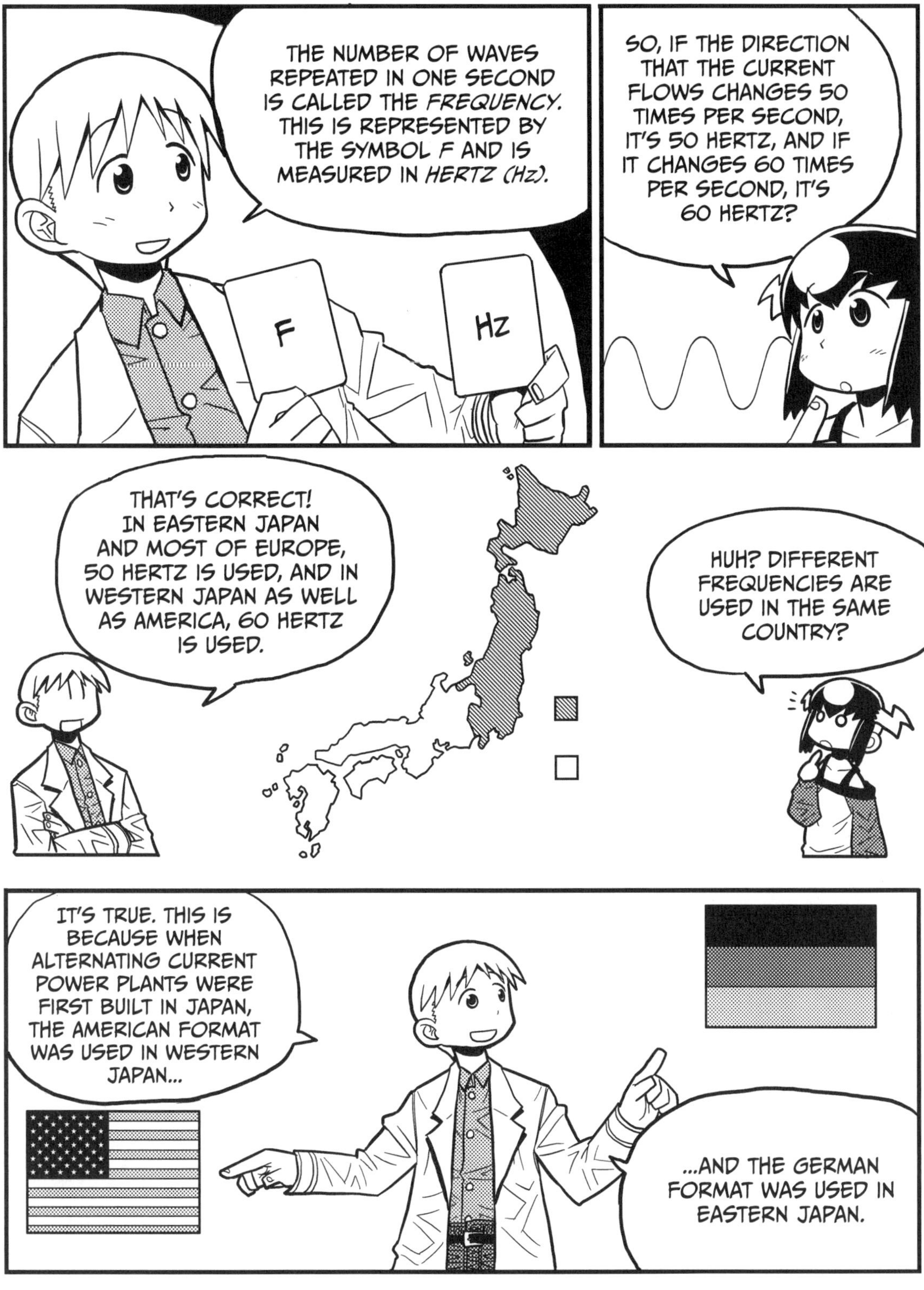
THE NUMBER OF WAVES REPEATED IN ONE SECOND IS CALLED THE *FREQUENCY*. THIS IS REPRESENTED BY THE SYMBOL *F* AND IS MEASURED IN *HERTZ (Hz)*.
F
Hz
SO, IF THE DIRECTION THAT THE CURRENT FLOWS CHANGES 50 TIMES PER SECOND, IT'S 50 HERTZ, AND IF IT CHANGES 60 TIMES PER SECOND, IT'S 60 HERTZ?
THAT'S CORRECT! IN EASTERN JAPAN AND MOST OF EUROPE, 50 HERTZ IS USED, AND IN WESTERN JAPAN AS WELL AS AMERICA, 60 HERTZ IS USED.
HUH? DIFFERENT FREQUENCIES ARE USED IN THE SAME COUNTRY?
IT'S TRUE. THIS IS BECAUSE WHEN ALTERNATING CURRENT POWER PLANTS WERE FIRST BUILT IN JAPAN, THE AMERICAN FORMAT WAS USED IN WESTERN JAPAN...
...AND THE GERMAN FORMAT WAS USED IN EASTERN JAPAN.

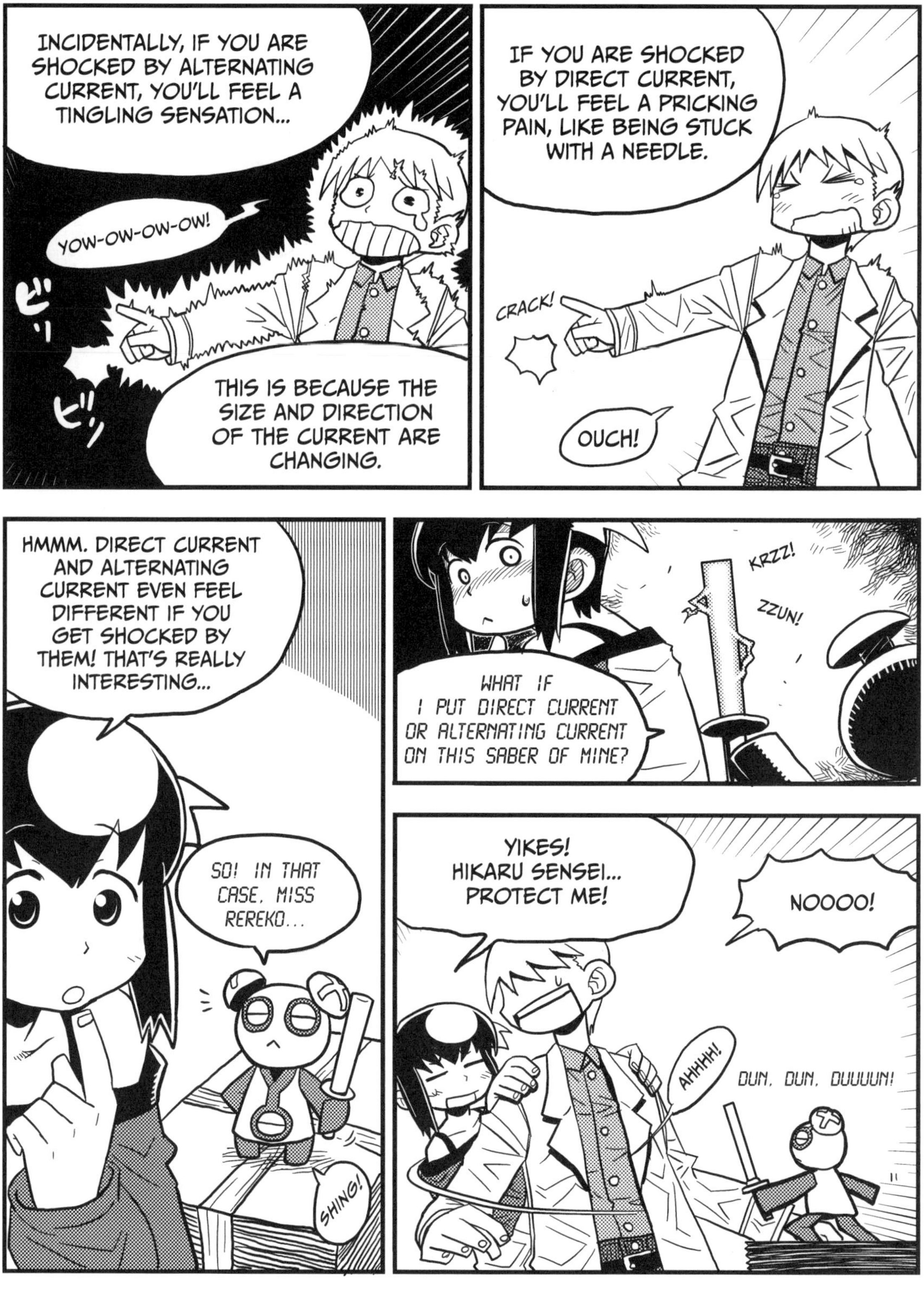
INCIDENTALLY, IF YOU ARE SHOCKED BY ALTERNATING CURRENT, YOU'LL FEEL A TINGLING SENSATION...
YOW-OW-OW-OW!
ビッ
ビッ
THIS IS BECAUSE THE SIZE AND DIRECTION OF THE CURRENT ARE CHANGING.
IF YOU ARE SHOCKED BY DIRECT CURRENT, YOU'LL FEEL A PRICKING PAIN, LIKE BEING STUCK WITH A NEEDLE.
CRACK!
OUCH!
HMMM. DIRECT CURRENT AND ALTERNATING CURRENT EVEN FEEL DIFFERENT IF YOU GET SHOCKED BY THEM! THAT'S REALLY INTERESTING...
SO! IN THAT CASE, MISS REREKO...
SHING!
KRZZ!
ZZUN!
WHAT IF I PUT DIRECT CURRENT OR ALTERNATING CURRENT ON THIS SABER OF MINE?
YIKES! HIKARU SENSEI... PROTECT ME!
NOOOO!
AHHHH!
DUN, DUN, DUUUUN!

OHM'S LAW AND METHODS OF CONNECTING ELECTRICAL COMPONENTS

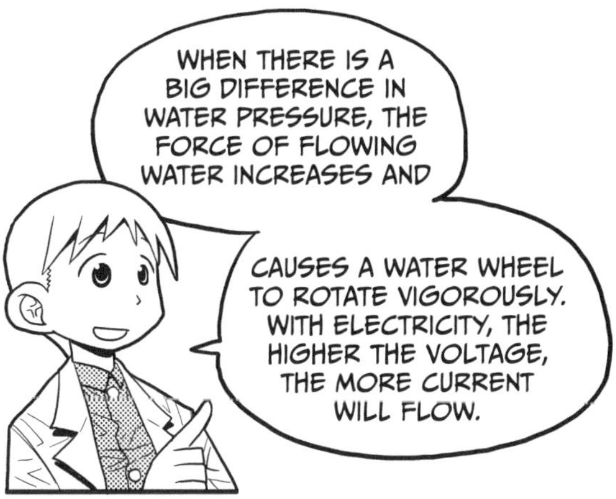

WHEN THERE IS A BIG DIFFERENCE IN WATER PRESSURE, THE FORCE OF FLOWING WATER INCREASES AND
CAUSES A WATER WHEEL TO ROTATE VIGOROUSLY. WITH ELECTRICITY, THE HIGHER THE VOLTAGE, THE MORE CURRENT WILL FLOW.

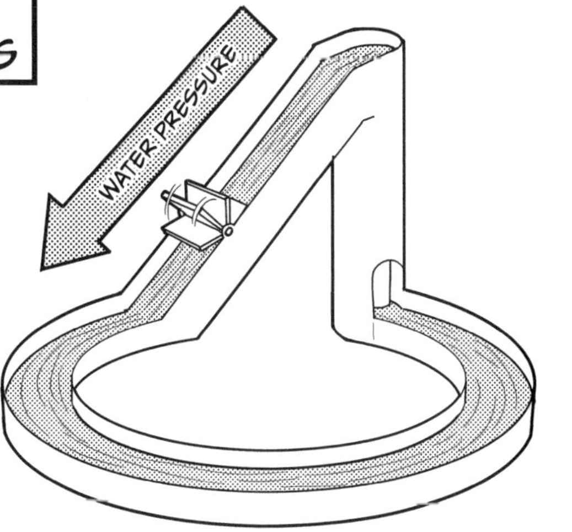

WATER PRESSURE

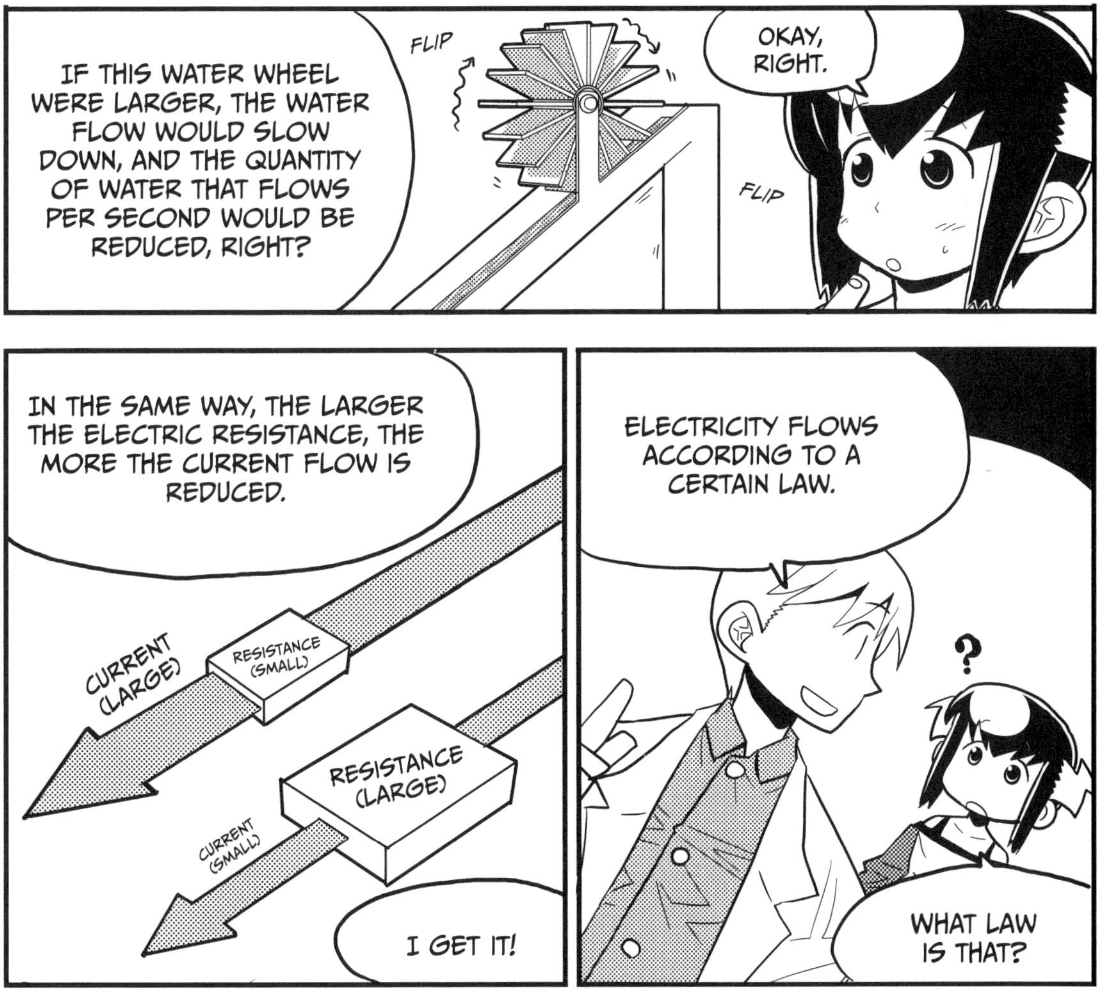

IF THIS WATER WHEEL WERE LARGER, THE WATER FLOW WOULD SLOW DOWN, AND THE QUANTITY OF WATER THAT FLOWS PER SECOND WOULD BE REDUCED, RIGHT?
FLIP
FLIP
OKAY, RIGHT.
IN THE SAME WAY, THE LARGER THE ELECTRIC RESISTANCE, THE MORE THE CURRENT FLOW IS REDUCED.
CURRENT (LARGE)
RESISTANCE (SMALL)
RESISTANCE (LARGE)
CURRENT (SMALL)
I GET IT!
ELECTRICITY FLOWS ACCORDING TO A CERTAIN LAW.
?
WHAT LAW IS THAT?

ELECTRIC CIRCUITS AND OHM'S LAW

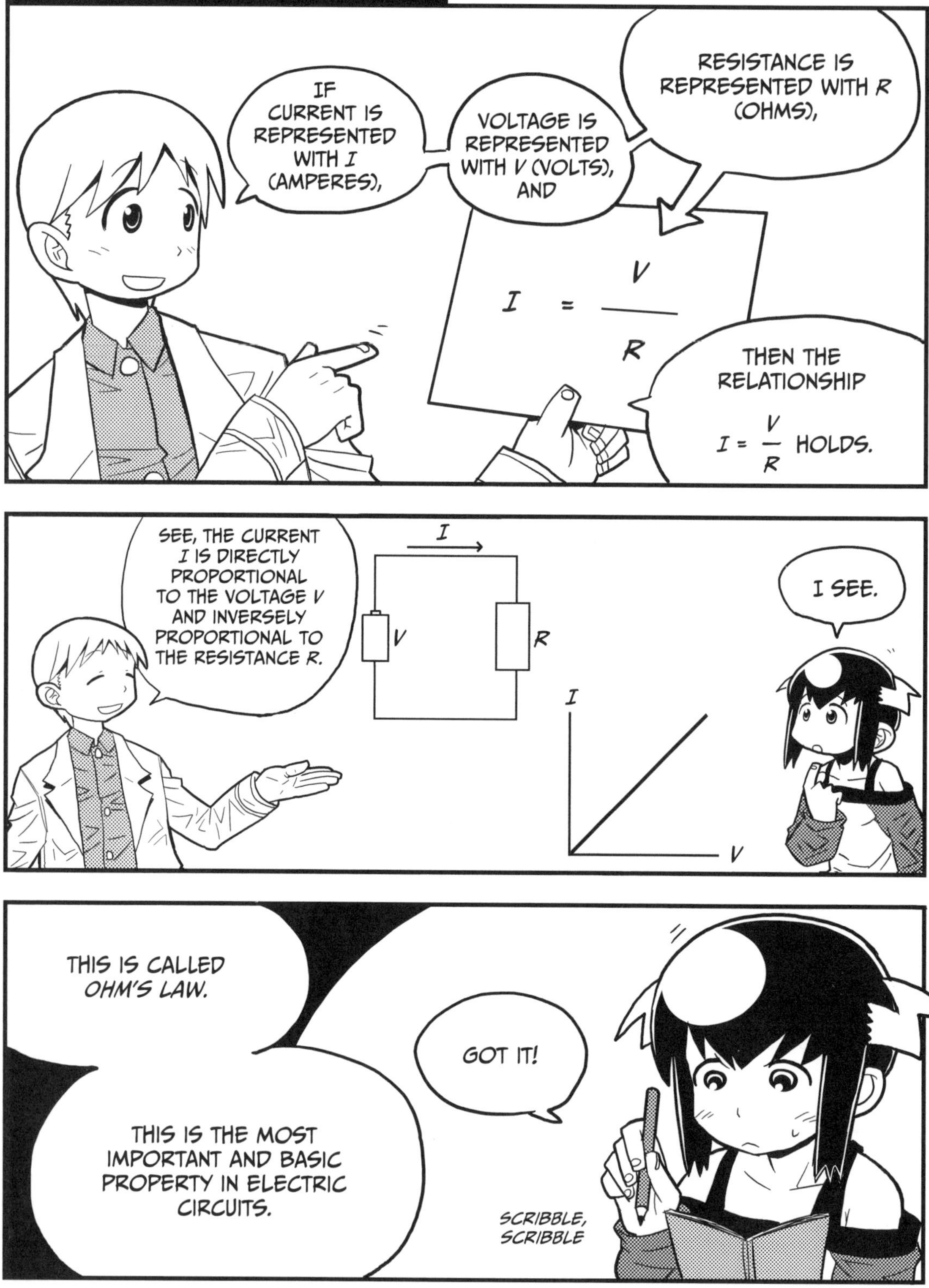
IF CURRENT IS REPRESENTED WITH I (AMPERES),
VOLTAGE IS REPRESENTED WITH V (VOLTS), AND
RESISTANCE IS REPRESENTED WITH R (OHMS),
I = V / R
THEN THE RELATIONSHIP I = V / R HOLDS.
SEE, THE CURRENT I IS DIRECTLY PROPORTIONAL TO THE VOLTAGE V AND INVERSELY PROPORTIONAL TO THE RESISTANCE R.
I
V
R
I
V
I SEE.
THIS IS CALLED OHM'S LAW.
THIS IS THE MOST IMPORTANT AND BASIC PROPERTY IN ELECTRIC CIRCUITS.
GOT IT!
SCRIBBLE, SCRIBBLE

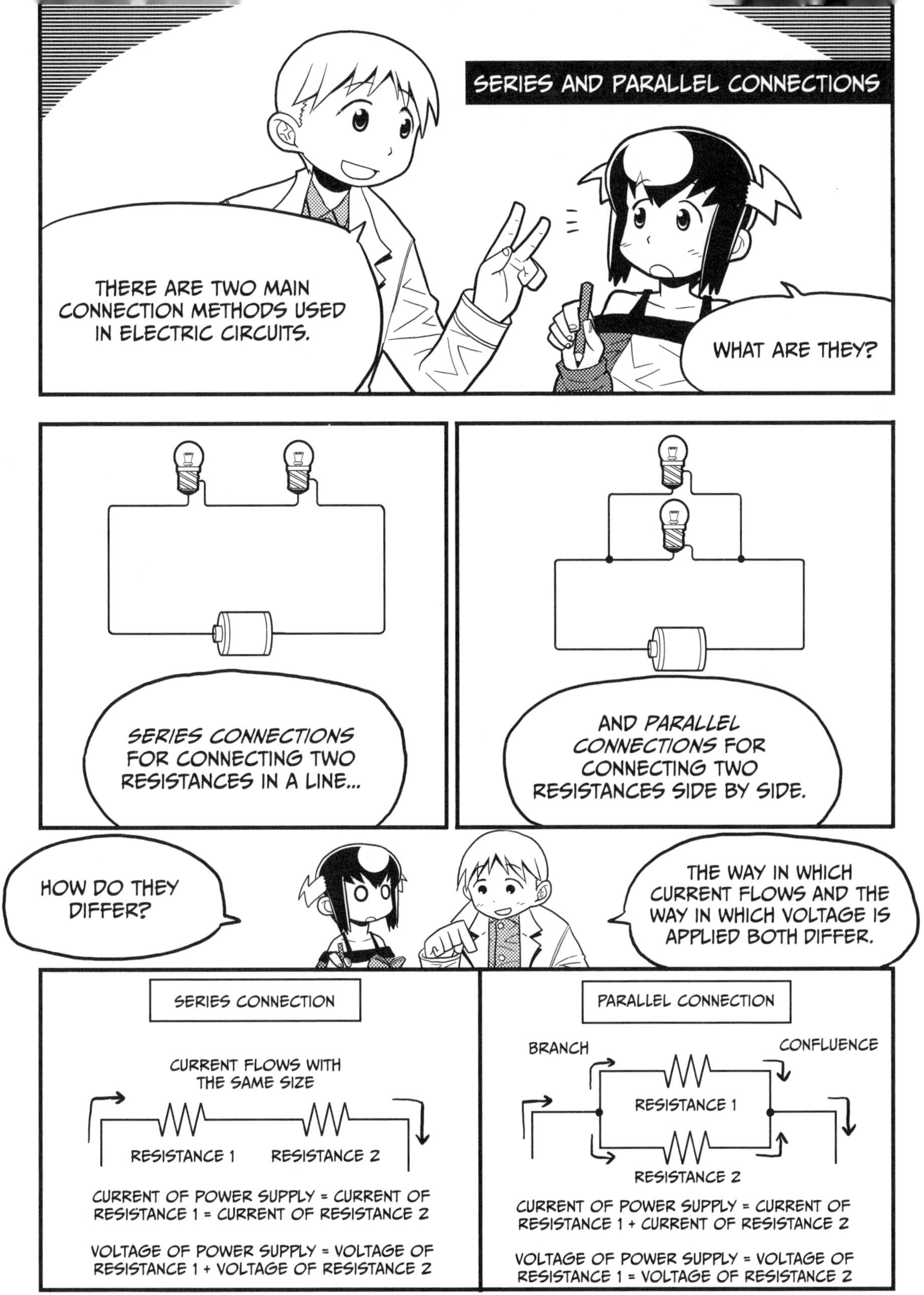
SERIES AND PARALLEL CONNECTIONS
THERE ARE TWO MAIN CONNECTION METHODS USED IN ELECTRIC CIRCUITS.
WHAT ARE THEY?
SERIES CONNECTIONS FOR CONNECTING TWO RESISTANCES IN A LINE...
AND PARALLEL CONNECTIONS FOR CONNECTING TWO RESISTANCES SIDE BY SIDE.
HOW DO THEY DIFFER?
THE WAY IN WHICH CURRENT FLOWS AND THE WAY IN WHICH VOLTAGE IS APPLIED BOTH DIFFER.
SERIES CONNECTION
CURRENT FLOWS WITH THE SAME SIZE
RESISTANCE 1
RESISTANCE 2
CURRENT OF POWER SUPPLY = CURRENT OF RESISTANCE 1 = CURRENT OF RESISTANCE 2
VOLTAGE OF POWER SUPPLY = VOLTAGE OF RESISTANCE 1 + VOLTAGE OF RESISTANCE 2
PARALLEL CONNECTION
BRANCH
CONFLUENCE
RESISTANCE 1
RESISTANCE 2
CURRENT OF POWER SUPPLY = CURRENT OF RESISTANCE 1 + CURRENT OF RESISTANCE 2
VOLTAGE OF POWER SUPPLY = VOLTAGE OF RESISTANCE 1 = VOLTAGE OF RESISTANCE 2

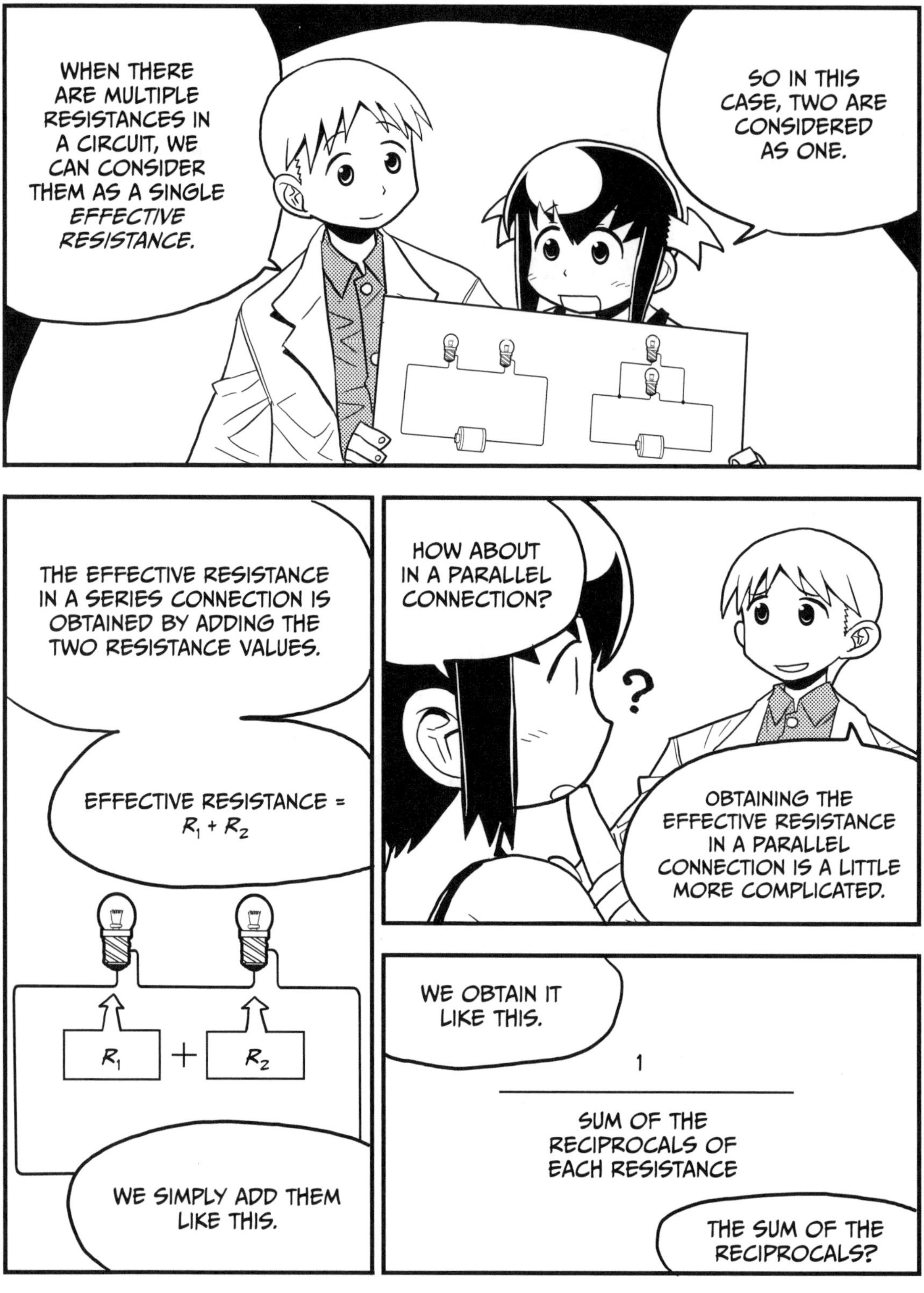
WHEN THERE ARE MULTIPLE RESISTANCES IN A CIRCUIT, WE CAN CONSIDER THEM AS A SINGLE *EFFECTIVE RESISTANCE*.
SO IN THIS CASE, TWO ARE CONSIDERED AS ONE.
THE EFFECTIVE RESISTANCE IN A SERIES CONNECTION IS OBTAINED BY ADDING THE TWO RESISTANCE VALUES.
EFFECTIVE RESISTANCE = $R_1 + R_2$
R_1
+
R_2
WE SIMPLY ADD THEM LIKE THIS.
HOW ABOUT IN A PARALLEL CONNECTION?
?
OBTAINING THE EFFECTIVE RESISTANCE IN A PARALLEL CONNECTION IS A LITTLE MORE COMPLICATED.
WE OBTAIN IT LIKE THIS.
1
SUM OF THE RECIPROCALS OF EACH RESISTANCE
THE SUM OF THE RECIPROCALS?

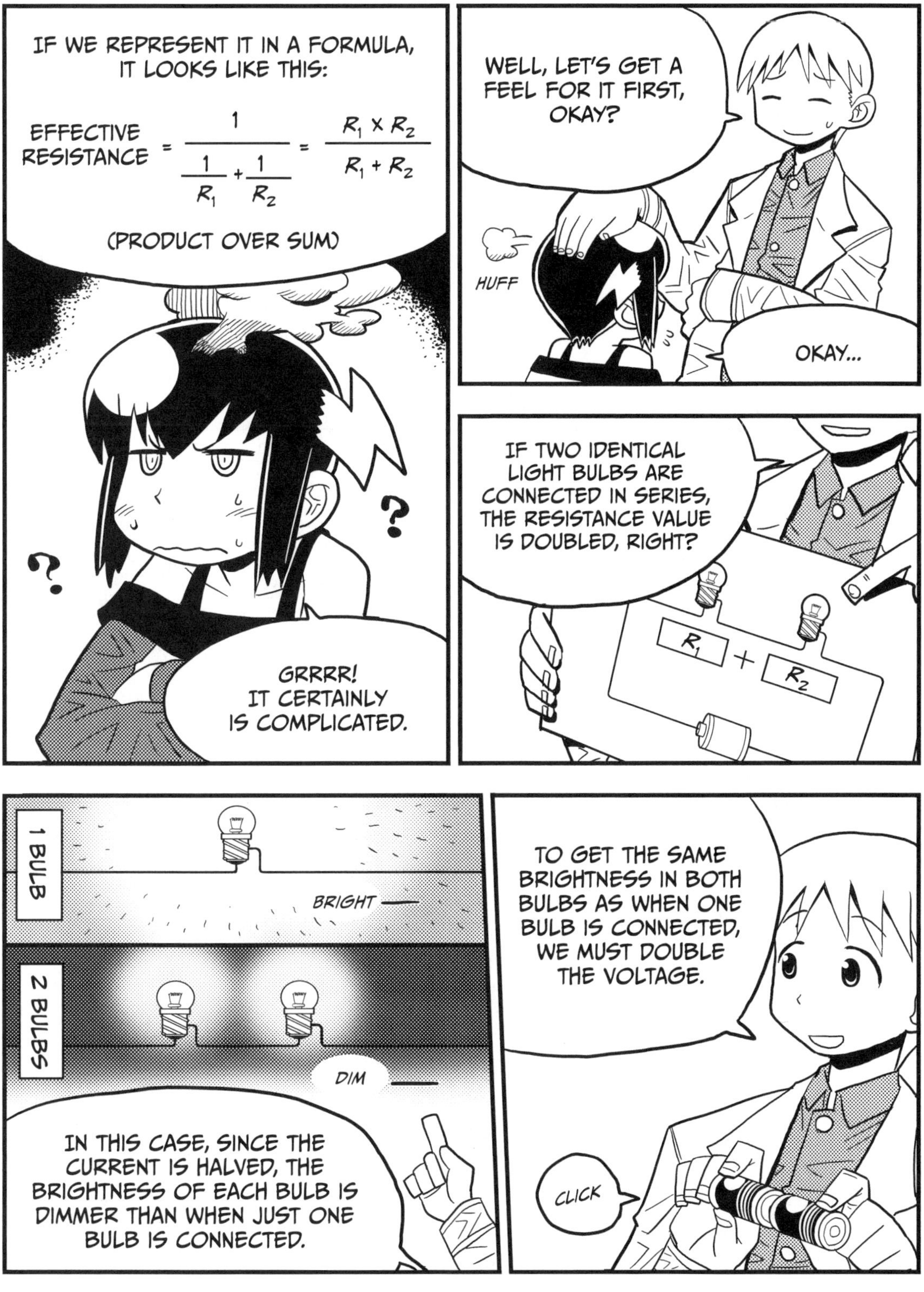
IF WE REPRESENT IT IN A FORMULA, IT LOOKS LIKE THIS:
$$\text{EFFECTIVE RESISTANCE} = \frac{1}{\frac{1}{R_1} + \frac{1}{R_2}} = \frac{R_1 \times R_2}{R_1 + R_2}$$
(PRODUCT OVER SUM)
GRRRR! IT CERTAINLY IS COMPLICATED.
WELL, LET'S GET A FEEL FOR IT FIRST, OKAY?
HUFF
OKAY...
IF TWO IDENTICAL LIGHT BULBS ARE CONNECTED IN SERIES, THE RESISTANCE VALUE IS DOUBLED, RIGHT?
R_1 + R_2
1 BULB
BRIGHT
2 BULBS
DIM
IN THIS CASE, SINCE THE CURRENT IS HALVED, THE BRIGHTNESS OF EACH BULB IS DIMMER THAN WHEN JUST ONE BULB IS CONNECTED.
TO GET THE SAME BRIGHTNESS IN BOTH BULBS AS WHEN ONE BULB IS CONNECTED, WE MUST DOUBLE THE VOLTAGE.
CLICK

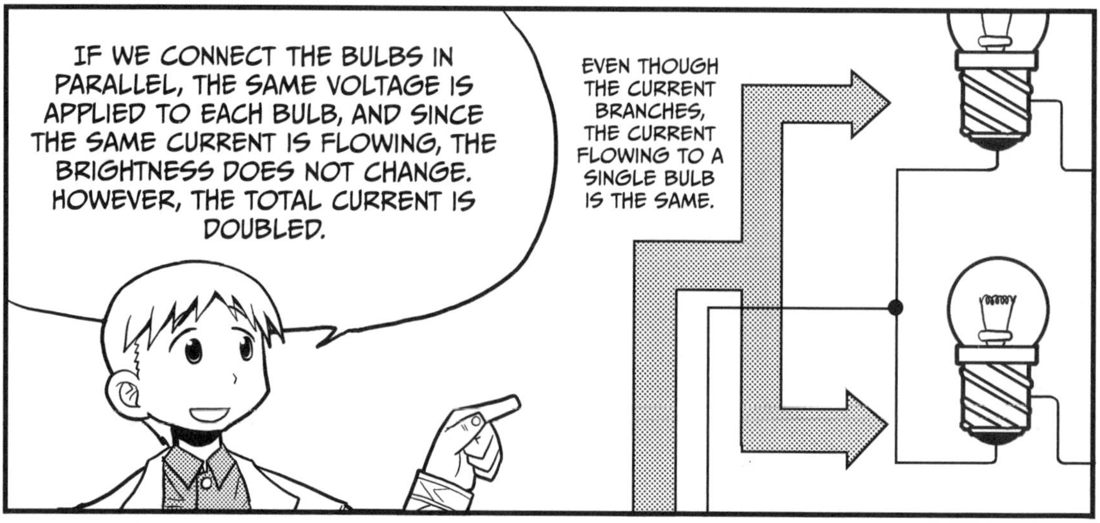
IF WE CONNECT THE BULBS IN PARALLEL, THE SAME VOLTAGE IS APPLIED TO EACH BULB, AND SINCE THE SAME CURRENT IS FLOWING, THE BRIGHTNESS DOES NOT CHANGE. HOWEVER, THE TOTAL CURRENT IS DOUBLED.
EVEN THOUGH THE CURRENT BRANCHES, THE CURRENT FLOWING TO A SINGLE BULB IS THE SAME.

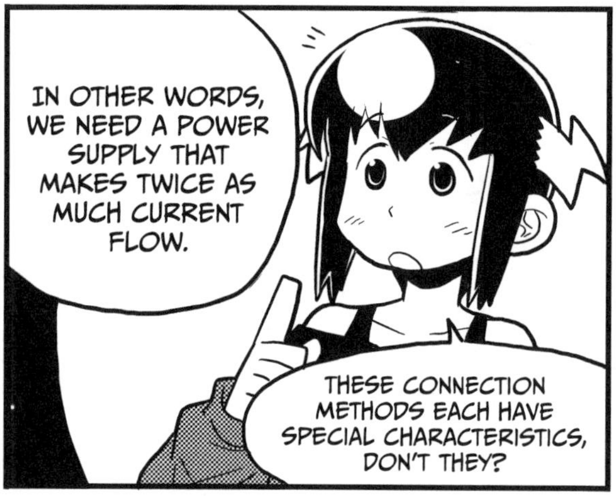
IN OTHER WORDS, WE NEED A POWER SUPPLY THAT MAKES TWICE AS MUCH CURRENT FLOW.
THESE CONNECTION METHODS EACH HAVE SPECIAL CHARACTERISTICS, DON'T THEY?

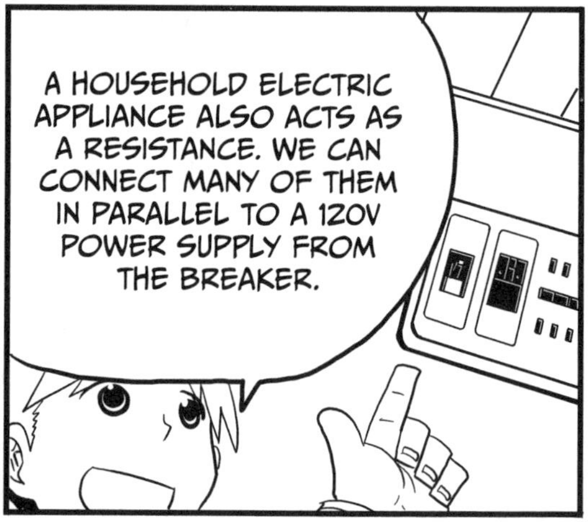
A HOUSEHOLD ELECTRIC APPLIANCE ALSO ACTS AS A RESISTANCE. WE CAN CONNECT MANY OF THEM IN PARALLEL TO A 120V POWER SUPPLY FROM THE BREAKER.

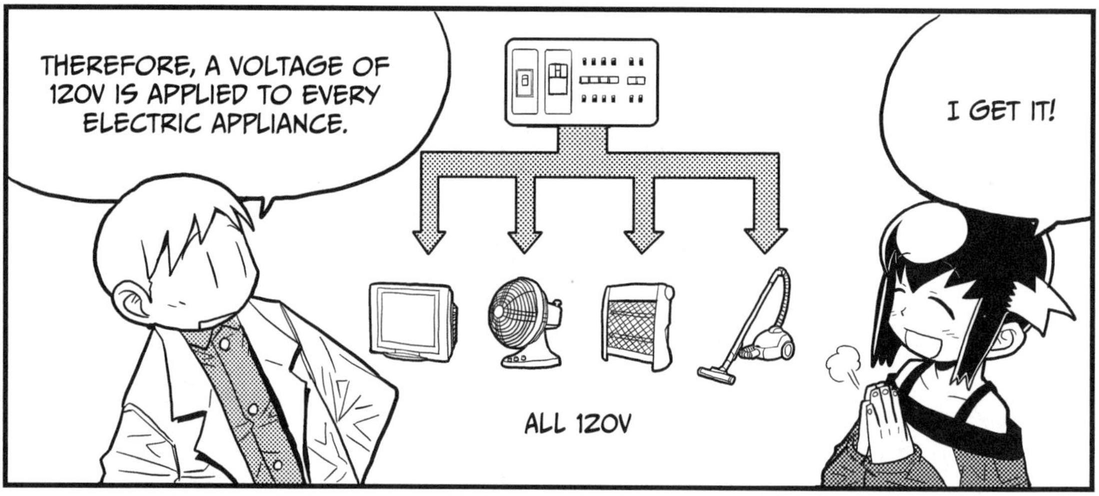
THEREFORE, A VOLTAGE OF 120V IS APPLIED TO EVERY ELECTRIC APPLIANCE.
I GET IT!
ALL 120V

ELECTRIC CIRCUITS AND CURRENT

The electrical parts that make up a flashlight are the batteries, a miniature light bulb, and a switch. A battery has the ability to make electrical current flow, so it is called the *power supply*. The light bulb is a part that emits light when current flows through it. The switch is a part that lets electricity flow or stops it according to the opening and closing of a contact.

When the switch is closed, current leaves the positive pole of the battery, passes through the light bulb and switch, and returns to the negative pole. The path through which current flows in this way is called an *electric circuit*, which always has a closed form (a *closed circuit*).

ELECTRIC PARTS IN A FLASHLIGHT
SWITCH
LIGHT BULB
BATTERIES
CIRCUIT DRAWING
SWITCH
CURRENT
CONTACT
LIGHT BULB
NEGATIVE POLE
POSITIVE POLE
BATTERIES

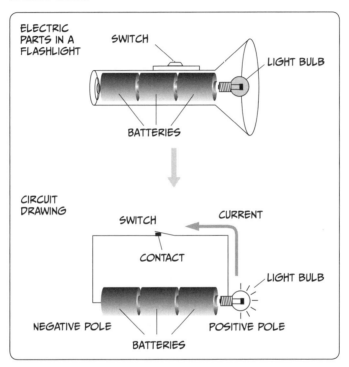

Electric circuit of a flashlight

GRAPHICAL SYMBOLS

A basic electric circuit consists of three elements: power supply voltage, current, and electric resistance. These elements are connected by electric wires.

The power supply voltage that does the work of making current flow is called the *electromotive force*. The element that converts electric energy to light or heat when current flows is called the *load* (loads can also convert electrical energy into other things like sound or motion). The load has a property that hinders the flow of current, and this is called *electric resistance* or simply *resistance*. Resistance is represented by the symbol *R* and measured in *ohms* (Ω), which come from the name of the German physicist Georg Simon Ohm.

Creating a realistic drawing of an electric circuit takes time and effort. Therefore, *graphical symbols* are generally used to draw a representation. Using standard graphical symbols enables anyone to easily understand a circuit diagram that was drawn by someone else.

CURRENT
I
POWER SUPPLY
VOLTAGE
V
+
-
RESISTANCE
R
ELECTRIC
WIRE
GRAPHICAL SYMBOLS
DIRECT CURRENT
POWER SUPPLY
RESISTANCE
ELECTRIC
WIRE
SWITCH

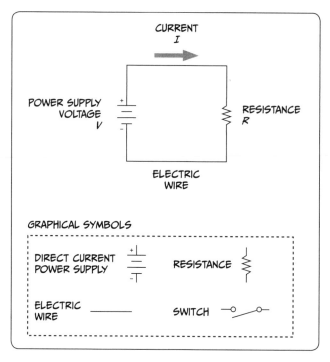

Electric circuit and graphical symbols

Appliances that use electric resistance include electric heaters and toasters. The electric heating element used in these appliances is the part that converts electrical energy to heat energy when current flows through the electric resistance. Note that the electric wire used in these appliances also has electric resistance; although it is only a small amount of resistance, when current flows through the electric wire, heat is generated.

DIRECT CURRENT CIRCUIT AND ALTERNATING CURRENT CIRCUIT

The direction the current flows in a circuit that has a battery as the power supply is fixed, and the size of the current is also constant. When the direction of the flow of current and the size of the current are fixed, we call the electricity *direct current (DC).* A circuit in which direct current flows is called a *direct current (DC) circuit.* A power supply that sends out direct current, such as a battery, is called a *DC power supply.* A size D or size AA battery has a DC 1.5V power supply voltage.

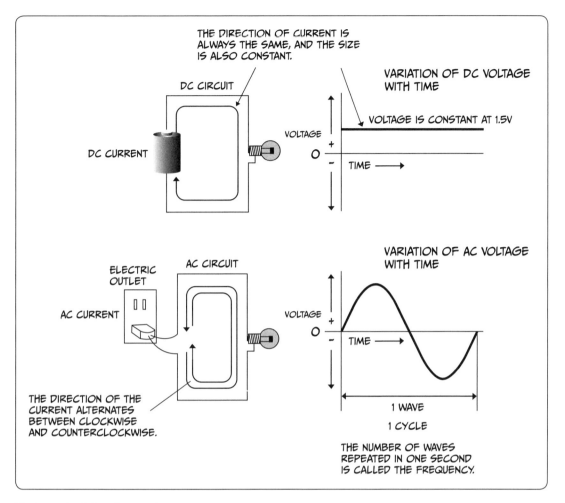

THE DIRECTION OF CURRENT IS ALWAYS THE SAME, AND THE SIZE IS ALSO CONSTANT.

DC CIRCUIT

DC CURRENT

VARIATION OF DC VOLTAGE WITH TIME

VOLTAGE IS CONSTANT AT 1.5V

VOLTAGE

O

TIME

ELECTRIC OUTLET

AC CIRCUIT

AC CURRENT

THE DIRECTION OF THE CURRENT ALTERNATES BETWEEN CLOCKWISE AND COUNTERCLOCKWISE.

VARIATION OF AC VOLTAGE WITH TIME

VOLTAGE

O

TIME

1 WAVE

1 CYCLE

THE NUMBER OF WAVES REPEATED IN ONE SECOND IS CALLED THE FREQUENCY.

Direct current and alternating current

On the other hand, the direction of flow and size of the current sent from the electric power company to a home changes cyclically. This kind of electricity is called *alternating current (AC)*, and a circuit in which alternating current flows is called an *alternating current (AC) circuit*. The direction that this electricity flows changes 50 or 60 times per second, and its size also varies cyclically with time. The number of waves repeated in one second is called the *frequency*, which is represented by *f* and measured in *hertz (Hz)*.

The size of AC voltage at any given time is called the *instantaneous voltage*, and the largest value among the instantaneous voltages is called the *peak voltage*. The size of the AC voltage that will perform the same amount of work as a DC voltage is called the *effective voltage*. The AC voltage that comes to an electric outlet in a home is generally 120V in the United States, but this is the effective voltage. The peak voltage is approximately 170V.

PEAK VOLTAGE: 170V
EFFECTIVE VOLTAGE: 120V
VOLTAGE
INSTANTANEOUS VOLTAGE
0
+
-
TIME

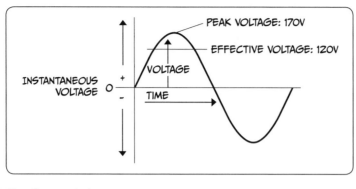

Alternating current value

OHM'S LAW

The current that flows in a circuit is directly proportional to the voltage and indirectly proportional to the resistance. This relationship is called *Ohm's law* and can be expressed in a formula as $I = V / R$. This is the most important and basic property in electric circuits.

For example, if a voltage of 120V is applied to a resistance of 120Ω, the current will be $I = V / R = 120 / 120$, and 1A of current will flow. Whenever you know two values among the current, voltage, and resistance in a circuit, you can use Ohm's law to calculate the value you don't know.

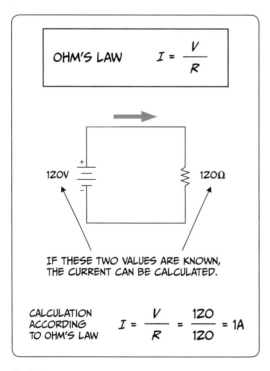

OHM'S LAW $I = \frac{V}{R}$
120V
120Ω
IF THESE TWO VALUES ARE KNOWN, THE CURRENT CAN BE CALCULATED.
CALCULATION ACCORDING TO OHM'S LAW $I = \frac{V}{R} = \frac{120}{120} = 1A$

Ohm's law

RESISTIVITY AND CONDUCTIVITY

Electric wire has a very low resistance and is used to connect circuit elements. When a small amount of current flows through wire, we can consider its resistance negligible. If a larger amount of current flows through a wire than can do so safely, heat will be generated.

Resistance (R) is a measure indicating the difficulty of the flow of current. The resistance (measured in ohms) of a conductor with length L meters and cross-sectional area A square meters can be represented by $R = \rho \times L / A$.

Resistivity measures how much a material opposes the flow of current and can be used to determine a wire's resistance. Resistivity, represented by the symbol ρ, is a material-specific resistance value and is measured in *ohm meters* (Ω *m*). From this equation it is apparent that for the same material, the size of the resistance is directly proportional to the length and inversely proportional to the cross-sectional area.

RESISTIVITY (IN ΩM) OF VARIOUS METALS AT ROOM TEMPERATURE (68°F)

Gold	2.22×10^{-8}
Silver	1.59×10^{-8}
Copper	1.69×10^{-8}
Aluminum	2.27×10^{-8}
Nichrome	107.5×10^{-8}

Conductance (G), in contrast to resistance, is a measure indicating the ease of the flow of current and is measured in *siemens (S)*. Conductivity, represented by the symbol σ, is the reciprocal of resistivity, and it is measured in *siemens per meter (S/m)*. (The siemens, named for German inventor Ernst Werner von Siemens, is an inverse ohm; it is also sometimes called a *mho* and can be represented by the symbol ℧ or Ω^{-1}).

CROSS-SECTIONAL AREA
A
LENGTH
L
CONDUCTOR
RESISTANCE OF THIS INTERVAL
$R = \rho \times \frac{L}{A}$
ρ IS RESISTIVITY (Ω M)
CONDUCTIVITY IS $\sigma = \frac{1}{\rho}$

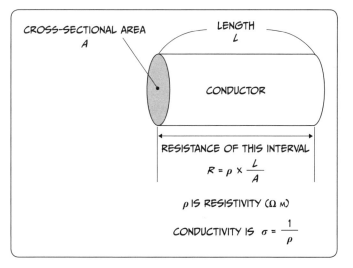

Resistivity and conductivity

EFFECTIVE RESISTANCE

There are two basic methods of connecting electrical components. Let's look at them both with respect to resistance. When there are multiple resistances in a circuit, we can consider them as a single *effective resistance*.

The method of connecting resistances in a line is called a *series connection*. We calculate the value of the effective resistance in a series connection by totaling the individual resistance values.

Effective resistance = $R_0 = R_1 + R_2 + \ldots + R_n$

In this connection, the size of the current that flows in each resistance is the same. The power supply voltage is voltage divided by each resistance.

EFFECTIVE RESISTANCE
$R_0 = R_1 + R_2 + R_3 + \ldots + R_N$
SERIES CONNECTION
R_1
R_2
R_3
R_N
V_1
V_2
V_3
V_N
POWER SUPPLY VOLTAGE (V)
CALCULATE POWER SUPPLY VOLTAGE BY DIVIDING VOLTAGE BY THE SUM OF THE RESISTANCES.

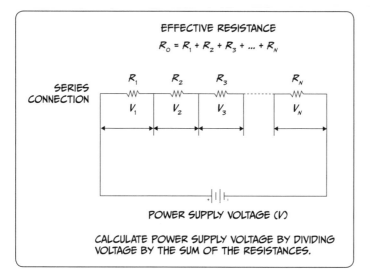

Series connection and effective resistance

If two light bulbs of the same size are connected in series to a power supply, the current will be halved, and the brightness of each bulb will be dimmer than it was when just a single bulb was connected, because the effective resistance is doubled. At this time, the voltage at both sides of each light bulb will be half the value of the power supply voltage.

ONE LIGHT BULB
CURRENT
TWO BULBS CONNECTED IN SERIES
CURRENT
BULBS NOW SHINE MORE DIMLY, BECAUSE CURRENT TO THE BULB IS HALF WHAT IT WAS.

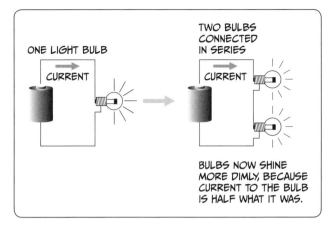

Series connection of light bulbs

The other basic method of connecting resistances is called a *parallel connection*. At this time, the value of the effective resistance can be obtained by calculating the reciprocal of the sum of the reciprocals of each resistance.

$$\text{Effective resistance} = R_0 = \frac{1}{\frac{1}{R_1} + \frac{1}{R_2} + \dots + \frac{1}{R_n}}$$

The total resistance when two resistances are connected in parallel can be obtained as follows.

$$\text{Effective resistance} = R_0 = \frac{R_1 \times R_2}{R_1 + R_2} \qquad \text{(Product over sum)}$$

In a parallel circuit, the voltage applied to each resistance is the same, because the current branches and flows to each resistance.

EFFECTIVE RESISTANCE
$R_0 = \frac{1}{\frac{1}{R_1} + \frac{1}{R_2} + \frac{1}{R_3} + \dots + \frac{1}{R_N}}$
VOLTAGE APPLIED TO EACH RESISTANCE IS THE SAME.
CURRENT BRANCHES TO EACH RESISTANCE
R_1
R_2
R_3
R_N
PARALLEL CONNECTION
CURRENT
POWER SUPPLY VOLTAGE

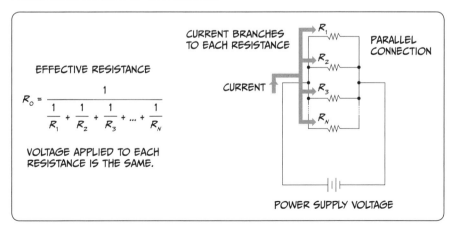

Parallel connection and effective resistance

If two light bulbs of the same size are connected in parallel to a power supply, the brightness of each bulb is the same as it is when there is only one bulb. Since the current flowing to each bulb is the same as the current flowing when only one bulb is connected, the total current is doubled.

ONE LIGHT BULB
CURRENT
TWO BULBS CONNECTED IN PARALLEL
CURRENT
BULBS NOW SHINE WITH THE SAME BRIGHTNESS, SINCE TOTAL CURRENT IS DOUBLED.

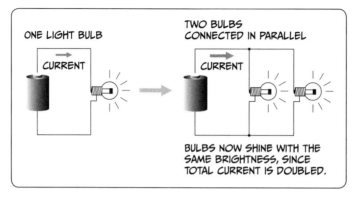

Series connection of light bulbs

The 120V electric appliances that we use in our homes are connected in parallel to a 120V power supply. If we increase the number of electric appliances connected to the power supply, the total current flowing also increases.

3
HOW DOES ELECTRICITY WORK?

WHY DOES ELECTRICITY PRODUCE HEAT?

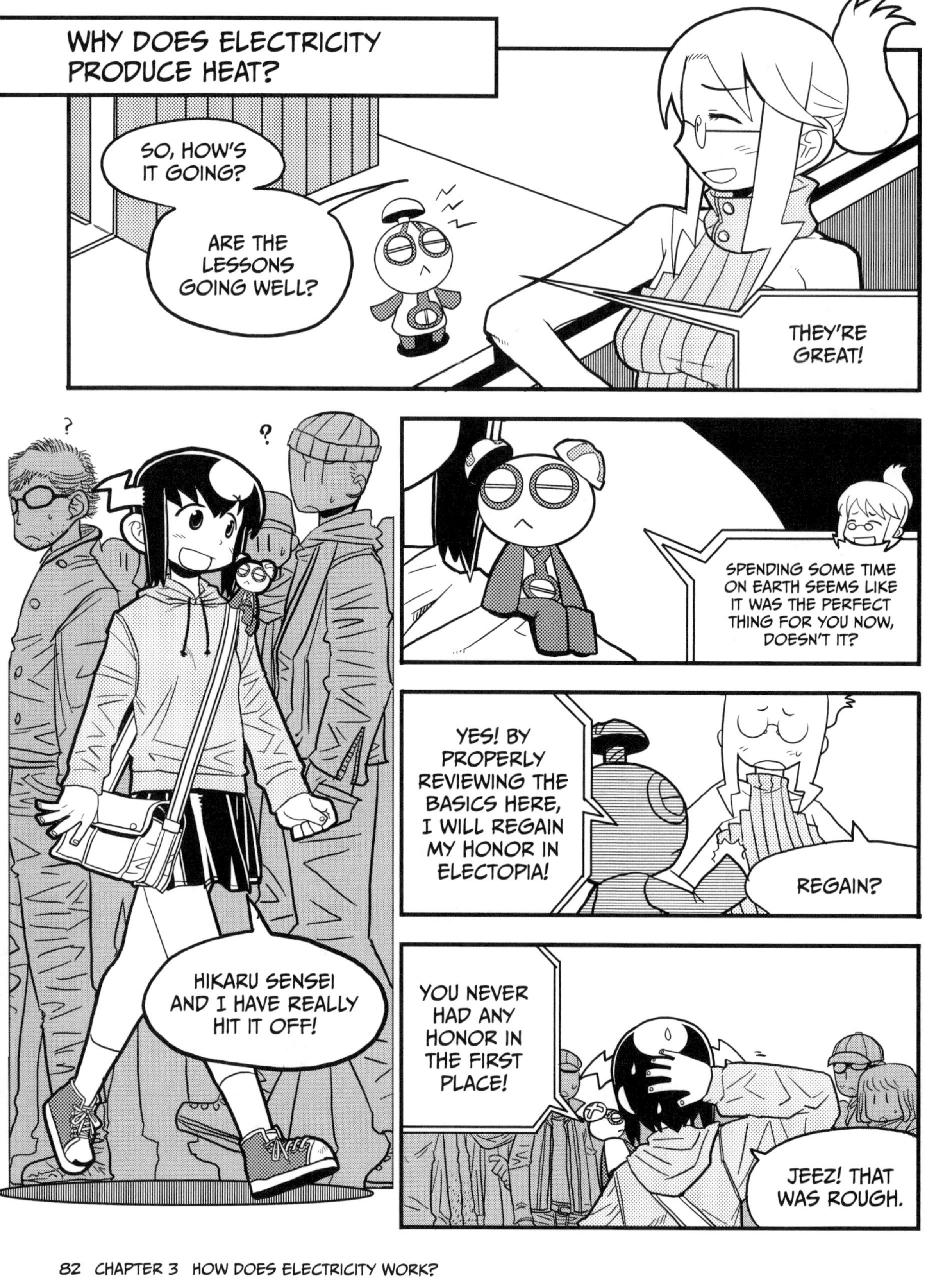
SO, HOW'S IT GOING?
ARE THE LESSONS GOING WELL?
THEY'RE GREAT!
HIKARU SENSEI AND I HAVE REALLY HIT IT OFF!
SPENDING SOME TIME ON EARTH SEEMS LIKE IT WAS THE PERFECT THING FOR YOU NOW, DOESN'T IT?
YES! BY PROPERLY REVIEWING THE BASICS HERE, I WILL REGAIN MY HONOR IN ELECTOPIA!
REGAIN?
YOU NEVER HAD ANY HONOR IN THE FIRST PLACE!
JEEZ! THAT WAS ROUGH.

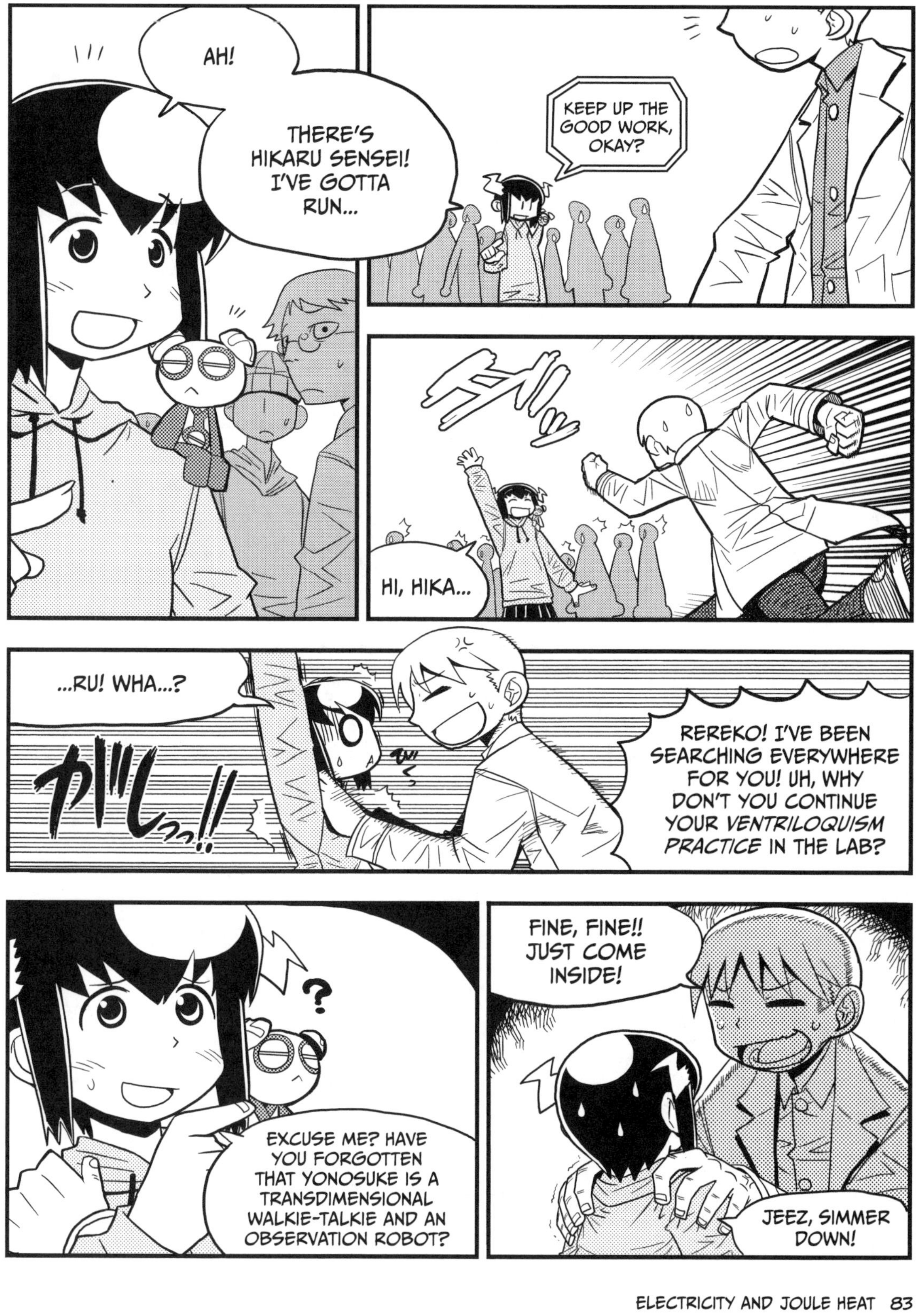
AH!
THERE'S HIKARU SENSEI! I'VE GOTTA RUN...
KEEP UP THE GOOD WORK, OKAY?
HI, HIKA...
...RU! WHA...?
REREKO! I'VE BEEN SEARCHING EVERYWHERE FOR YOU! UH, WHY DON'T YOU CONTINUE YOUR *VENTRILOQUISM PRACTICE* IN THE LAB?
EXCUSE ME? HAVE YOU FORGOTTEN THAT YONOSUKE IS A TRANSDIMENSIONAL WALKIE-TALKIE AND AN OBSERVATION ROBOT?
FINE, FINE!! JUST COME INSIDE!
JEEZ, SIMMER DOWN!

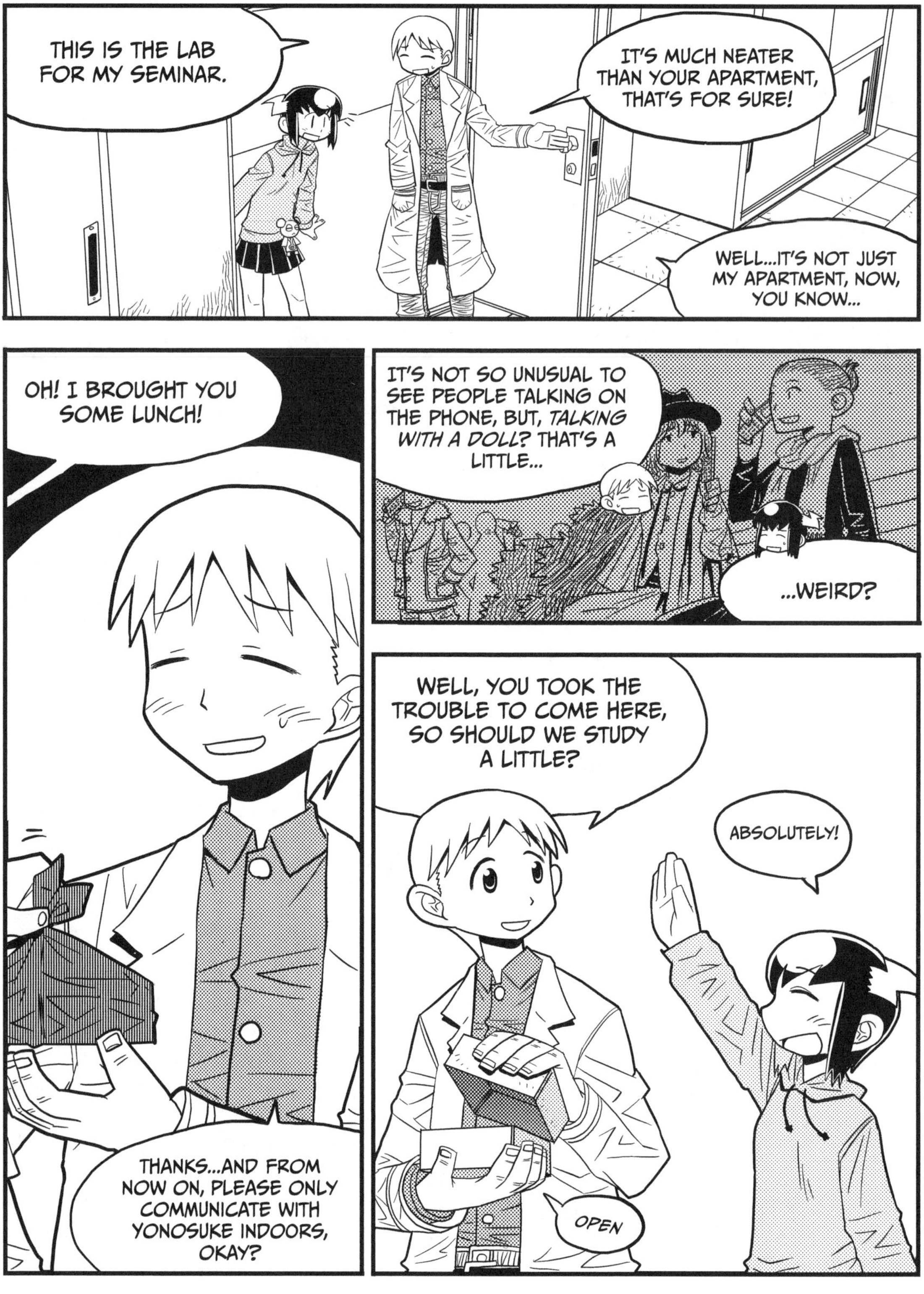
THIS IS THE LAB FOR MY SEMINAR.
IT'S MUCH NEATER THAN YOUR APARTMENT, THAT'S FOR SURE!
WELL...IT'S NOT JUST MY APARTMENT, NOW, YOU KNOW...
OH! I BROUGHT YOU SOME LUNCH!
THANKS...AND FROM NOW ON, PLEASE ONLY COMMUNICATE WITH YONOSUKE INDOORS, OKAY?
IT'S NOT SO UNUSUAL TO SEE PEOPLE TALKING ON THE PHONE, BUT, *TALKING WITH A DOLL?* THAT'S A LITTLE...
...WEIRD?
WELL, YOU TOOK THE TROUBLE TO COME HERE, SO SHOULD WE STUDY A LITTLE?
ABSOLUTELY!
OPEN

ELECTRICITY AND JOULE HEAT

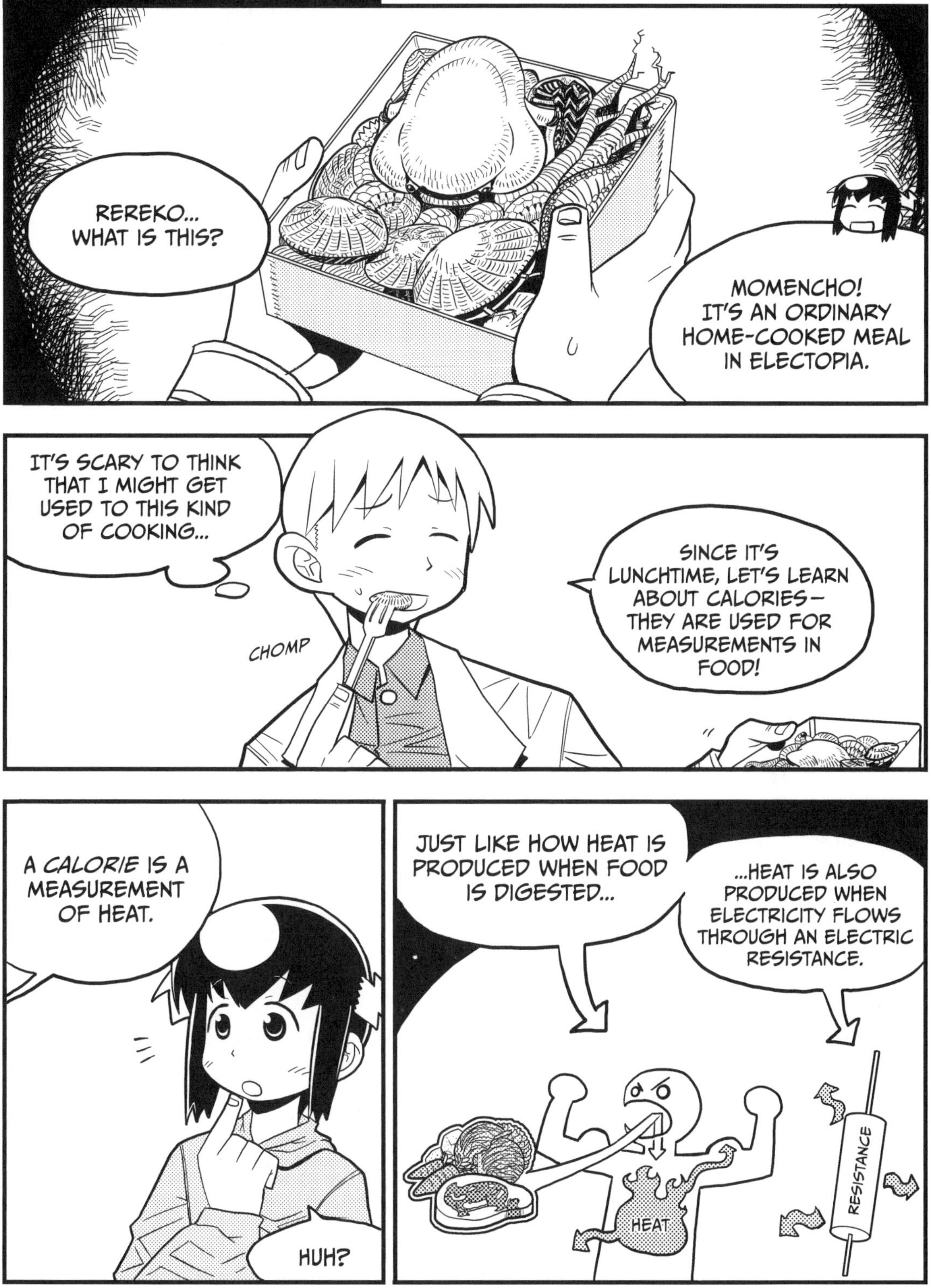

RereKO... WHAT IS THIS?
MOMENCHO! IT'S AN ORDINARY HOME-COOKED MEAL IN ELECTOPIA.
IT'S SCARY TO THINK THAT I MIGHT GET USED TO THIS KIND OF COOKING...
CHOMP
SINCE IT'S LUNCHTIME, LET'S LEARN ABOUT CALORIES— THEY ARE USED FOR MEASUREMENTS IN FOOD!
A *CALORIE* IS A MEASUREMENT OF HEAT.
HUH?
JUST LIKE HOW HEAT IS PRODUCED WHEN FOOD IS DIGESTED...
...HEAT IS ALSO PRODUCED WHEN ELECTRICITY FLOWS THROUGH AN ELECTRIC RESISTANCE.
HEAT
RESISTANCE

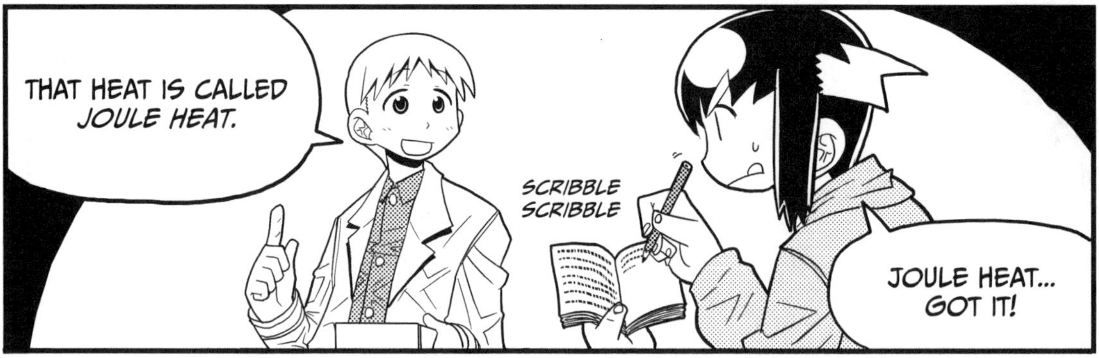
THAT HEAT IS CALLED *JOULE HEAT.*
SCRIBBLE SCRIBBLE
JOULE HEAT... GOT IT!

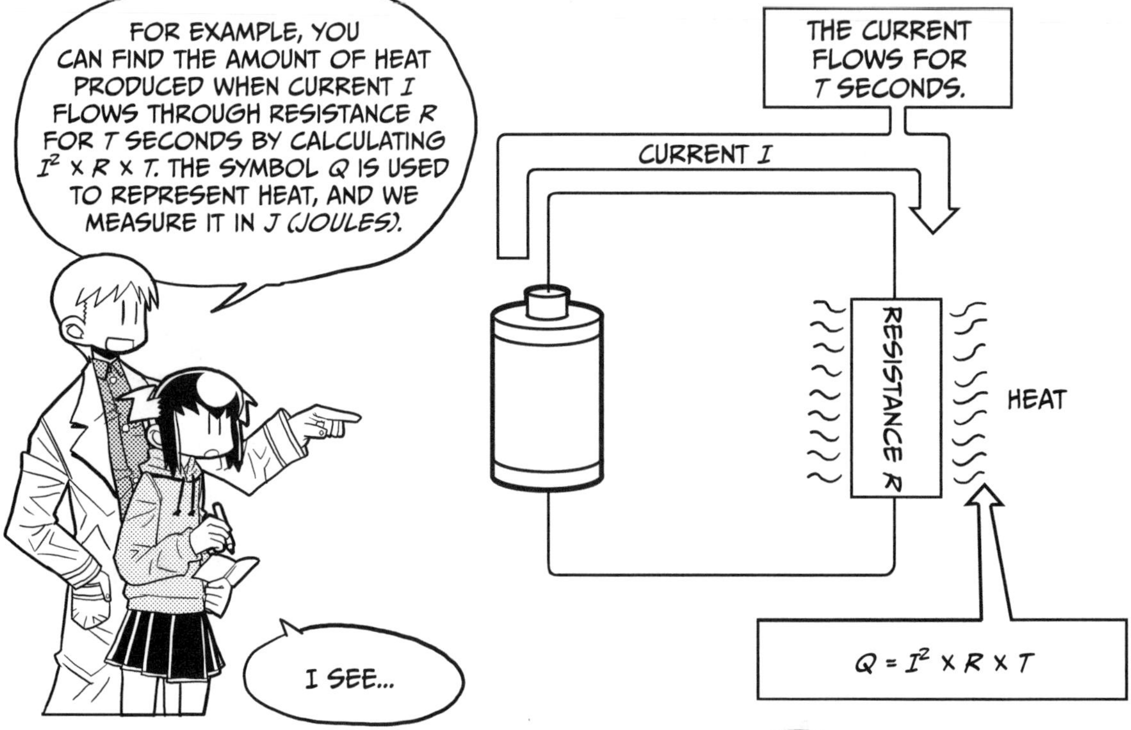
FOR EXAMPLE, YOU CAN FIND THE AMOUNT OF HEAT PRODUCED WHEN CURRENT I FLOWS THROUGH RESISTANCE R FOR T SECONDS BY CALCULATING $I^2 \times R \times T$. THE SYMBOL Q IS USED TO REPRESENT HEAT, AND WE MEASURE IT IN J (JOULES).
THE CURRENT FLOWS FOR T SECONDS.
CURRENT I
RESISTANCE R
HEAT
I SEE...
$Q = I^2 \times R \times T$

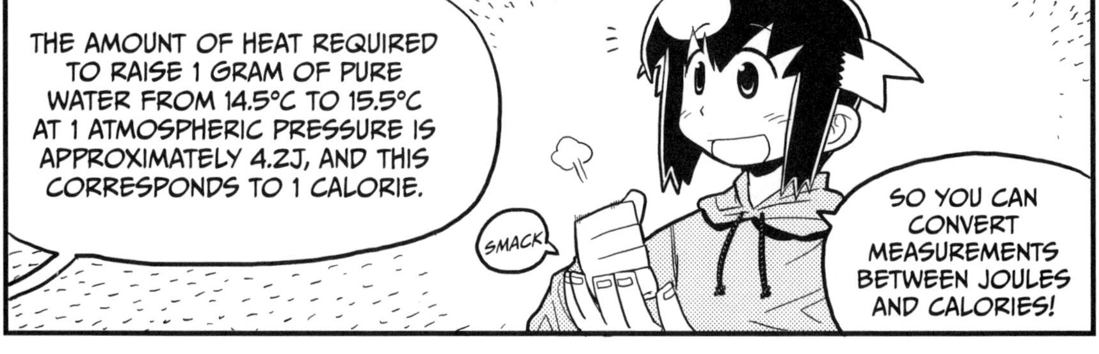
THE AMOUNT OF HEAT REQUIRED TO RAISE 1 GRAM OF PURE WATER FROM 14.5°C TO 15.5°C AT 1 ATMOSPHERIC PRESSURE IS APPROXIMATELY 4.2J, AND THIS CORRESPONDS TO 1 CALORIE.
SMACK
SO YOU CAN CONVERT MEASUREMENTS BETWEEN JOULES AND CALORIES!

HOW IS HEAT GENERATED BY CURRENT?

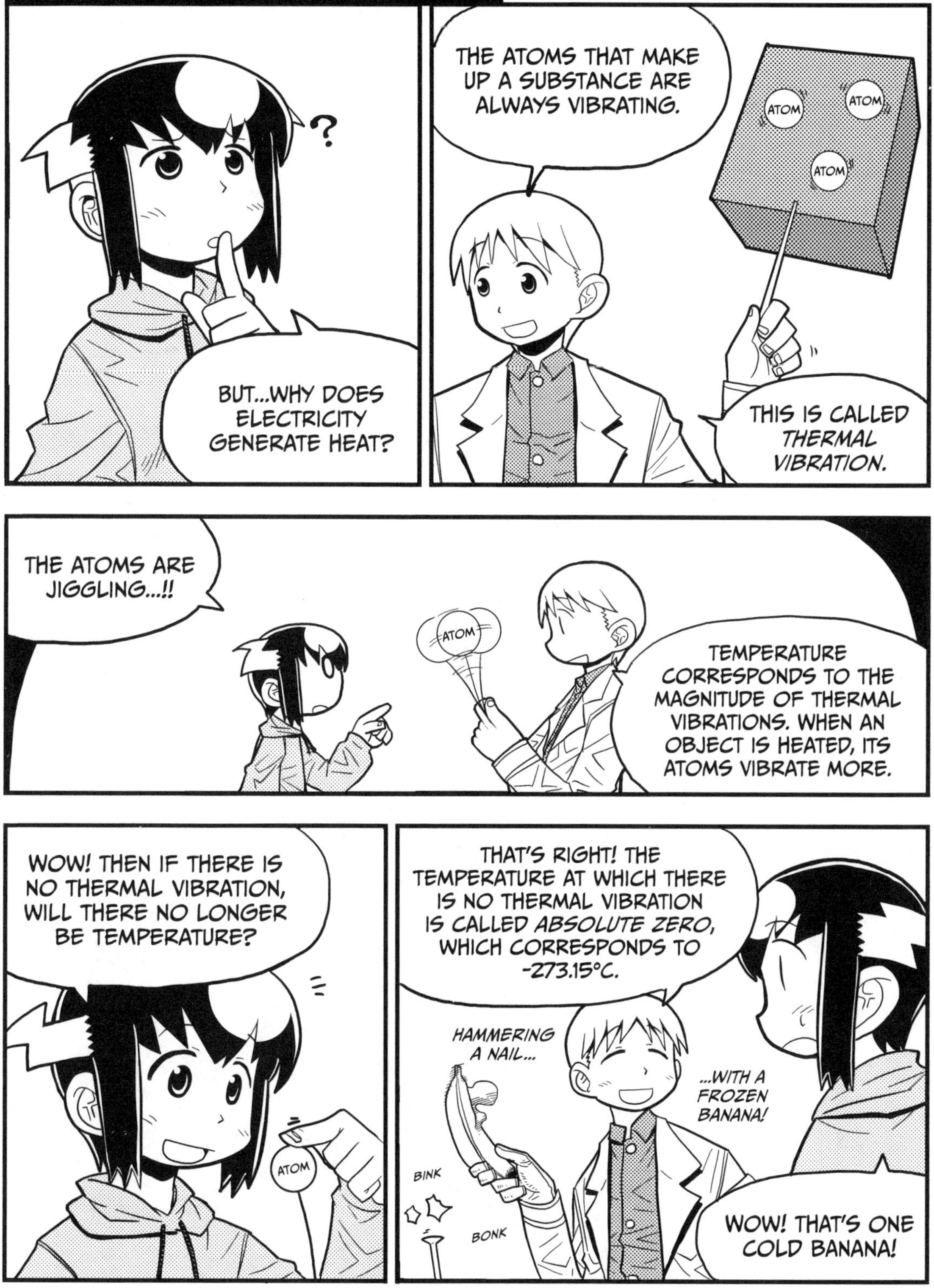
?
BUT...WHY DOES ELECTRICITY GENERATE HEAT?
THE ATOMS THAT MAKE UP A SUBSTANCE ARE ALWAYS VIBRATING.
ATOM
ATOM
ATOM
THIS IS CALLED *THERMAL VIBRATION.*
THE ATOMS ARE JIGGLING...!!
ATOM
TEMPERATURE CORRESPONDS TO THE MAGNITUDE OF THERMAL VIBRATIONS. WHEN AN OBJECT IS HEATED, ITS ATOMS VIBRATE MORE.
WOW! THEN IF THERE IS NO THERMAL VIBRATION, WILL THERE NO LONGER BE TEMPERATURE?
ATOM
THAT'S RIGHT! THE TEMPERATURE AT WHICH THERE IS NO THERMAL VIBRATION IS CALLED *ABSOLUTE ZERO,* WHICH CORRESPONDS TO -273.15°C.
HAMMERING A NAIL...
...WITH A FROZEN BANANA!
BINK
BONK
WOW! THAT'S ONE COLD BANANA!

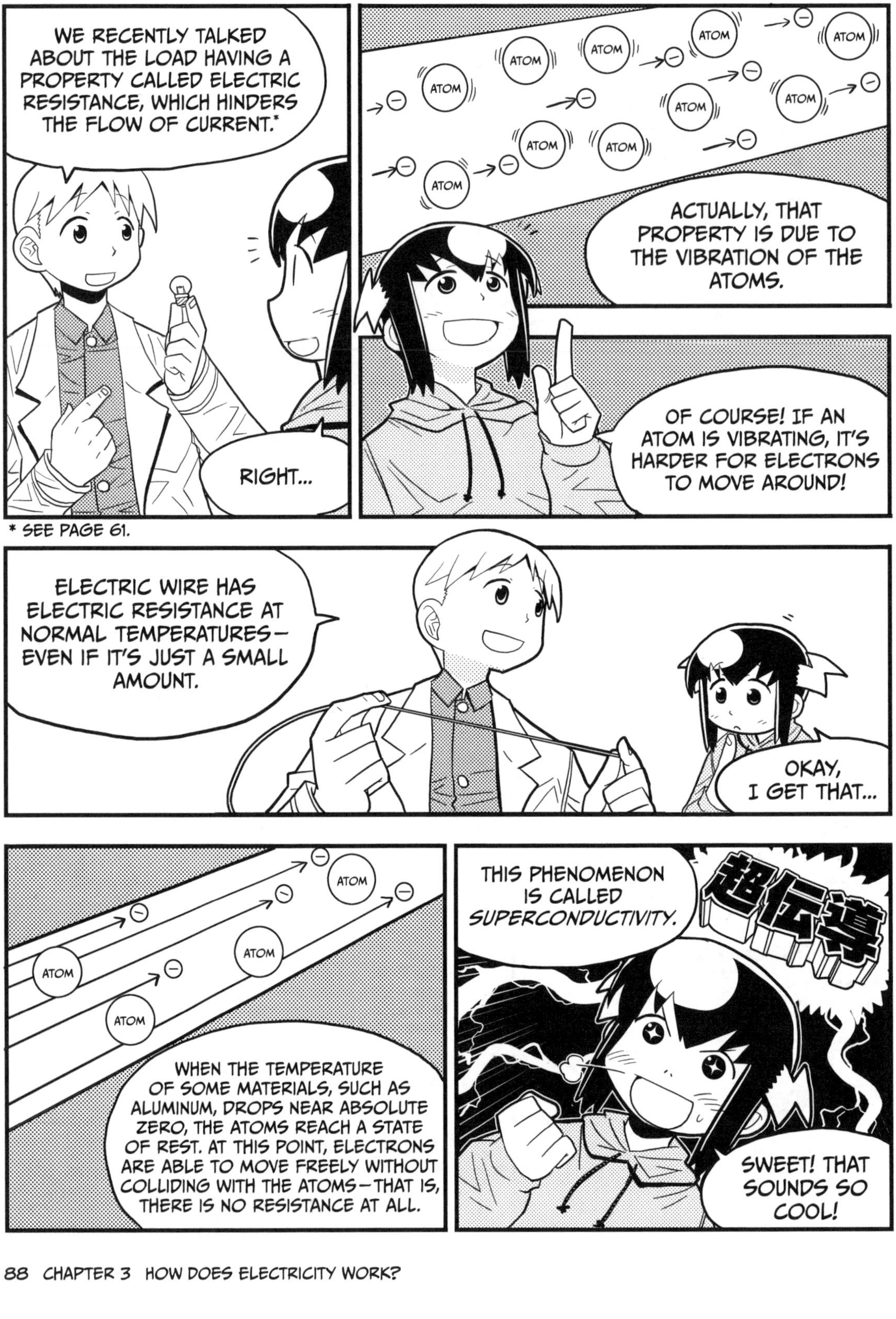
WE RECENTLY TALKED ABOUT THE LOAD HAVING A PROPERTY CALLED ELECTRIC RESISTANCE, WHICH HINDERS THE FLOW OF CURRENT.*
RIGHT...
ATOM
ATOM
ATOM
ATOM
ATOM
ATOM
ATOM
ATOM
ATOM
ATOM
ACTUALLY, THAT PROPERTY IS DUE TO THE VIBRATION OF THE ATOMS.
OF COURSE! IF AN ATOM IS VIBRATING, IT'S HARDER FOR ELECTRONS TO MOVE AROUND!
ELECTRIC WIRE HAS ELECTRIC RESISTANCE AT NORMAL TEMPERATURES—EVEN IF IT'S JUST A SMALL AMOUNT.
OKAY, I GET THAT...
ATOM
ATOM
ATOM
ATOM
WHEN THE TEMPERATURE OF SOME MATERIALS, SUCH AS ALUMINUM, DROPS NEAR ABSOLUTE ZERO, THE ATOMS REACH A STATE OF REST. AT THIS POINT, ELECTRONS ARE ABLE TO MOVE FREELY WITHOUT COLLIDING WITH THE ATOMS—THAT IS, THERE IS NO RESISTANCE AT ALL.
THIS PHENOMENON IS CALLED *SUPERCONDUCTIVITY*.
超伝導
SWEET! THAT SOUNDS SO COOL!

* SEE PAGE 61.

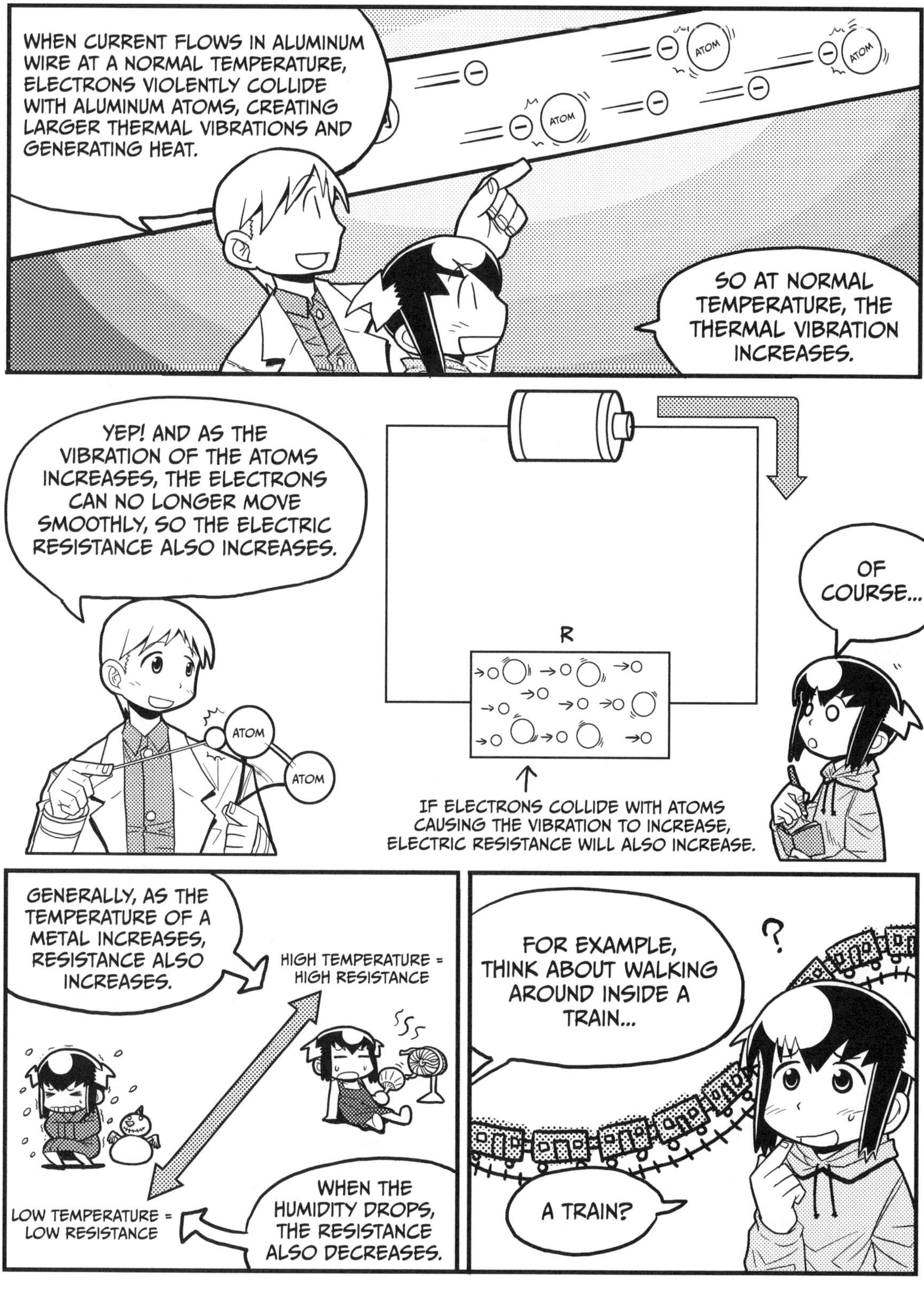
WHEN CURRENT FLOWS IN ALUMINUM WIRE AT A NORMAL TEMPERATURE, ELECTRONS VIOLENTLY COLLIDE WITH ALUMINUM ATOMS, CREATING LARGER THERMAL VIBRATIONS AND GENERATING HEAT.
ATOM
ATOM
ATOM
SO AT NORMAL TEMPERATURE, THE THERMAL VIBRATION INCREASES.
YEP! AND AS THE VIBRATION OF THE ATOMS INCREASES, THE ELECTRONS CAN NO LONGER MOVE SMOOTHLY, SO THE ELECTRIC RESISTANCE ALSO INCREASES.
ATOM
ATOM
R
IF ELECTRONS COLLIDE WITH ATOMS CAUSING THE VIBRATION TO INCREASE, ELECTRIC RESISTANCE WILL ALSO INCREASE.
OF COURSE...
GENERALLY, AS THE TEMPERATURE OF A METAL INCREASES, RESISTANCE ALSO INCREASES.
HIGH TEMPERATURE = HIGH RESISTANCE
LOW TEMPERATURE = LOW RESISTANCE
WHEN THE HUMIDITY DROPS, THE RESISTANCE ALSO DECREASES.
?
FOR EXAMPLE, THINK ABOUT WALKING AROUND INSIDE A TRAIN...
A TRAIN?

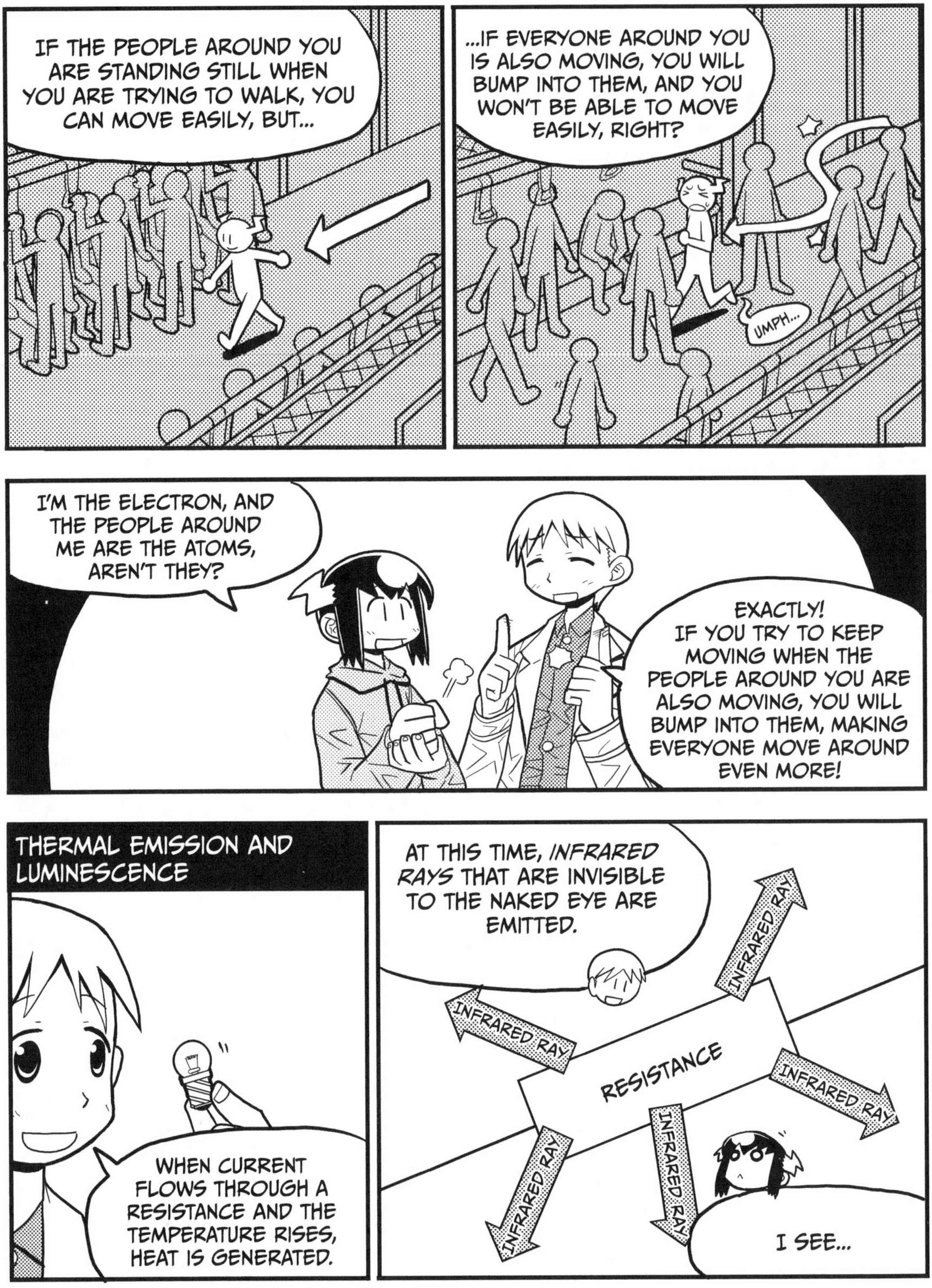
IF THE PEOPLE AROUND YOU ARE STANDING STILL WHEN YOU ARE TRYING TO WALK, YOU CAN MOVE EASILY, BUT...
...IF EVERYONE AROUND YOU IS ALSO MOVING, YOU WILL BUMP INTO THEM, AND YOU WON'T BE ABLE TO MOVE EASILY, RIGHT?
UMPH...
I'M THE ELECTRON, AND THE PEOPLE AROUND ME ARE THE ATOMS, AREN'T THEY?
EXACTLY! IF YOU TRY TO KEEP MOVING WHEN THE PEOPLE AROUND YOU ARE ALSO MOVING, YOU WILL BUMP INTO THEM, MAKING EVERYONE MOVE AROUND EVEN MORE!
THERMAL EMISSION AND LUMINESCENCE
WHEN CURRENT FLOWS THROUGH A RESISTANCE AND THE TEMPERATURE RISES, HEAT IS GENERATED.
AT THIS TIME, *INFRARED RAYS* THAT ARE INVISIBLE TO THE NAKED EYE ARE EMITTED.
INFRARED RAY
INFRARED RAY
RESISTANCE
INFRARED RAY
INFRARED RAY
INFRARED RAY
I SEE...

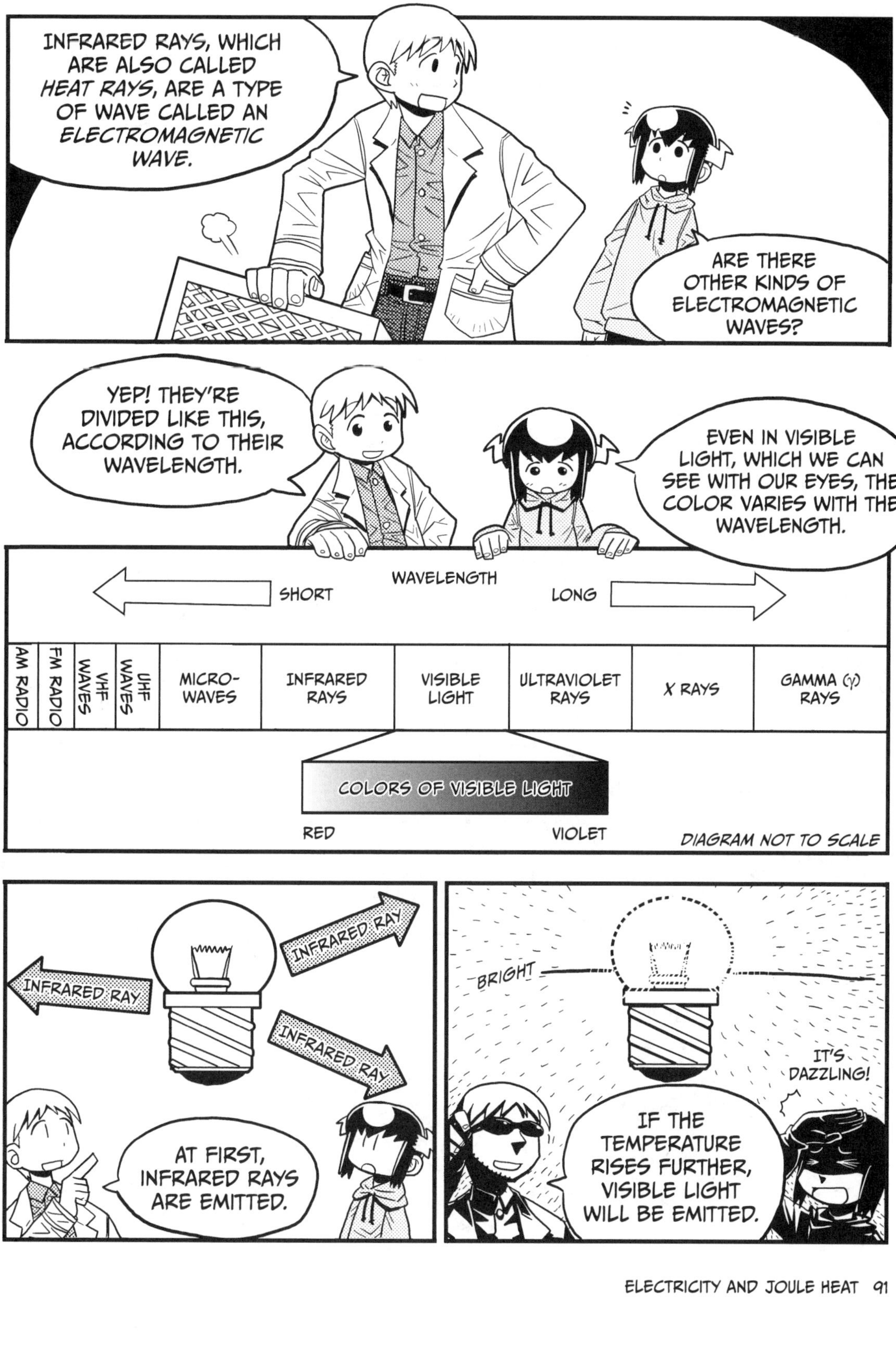
INFRARED RAYS, WHICH ARE ALSO CALLED *HEAT RAYS*, ARE A TYPE OF WAVE CALLED AN *ELECTROMAGNETIC WAVE*.
ARE THERE OTHER KINDS OF ELECTROMAGNETIC WAVES?
YEP! THEY'RE DIVIDED LIKE THIS, ACCORDING TO THEIR WAVELENGTH.
EVEN IN VISIBLE LIGHT, WHICH WE CAN SEE WITH OUR EYES, THE COLOR VARIES WITH THE WAVELENGTH.
WAVELENGTH
SHORT
LONG
AM RADIO
FM RADIO
VHF WAVES
UHF WAVES
MICRO-WAVES
INFRARED RAYS
VISIBLE LIGHT
ULTRAVIOLET RAYS
X RAYS
GAMMA (γ) RAYS
COLORS OF VISIBLE LIGHT
RED
VIOLET
DIAGRAM NOT TO SCALE
INFRARED RAY
INFRARED RAY
INFRARED RAY
AT FIRST, INFRARED RAYS ARE EMITTED.
BRIGHT
IF THE TEMPERATURE RISES FURTHER, VISIBLE LIGHT WILL BE EMITTED.
IT'S DAZZLING!

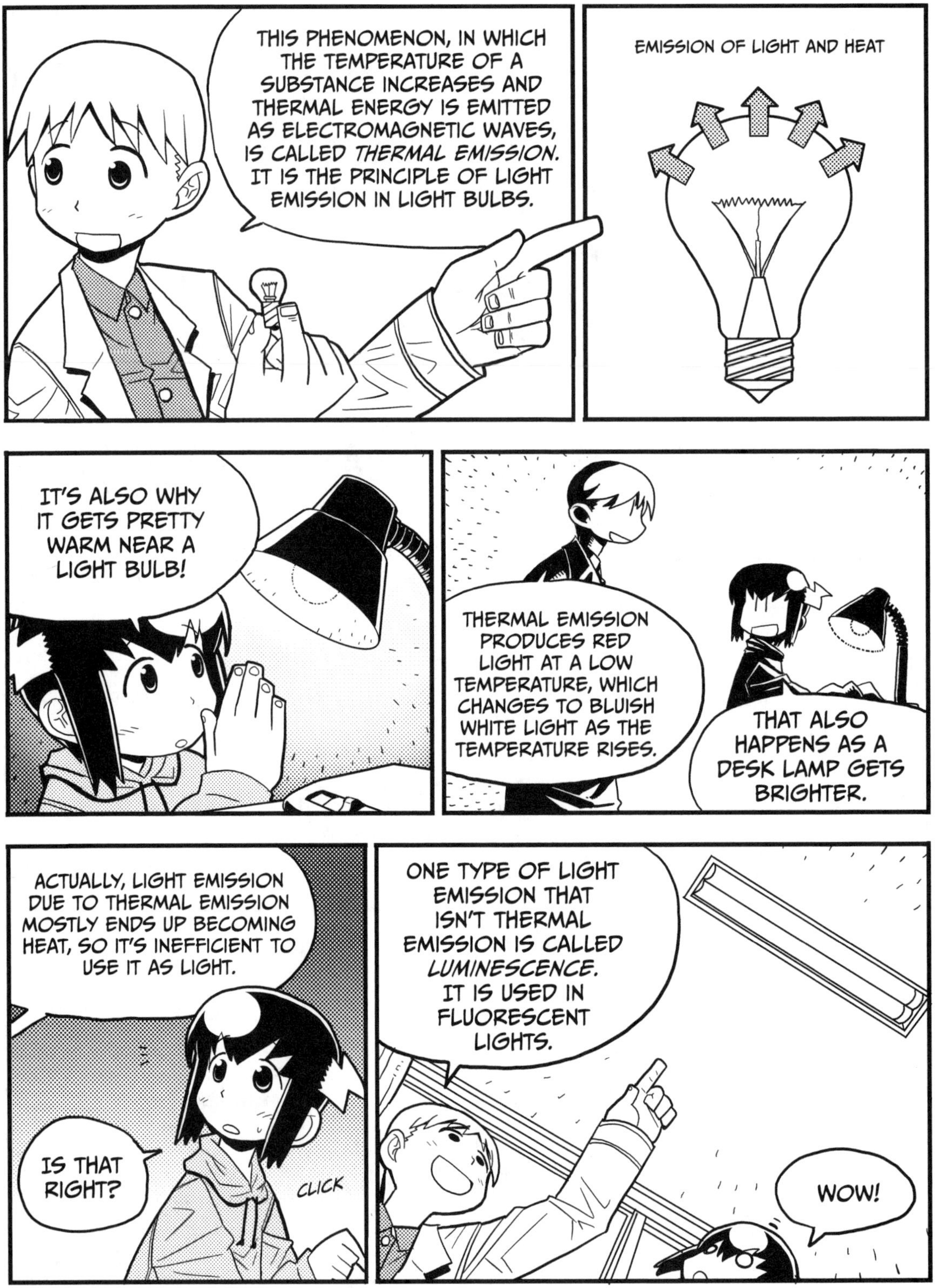
THIS PHENOMENON, IN WHICH THE TEMPERATURE OF A SUBSTANCE INCREASES AND THERMAL ENERGY IS EMITTED AS ELECTROMAGNETIC WAVES, IS CALLED *THERMAL EMISSION*. IT IS THE PRINCIPLE OF LIGHT EMISSION IN LIGHT BULBS.
EMISSION OF LIGHT AND HEAT
IT'S ALSO WHY IT GETS PRETTY WARM NEAR A LIGHT BULB!
THERMAL EMISSION PRODUCES RED LIGHT AT A LOW TEMPERATURE, WHICH CHANGES TO BLUISH WHITE LIGHT AS THE TEMPERATURE RISES.
THAT ALSO HAPPENS AS A DESK LAMP GETS BRIGHTER.
ACTUALLY, LIGHT EMISSION DUE TO THERMAL EMISSION MOSTLY ENDS UP BECOMING HEAT, SO IT'S INEFFICIENT TO USE IT AS LIGHT.
IS THAT RIGHT?
CLICK
ONE TYPE OF LIGHT EMISSION THAT ISN'T THERMAL EMISSION IS CALLED *LUMINESCENCE*. IT IS USED IN FLUORESCENT LIGHTS.
WOW!

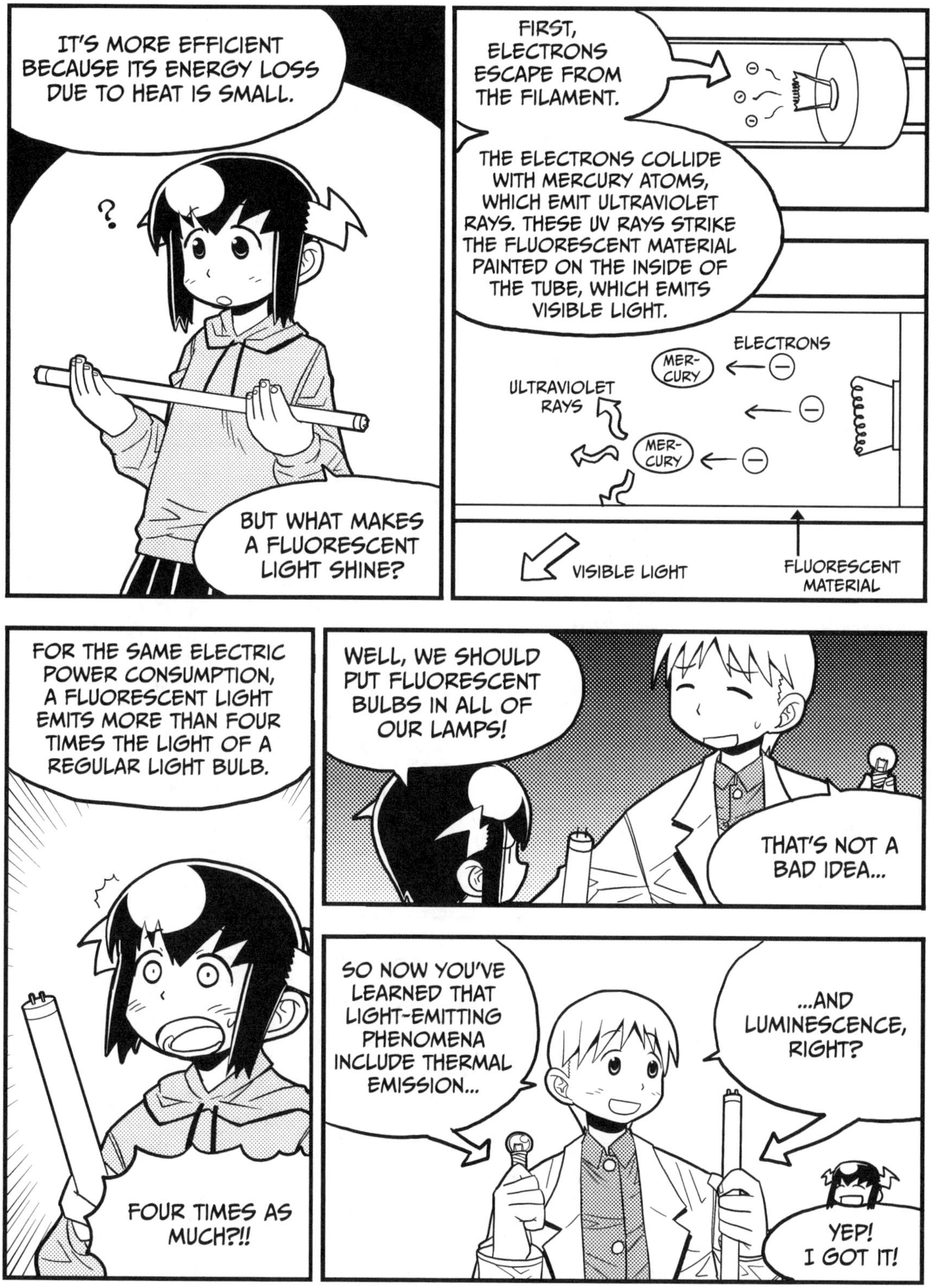
IT'S MORE EFFICIENT BECAUSE ITS ENERGY LOSS DUE TO HEAT IS SMALL.
?
BUT WHAT MAKES A FLUORESCENT LIGHT SHINE?
FIRST, ELECTRONS ESCAPE FROM THE FILAMENT.
THE ELECTRONS COLLIDE WITH MERCURY ATOMS, WHICH EMIT ULTRAVIOLET RAYS. THESE UV RAYS STRIKE THE FLUORESCENT MATERIAL PAINTED ON THE INSIDE OF THE TUBE, WHICH EMITS VISIBLE LIGHT.
ELECTRONS
MER-CURY
ULTRAVIOLET RAYS
MER-CURY
VISIBLE LIGHT
FLUORESCENT MATERIAL
FOR THE SAME ELECTRIC POWER CONSUMPTION, A FLUORESCENT LIGHT EMITS MORE THAN FOUR TIMES THE LIGHT OF A REGULAR LIGHT BULB.
FOUR TIMES AS MUCH?!!
WELL, WE SHOULD PUT FLUORESCENT BULBS IN ALL OF OUR LAMPS!
THAT'S NOT A BAD IDEA...
SO NOW YOU'VE LEARNED THAT LIGHT-EMITTING PHENOMENA INCLUDE THERMAL EMISSION...
...AND LUMINESCENCE, RIGHT?
YEP! I GOT IT!

CURRENT AND MAGNETIC FIELDS

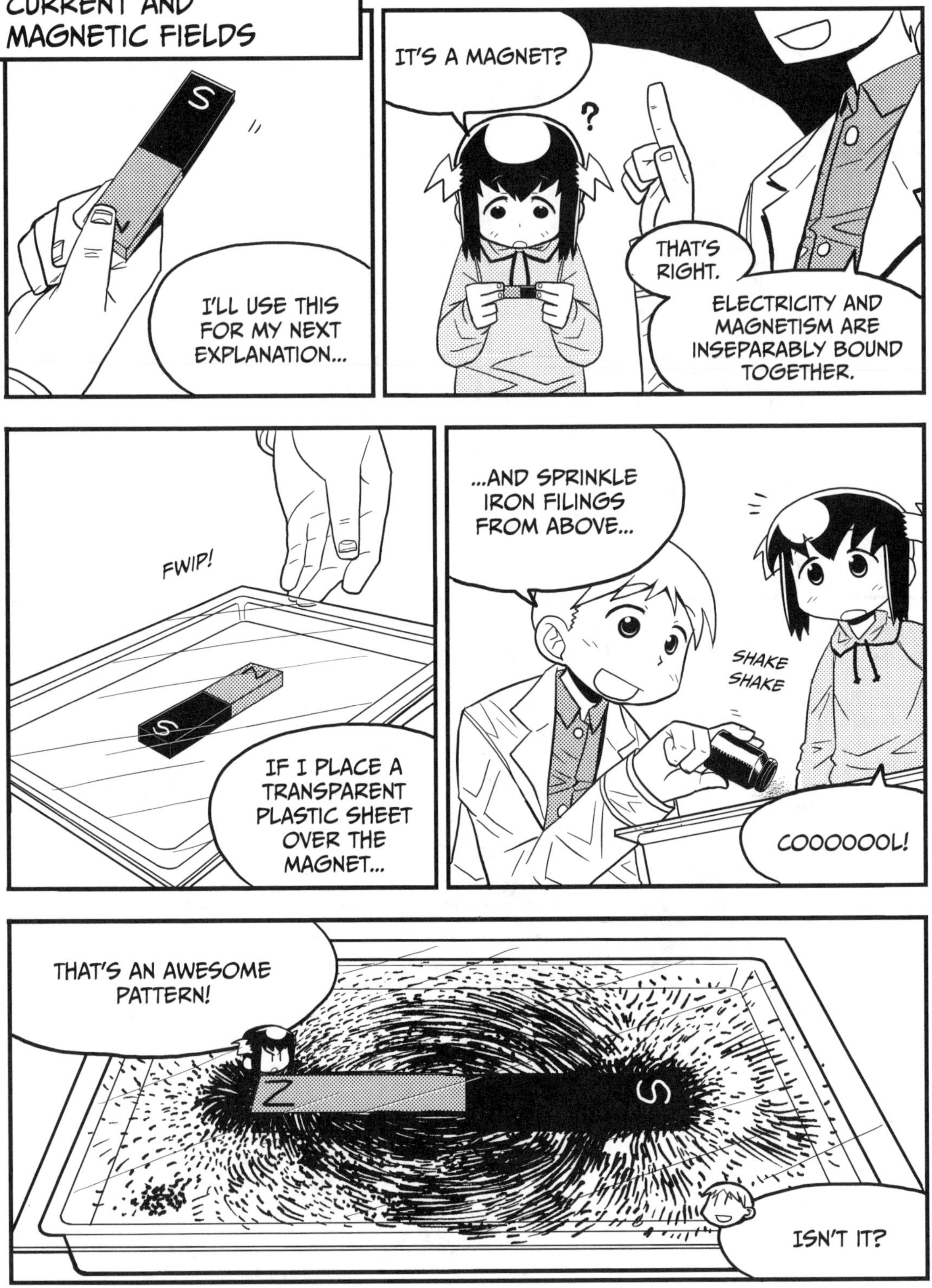
I'LL USE THIS FOR MY NEXT EXPLANATION...
IT'S A MAGNET?
THAT'S RIGHT.
ELECTRICITY AND MAGNETISM ARE INSEPARABLY BOUND TOGETHER.
FWIP!
IF I PLACE A TRANSPARENT PLASTIC SHEET OVER THE MAGNET...
...AND SPRINKLE IRON FILINGS FROM ABOVE...
SHAKE SHAKE
COOOOOOL!
THAT'S AN AWESOME PATTERN!
ISN'T IT?

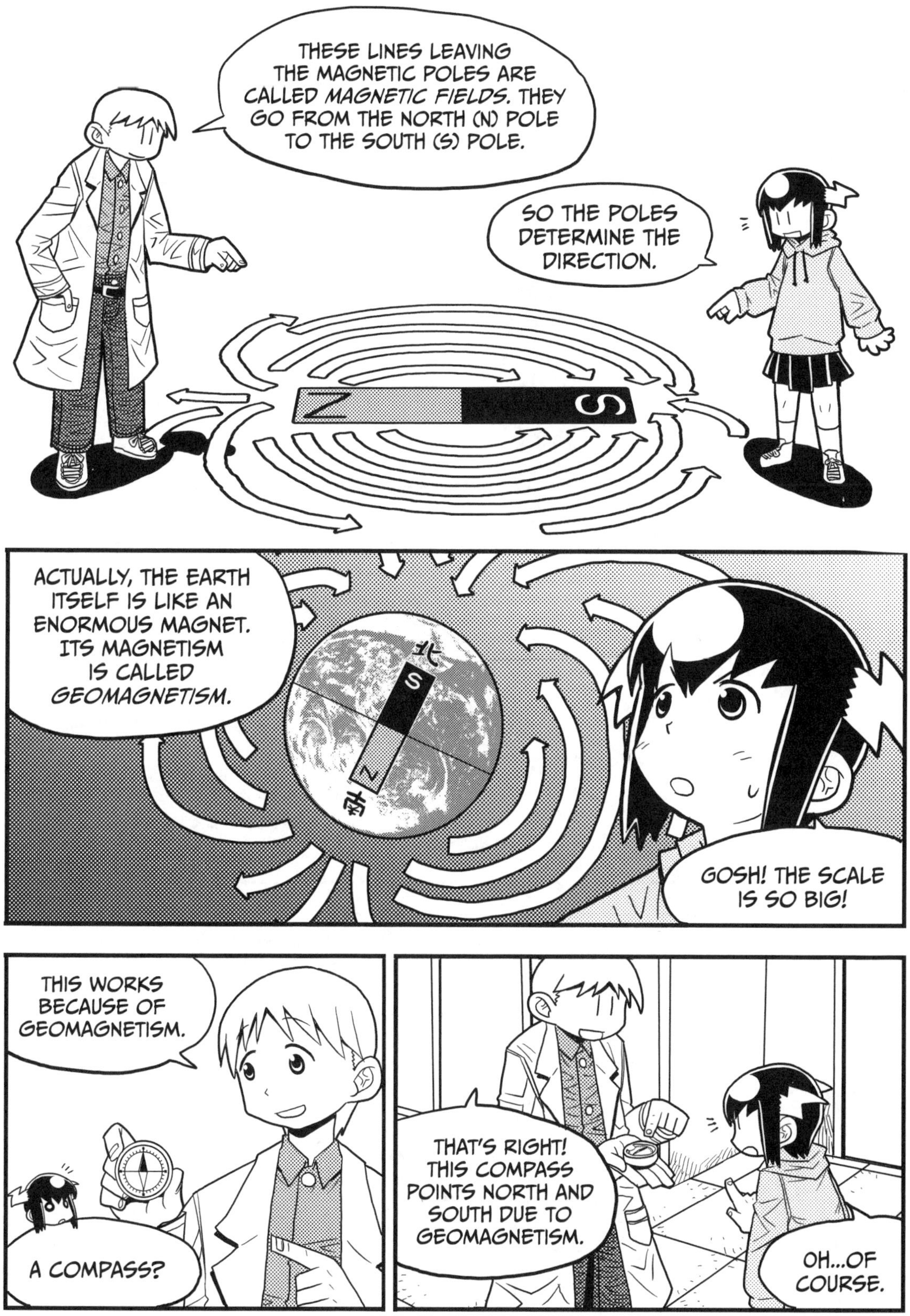
THESE LINES LEAVING THE MAGNETIC POLES ARE CALLED *MAGNETIC FIELDS*. THEY GO FROM THE NORTH (N) POLE TO THE SOUTH (S) POLE.
SO THE POLES DETERMINE THE DIRECTION.
N
S
ACTUALLY, THE EARTH ITSELF IS LIKE AN ENORMOUS MAGNET. ITS MAGNETISM IS CALLED *GEOMAGNETISM*.
北
S
N
南
GOSH! THE SCALE IS SO BIG!
THIS WORKS BECAUSE OF GEOMAGNETISM.
A COMPASS?
THAT'S RIGHT! THIS COMPASS POINTS NORTH AND SOUTH DUE TO GEOMAGNETISM.
OH...OF COURSE.

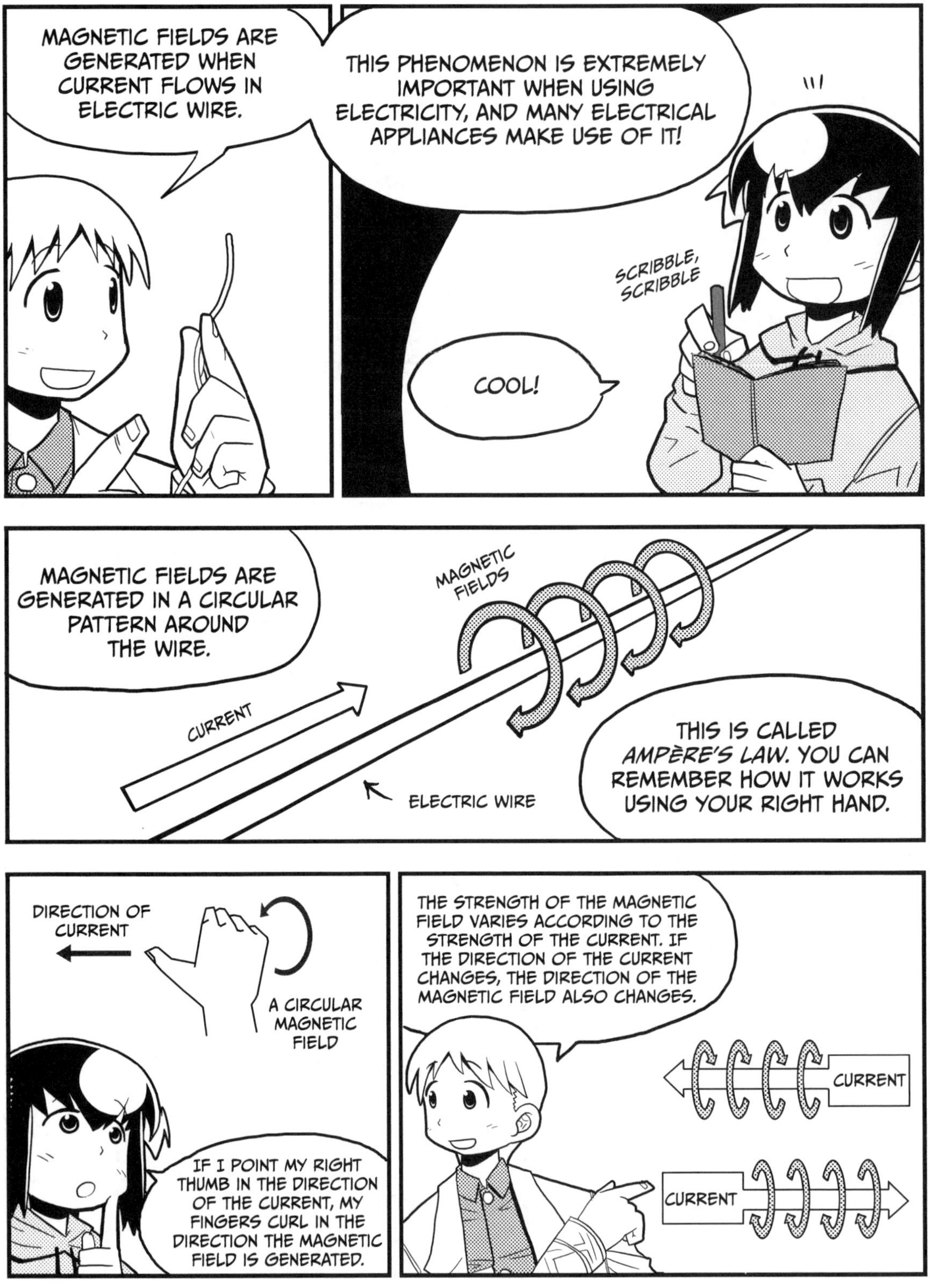
MAGNETIC FIELDS ARE GENERATED WHEN CURRENT FLOWS IN ELECTRIC WIRE.
THIS PHENOMENON IS EXTREMELY IMPORTANT WHEN USING ELECTRICITY, AND MANY ELECTRICAL APPLIANCES MAKE USE OF IT!
SCRIBBLE, SCRIBBLE
COOL!
MAGNETIC FIELDS ARE GENERATED IN A CIRCULAR PATTERN AROUND THE WIRE.
MAGNETIC FIELDS
CURRENT
ELECTRIC WIRE
THIS IS CALLED AMPÈRE'S LAW. YOU CAN REMEMBER HOW IT WORKS USING YOUR RIGHT HAND.
DIRECTION OF CURRENT
A CIRCULAR MAGNETIC FIELD
IF I POINT MY RIGHT THUMB IN THE DIRECTION OF THE CURRENT, MY FINGERS CURL IN THE DIRECTION THE MAGNETIC FIELD IS GENERATED.
THE STRENGTH OF THE MAGNETIC FIELD VARIES ACCORDING TO THE STRENGTH OF THE CURRENT. IF THE DIRECTION OF THE CURRENT CHANGES, THE DIRECTION OF THE MAGNETIC FIELD ALSO CHANGES.
CURRENT
CURRENT

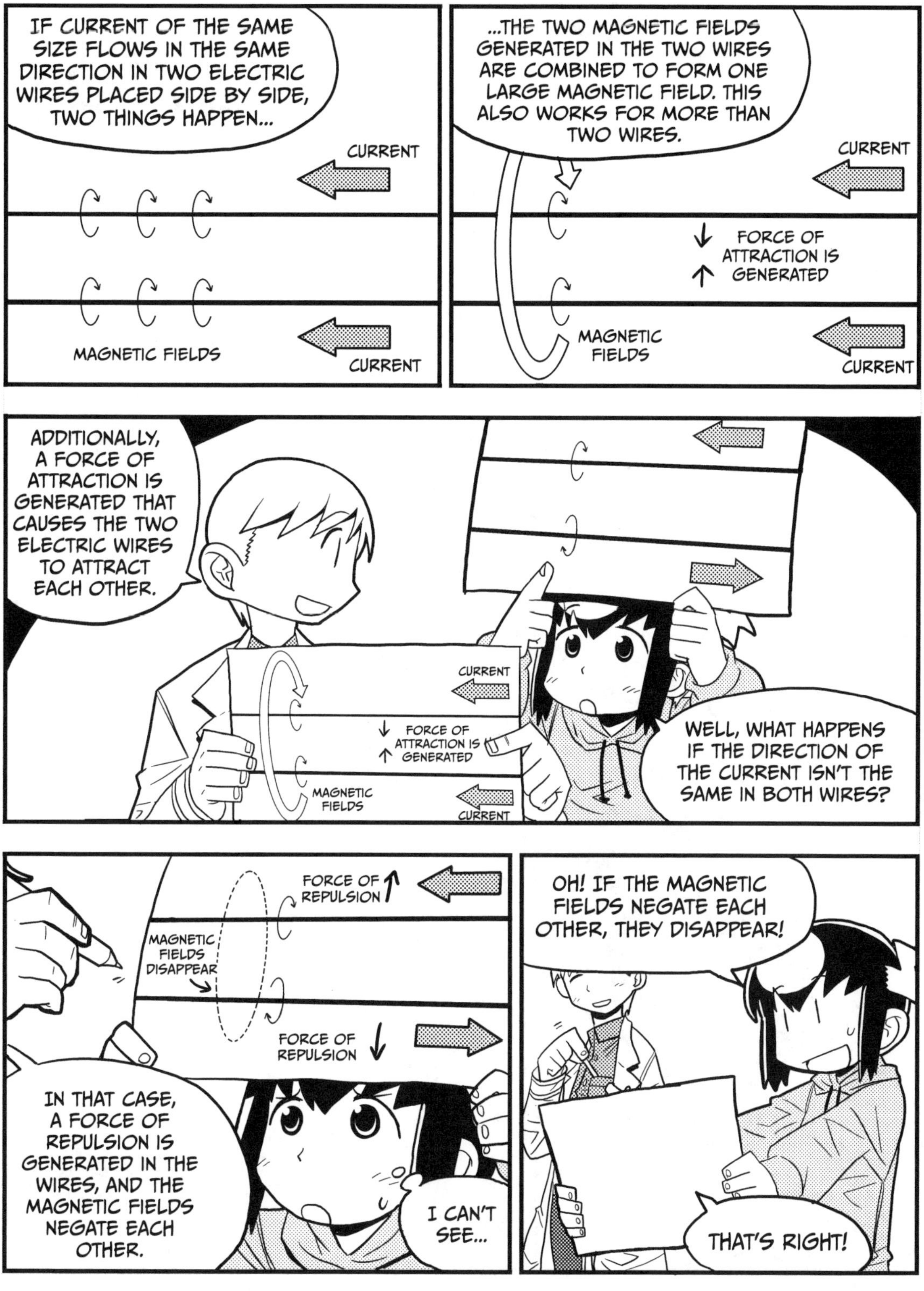
IF CURRENT OF THE SAME SIZE FLOWS IN THE SAME DIRECTION IN TWO ELECTRIC WIRES PLACED SIDE BY SIDE, TWO THINGS HAPPEN...
CURRENT
MAGNETIC FIELDS
CURRENT
...THE TWO MAGNETIC FIELDS GENERATED IN THE TWO WIRES ARE COMBINED TO FORM ONE LARGE MAGNETIC FIELD. THIS ALSO WORKS FOR MORE THAN TWO WIRES.
CURRENT
FORCE OF ATTRACTION IS GENERATED
MAGNETIC FIELDS
CURRENT
ADDITIONALLY, A FORCE OF ATTRACTION IS GENERATED THAT CAUSES THE TWO ELECTRIC WIRES TO ATTRACT EACH OTHER.
CURRENT
FORCE OF ATTRACTION IS GENERATED
MAGNETIC FIELDS
CURRENT
WELL, WHAT HAPPENS IF THE DIRECTION OF THE CURRENT ISN'T THE SAME IN BOTH WIRES?
FORCE OF REPULSION
MAGNETIC FIELDS DISAPPEAR
FORCE OF REPULSION
IN THAT CASE, A FORCE OF REPULSION IS GENERATED IN THE WIRES, AND THE MAGNETIC FIELDS NEGATE EACH OTHER.
I CAN'T SEE...
OH! IF THE MAGNETIC FIELDS NEGATE EACH OTHER, THEY DISAPPEAR!
THAT'S RIGHT!

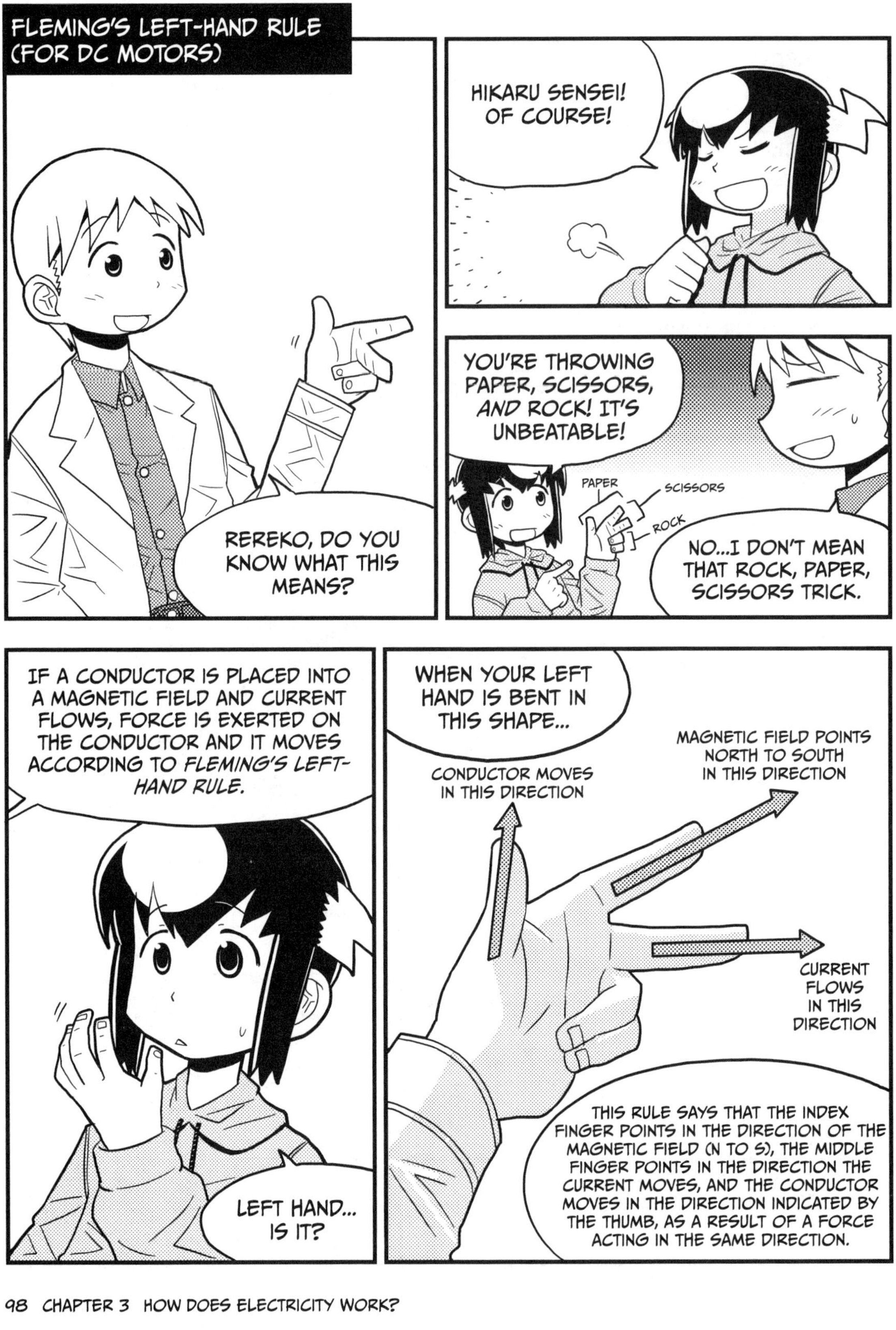
FLEMING'S LEFT-HAND RULE (FOR DC MOTORS)
REREKO, DO YOU KNOW WHAT THIS MEANS?
HIKARU SENSEI! OF COURSE!
YOU'RE THROWING PAPER, SCISSORS, *AND* ROCK! IT'S UNBEATABLE!
PAPER
SCISSORS
ROCK
NO...I DON'T MEAN THAT ROCK, PAPER, SCISSORS TRICK.
IF A CONDUCTOR IS PLACED INTO A MAGNETIC FIELD AND CURRENT FLOWS, FORCE IS EXERTED ON THE CONDUCTOR AND IT MOVES ACCORDING TO *FLEMING'S LEFT-HAND RULE*.
LEFT HAND... IS IT?
WHEN YOUR LEFT HAND IS BENT IN THIS SHAPE...
CONDUCTOR MOVES IN THIS DIRECTION
MAGNETIC FIELD POINTS NORTH TO SOUTH IN THIS DIRECTION
CURRENT FLOWS IN THIS DIRECTION
THIS RULE SAYS THAT THE INDEX FINGER POINTS IN THE DIRECTION OF THE MAGNETIC FIELD (N TO S), THE MIDDLE FINGER POINTS IN THE DIRECTION THE CURRENT MOVES, AND THE CONDUCTOR MOVES IN THE DIRECTION INDICATED BY THE THUMB, AS A RESULT OF A FORCE ACTING IN THE SAME DIRECTION.

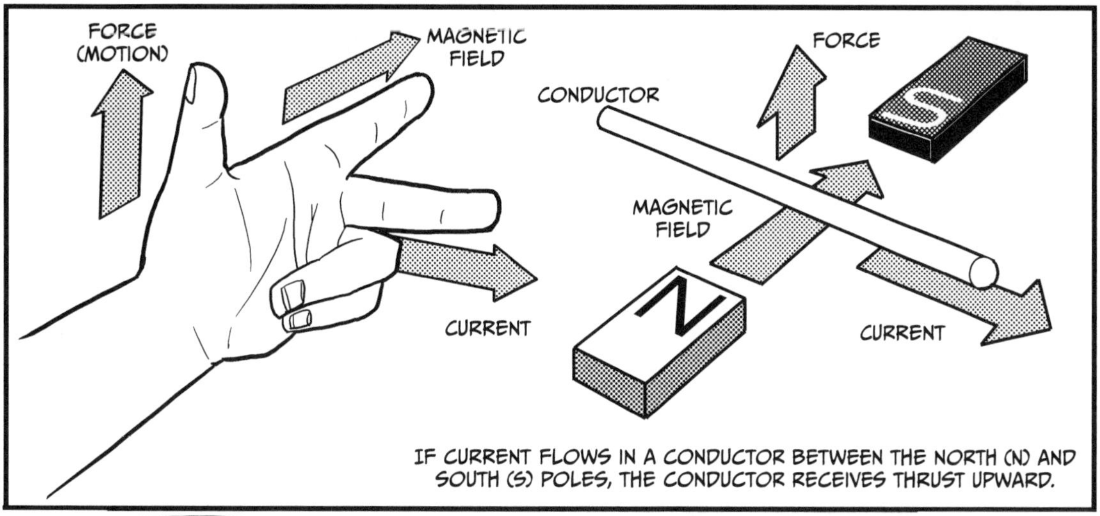
FORCE (MOTION)
MAGNETIC FIELD
CURRENT
CONDUCTOR
FORCE
S
MAGNETIC FIELD
N
CURRENT
IF CURRENT FLOWS IN A CONDUCTOR BETWEEN THE NORTH (N) AND SOUTH (S) POLES, THE CONDUCTOR RECEIVES THRUST UPWARD.

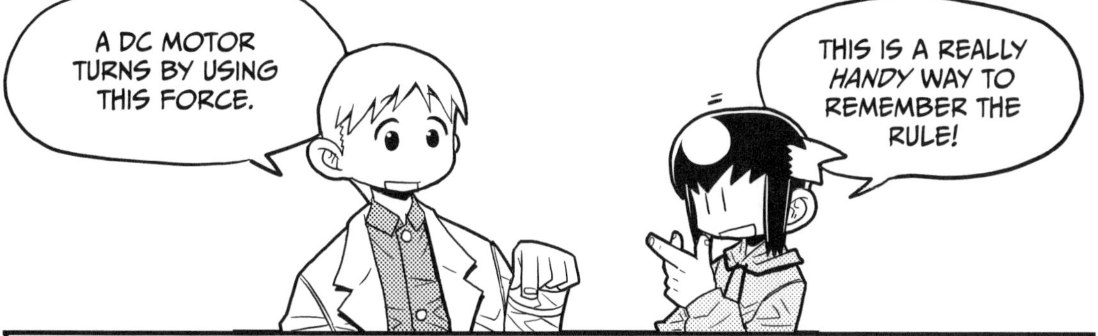
A DC MOTOR TURNS BY USING THIS FORCE.
THIS IS A REALLY *HANDY* WAY TO REMEMBER THE RULE!

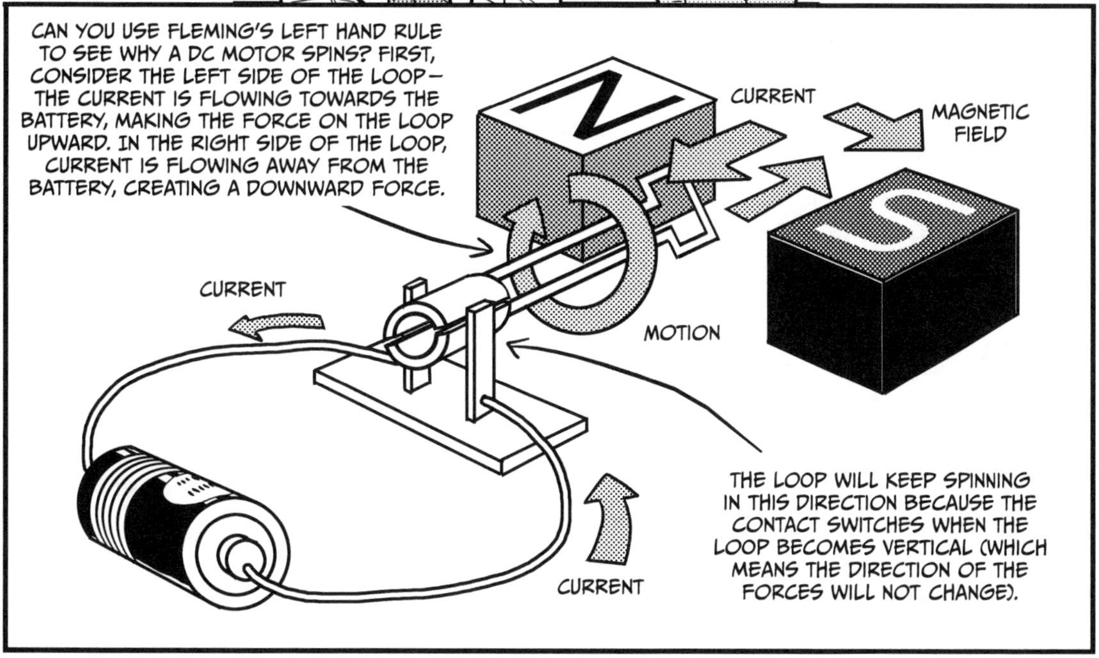
CAN YOU USE FLEMING'S LEFT HAND RULE TO SEE WHY A DC MOTOR SPINS? FIRST, CONSIDER THE LEFT SIDE OF THE LOOP— THE CURRENT IS FLOWING TOWARDS THE BATTERY, MAKING THE FORCE ON THE LOOP UPWARD. IN THE RIGHT SIDE OF THE LOOP, CURRENT IS FLOWING AWAY FROM THE BATTERY, CREATING A DOWNWARD FORCE.
N
CURRENT
MAGNETIC FIELD
S
CURRENT
MOTION
THE LOOP WILL KEEP SPINNING IN THIS DIRECTION BECAUSE THE CONTACT SWITCHES WHEN THE LOOP BECOMES VERTICAL (WHICH MEANS THE DIRECTION OF THE FORCES WILL NOT CHANGE).
CURRENT

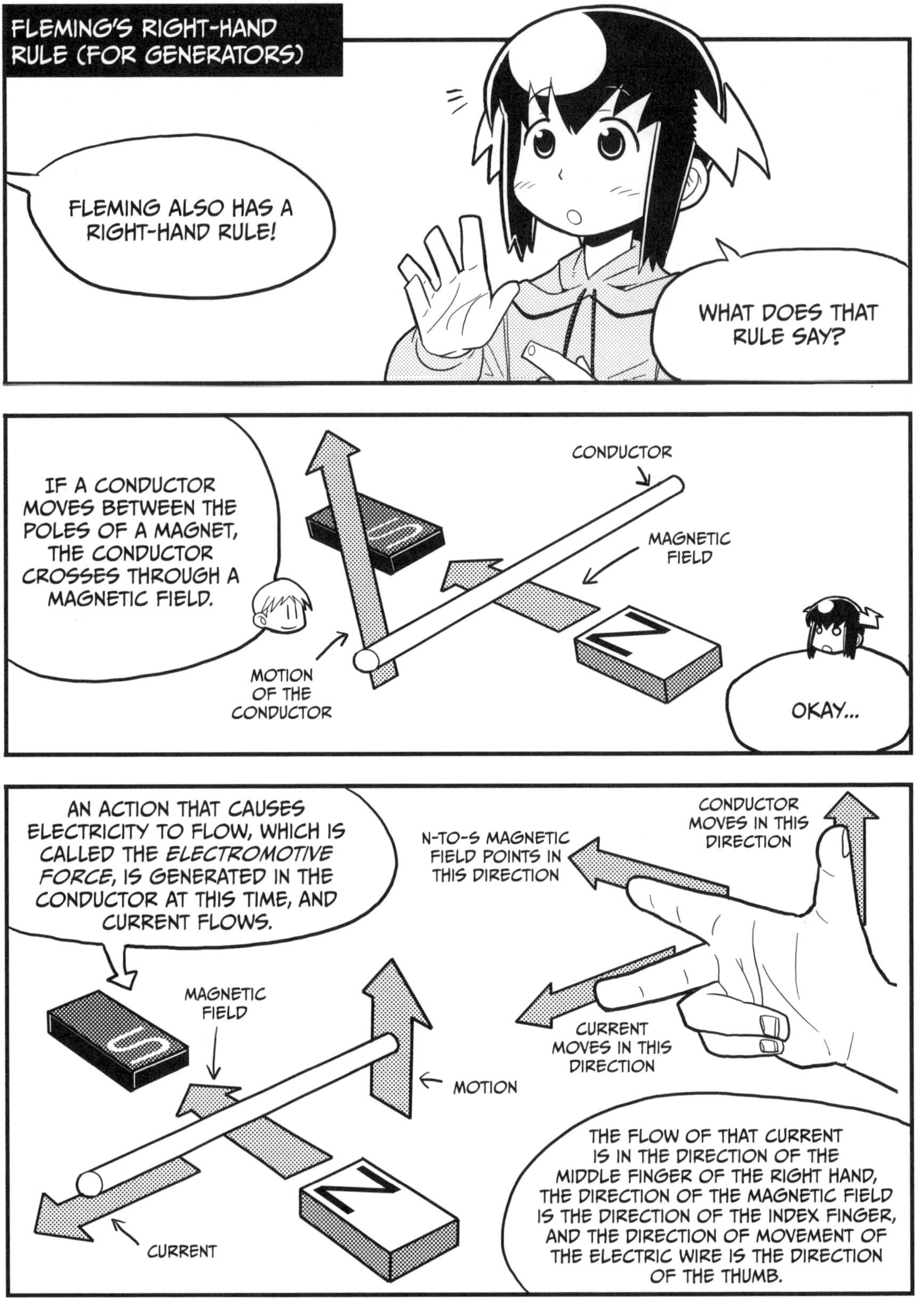
FLEMING'S RIGHT-HAND RULE (FOR GENERATORS)
FLEMING ALSO HAS A RIGHT-HAND RULE!
WHAT DOES THAT RULE SAY?
IF A CONDUCTOR MOVES BETWEEN THE POLES OF A MAGNET, THE CONDUCTOR CROSSES THROUGH A MAGNETIC FIELD.
CONDUCTOR
MAGNETIC FIELD
S
N
MOTION OF THE CONDUCTOR
OKAY...
AN ACTION THAT CAUSES ELECTRICITY TO FLOW, WHICH IS CALLED THE *ELECTROMOTIVE FORCE*, IS GENERATED IN THE CONDUCTOR AT THIS TIME, AND CURRENT FLOWS.
N-TO-S MAGNETIC FIELD POINTS IN THIS DIRECTION
CONDUCTOR MOVES IN THIS DIRECTION
MAGNETIC FIELD
S
MOTION
CURRENT MOVES IN THIS DIRECTION
N
CURRENT
THE FLOW OF THAT CURRENT IS IN THE DIRECTION OF THE MIDDLE FINGER OF THE RIGHT HAND, THE DIRECTION OF THE MAGNETIC FIELD IS THE DIRECTION OF THE INDEX FINGER, AND THE DIRECTION OF MOVEMENT OF THE ELECTRIC WIRE IS THE DIRECTION OF THE THUMB.

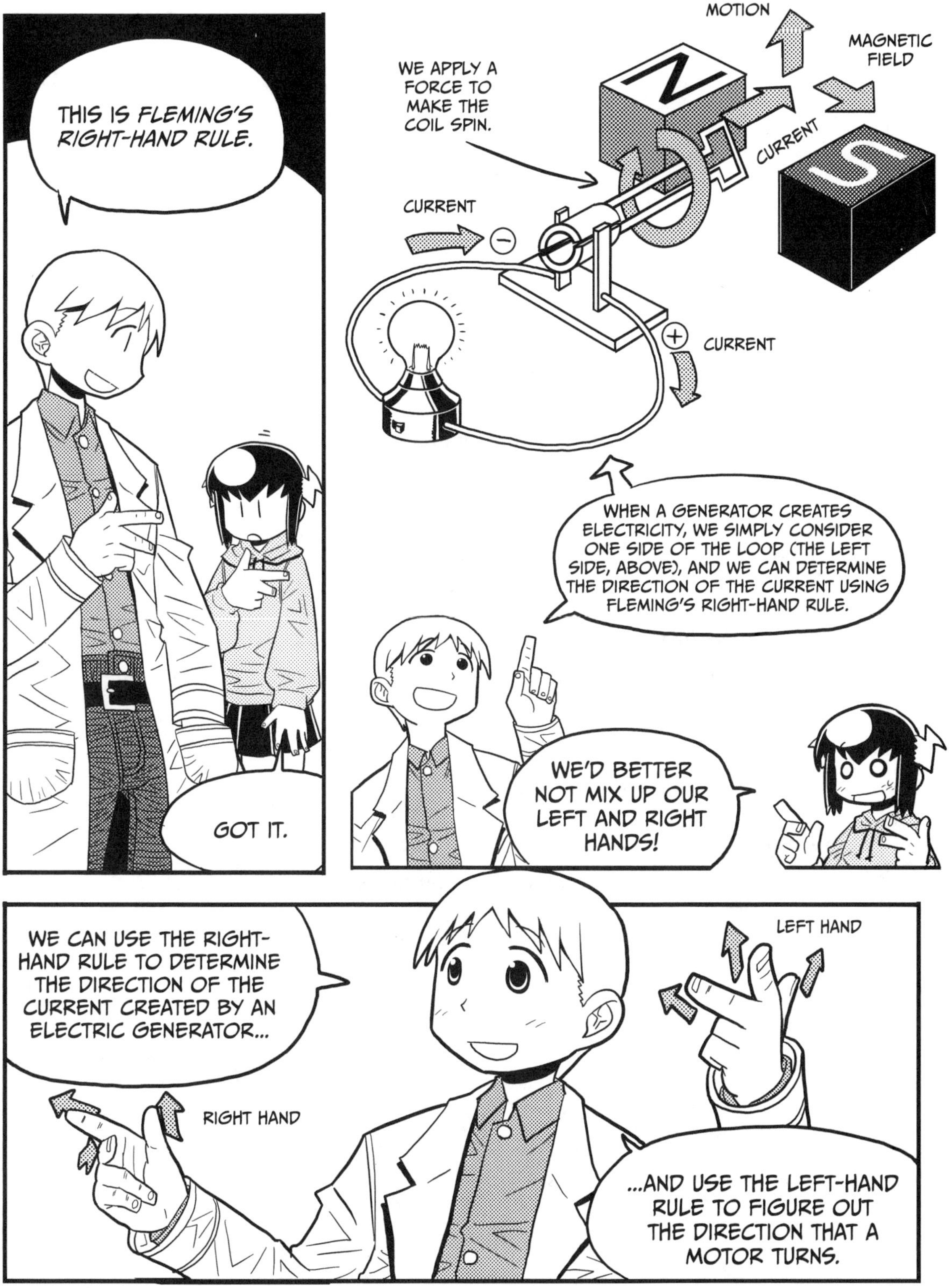
THIS IS FLEMING'S RIGHT-HAND RULE.
GOT IT.
MOTION
MAGNETIC FIELD
WE APPLY A FORCE TO MAKE THE COIL SPIN.
CURRENT
CURRENT
CURRENT
WHEN A GENERATOR CREATES ELECTRICITY, WE SIMPLY CONSIDER ONE SIDE OF THE LOOP (THE LEFT SIDE, ABOVE), AND WE CAN DETERMINE THE DIRECTION OF THE CURRENT USING FLEMING'S RIGHT-HAND RULE.
WE'D BETTER NOT MIX UP OUR LEFT AND RIGHT HANDS!
WE CAN USE THE RIGHT-HAND RULE TO DETERMINE THE DIRECTION OF THE CURRENT CREATED BY AN ELECTRIC GENERATOR...
LEFT HAND
RIGHT HAND
...AND USE THE LEFT-HAND RULE TO FIGURE OUT THE DIRECTION THAT A MOTOR TURNS.

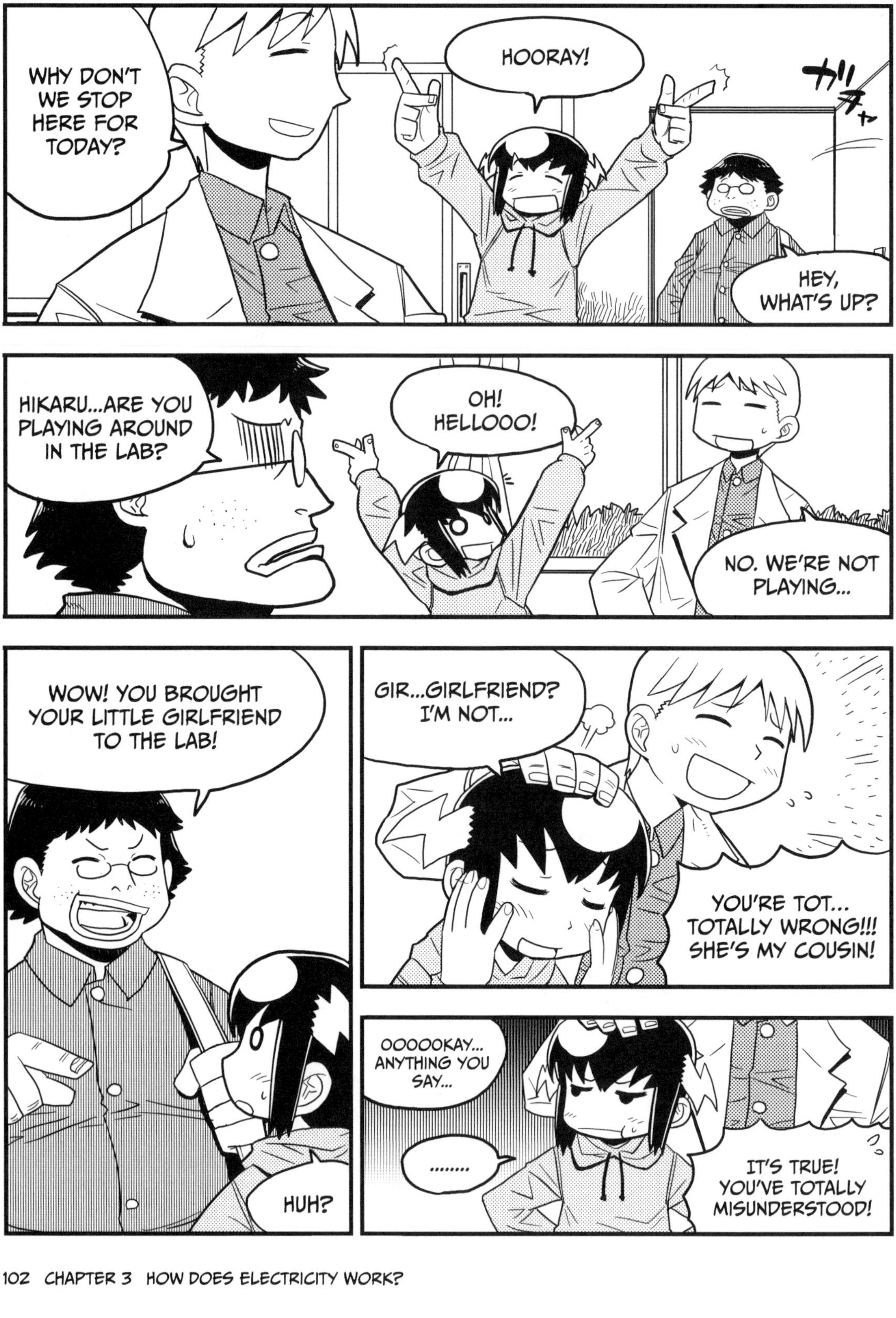
WHY DON'T WE STOP HERE FOR TODAY?
HOORAY!
HEY, WHAT'S UP?
HIKARU...ARE YOU PLAYING AROUND IN THE LAB?
OH! HELLOOO!
NO. WE'RE NOT PLAYING...
WOW! YOU BROUGHT YOUR LITTLE GIRLFRIEND TO THE LAB!
HUH?
GIR...GIRLFRIEND? I'M NOT...
YOU'RE TOT... TOTALLY WRONG!!! SHE'S MY COUSIN!
OOOOOKAY... ANYTHING YOU SAY...
....
IT'S TRUE! YOU'VE TOTALLY MISUNDERSTOOD!

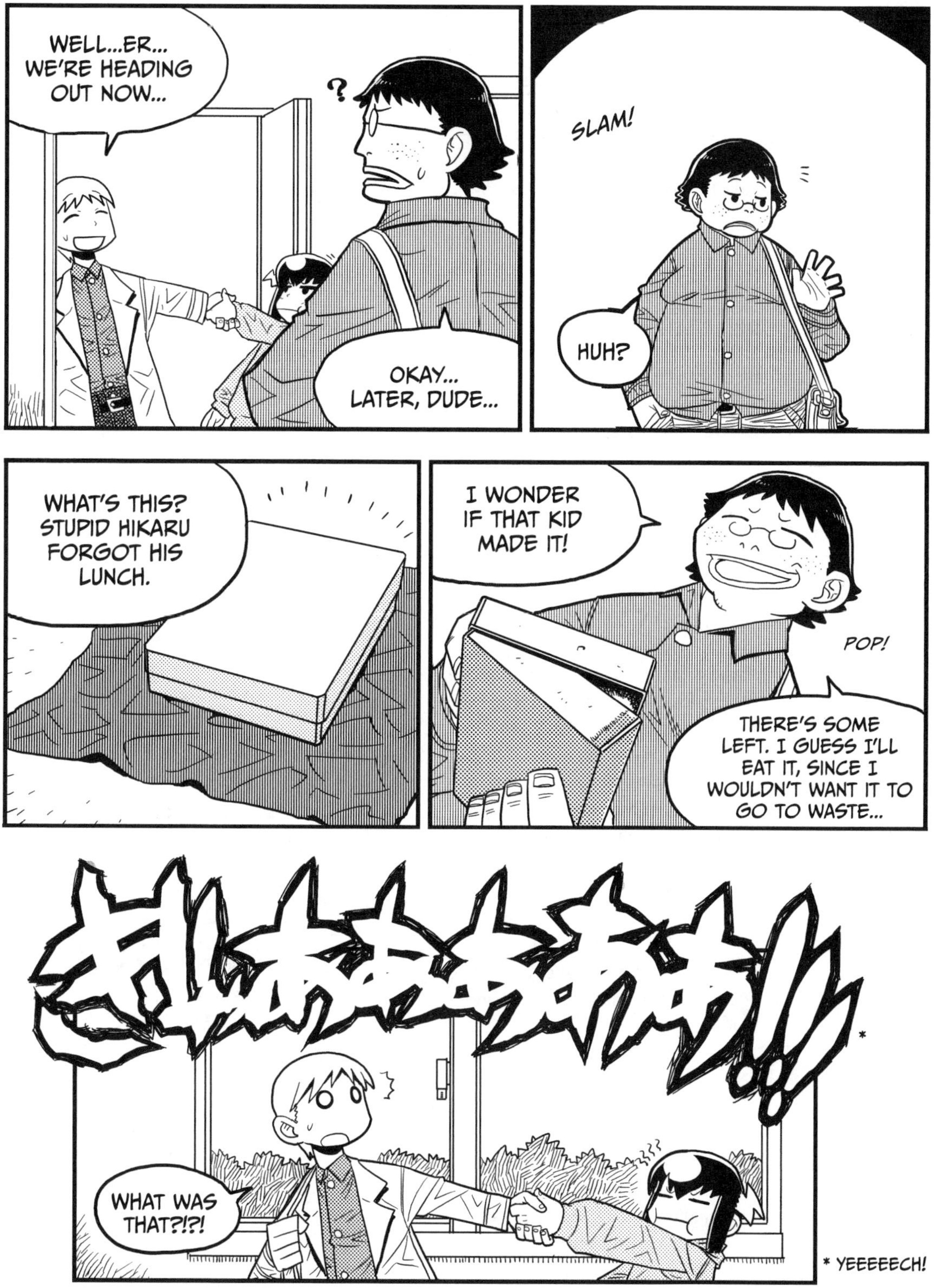
WELL...ER... WE'RE HEADING OUT NOW...
?
OKAY... LATER, DUDE...
SLAM!
HUH?
WHAT'S THIS? STUPID HIKARU FORGOT HIS LUNCH.
I WONDER IF THAT KID MADE IT!
POP!
THERE'S SOME LEFT. I GUESS I'LL EAT IT, SINCE I WOULDN'T WANT IT TO GO TO WASTE...
WHAT WAS THAT?!?!
* YEEEEECH!

JOULE HEAT

Heat that is produced when current flows through an electric resistance is called *joule heat*. For example, the amount of heat produced when current *I* flows through resistance *R* for *t* seconds can be obtained by calculating $I^2 \times R \times t$. The amount of heat is represented by the symbol *Q* and is measured in *joules (J)*, which are named after the English physicist James Prescott Joule. One joule corresponds to the electric power consumption of 1Ws (watt second)—and one joule is equivalent to a kg × m^2 / s^2. The amount of heat required to raise 1 gram of pure water from 14.5°C to 15.5°C at 1 atmosphere of pressure is approximately 4.2J, and this is equivalent to 1 *calorie (cal)*.

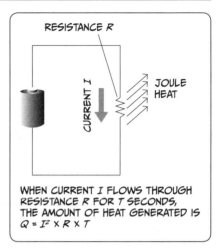

RESISTANCE *R*

CURRENT *I*

JOULE HEAT

WHEN CURRENT *I* FLOWS THROUGH RESISTANCE *R* FOR *T* SECONDS, THE AMOUNT OF HEAT GENERATED IS $Q = I^2 \times R \times T$

Resistance and joule heat

THERMAL VIBRATION

What is heat? The atoms that make up a substance are always vibrating, and this is called *thermal vibration*. The magnitude of the thermal vibration in a substance is directly related to the magnitude of the temperature of that substance—this thermal vibration of atoms is the true nature of heat.

If the atoms in a substance are not vibrating, that substance will have no temperature—that temperature is called *absolute zero*, which is equal to −273.15°C.

Even when copper wire, which is used for electric wire because of its low resistance, is at normal temperatures, the vibration of the copper atoms resists the movement of electrons, creating additional heat and additional resistance.

However, if the temperature of a material drops to near absolute zero, the vibrations of the atoms become very small. In such a state, electrons can travel much more easily—in other words, the material's resistance decreases. In some materials, such as aluminum, if the temperature becomes low enough, the electrons can move without

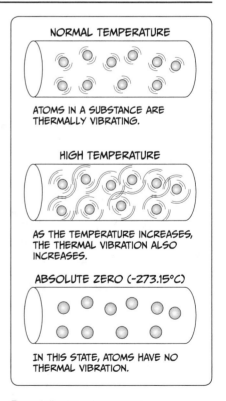

NORMAL TEMPERATURE

ATOMS IN A SUBSTANCE ARE THERMALLY VIBRATING.

HIGH TEMPERATURE

AS THE TEMPERATURE INCREASES, THE THERMAL VIBRATION ALSO INCREASES.

ABSOLUTE ZERO (−273.15°C)

IN THIS STATE, ATOMS HAVE NO THERMAL VIBRATION.

Thermal vibration and temperature

being obstructed by the atoms at all! When a material's resistance becomes zero, we call the phenomenon *superconductivity*.

Many metals are found to naturally superconduct when they are cold enough, but most need to be near absolute zero. However, since it is extremely difficult to actually lower a substance's temperature near absolute zero, research is being conducted on superconductivity phenomena that occur at temperatures much higher than absolute zero, a field called *high-temperature superconductors*. Someday, materials like these may be used to send electricity to homes everywhere without current loss due to joule heating.

CURRENT
ELECTRON
HIGH TEMPERATURE
IF THERMAL VIBRATION INCREASES, THE MOVEMENT OF ELECTRONS IS OBSTRUCTED.
CURRENT
ABSOLUTE ZERO
NEAR ABSOLUTE ZERO, THERE IS NO THERMAL VIBRATION TO OBSTRUCT THE MOVEMENT OF ELECTRONS.

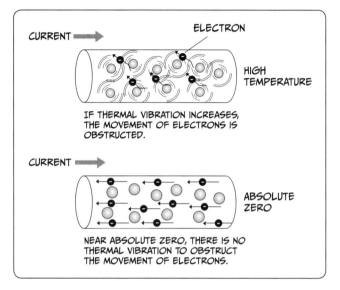

Superconductivity and current

In wires at normal temperatures, electrons will violently collide with other atoms, which creates even more thermal vibrations—that is, more heat. As a wire heats up, its resistance increases. Conversely, as its temperature decreases, electric resistance decreases.

CURRENT
HEAT
HEAT
ATOM
HEAT
HEAT
ELECTRON
THE THERMAL VIBRATION OF ATOMS INCREASES FURTHER DUE TO COLLISIONS WITH ELECTRONS, AND HEAT IS GENERATED.

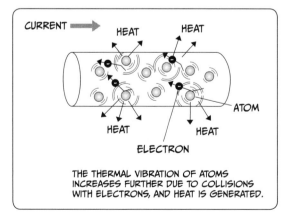

Collisions with electrons and generation of heat

ELECTROMAGNETIC WAVES

When current flows through a resistance and the temperature rises, heat is generated. At first, infrared rays that are invisible to the naked eye are emitted. *Infrared rays*, which are also called *heat rays*, are a type of *electromagnetic wave*—a wave that has thermal energy. Electromagnetic waves (in order of decreasing wavelength) include radio waves, infrared rays, visible light, ultraviolet rays, and X rays, among others. Radio waves are used for TV or radio broadcasting and communication for ships. The color of visible light varies with the wavelength—red light has the longest wavelength, and violet light has the shortest.

WAVELENGTH IS LONG.
FREQUENCY IS LOW.
WAVELENGTH IS SHORT.
FREQUENCY IS HIGH.
RADIATION:
RADIO WAVE
INFRARED RAY
VISIBLE LIGHT
RED
ORANGE
YELLOW
GREEN
BLUE
VIOLET
ULTRA-VIOLET RAY
X RAY
γ RAY
WAVELENGTH:
1MM
720NM
400NM
10NM

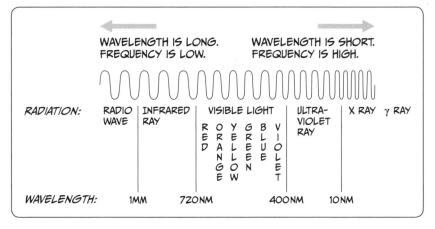

Wavelength and classification of electromagnetic waves

After infrared rays are emitted from a substance, visible light will be emitted if the temperature continues to rise. This phenomenon in which thermal energy is emitted as electromagnetic waves as the temperature of a substance increases is called *thermal emission*. This is the principle of light emission in light bulbs. Thermal emission produces red light at a low temperature, which changes to bluish white light as the temperature rises.

Light emission due to thermal emission mostly ends up becoming heat, so it is inefficient to use it as light. Light emission in which the emitter does not need to be heated is called *luminescence*; it is the principle used in fluorescent lights. In a fluorescent light, electrons that escape from the filament collide with mercury vapor inside the fluorescent tube; the ultraviolet rays that are generated at that time excite the fluorescent substance on the inner surface of the fluorescent tube and become visible light. The light emission of a fluorescent light is very efficient—for the same electric power consumption, it emits more than four times the light that a regular light bulb emits.

Light-emitting phenomena include thermal emission and luminescence, as shown here.

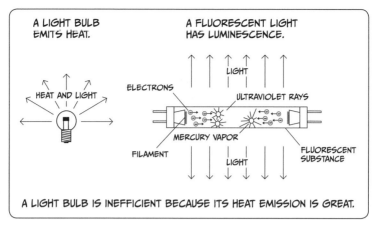

A LIGHT BULB
EMITS HEAT.

A FLUORESCENT LIGHT
HAS LUMINESCENCE.

HEAT AND LIGHT

ELECTRONS

LIGHT

ULTRAVIOLET RAYS

MERCURY VAPOR

FILAMENT

LIGHT

FLUORESCENT
SUBSTANCE

A LIGHT BULB IS INEFFICIENT BECAUSE ITS HEAT EMISSION IS GREAT.

Light emission of a light bulb and a fluorescent tube

ELECTRICITY AND MAGNETISM

If iron filings are sprinkled on a sheet of paper placed over a bar magnet, a pattern of lines is produced. These lines originate from the north (N) pole and lead toward the south (S) pole; they are called a *magnetic field*.

Magnetic fields are also generated when current flows. This phenomenon is extremely important when using electricity, and many common electrical appliances make use of it.

When current flows in an electric wire, a magnetic field with a circular pattern is generated around that wire. This is called *Ampère's law*. The magnitude of this magnetic field varies according to the strength of the current; if the direction of the current changes, the direction of the magnetic field also changes.

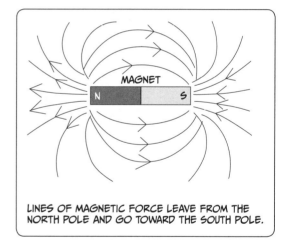

LINES OF MAGNETIC FORCE LEAVE FROM THE
NORTH POLE AND GO TOWARD THE SOUTH POLE.

A magnet and magnetic fields

If current of the same magnitude flows in the same direction in two electric wires placed side by side, the magnetic fields generated in each wire are combined to generate a magnetic field of twice the current around both conductors. At this time, a force of attraction is generated between the two electric wires. If current is flowing in opposite directions in two wires, a force of repulsion is generated between the wires. In this case, the magnetic fields around the wires negate each other and become smaller.

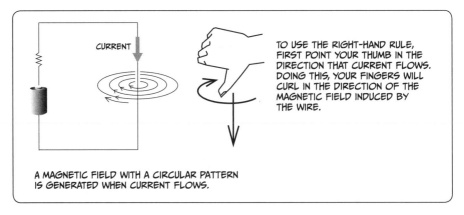

CURRENT

TO USE THE RIGHT-HAND RULE, FIRST POINT YOUR THUMB IN THE DIRECTION THAT CURRENT FLOWS. DOING THIS, YOUR FINGERS WILL CURL IN THE DIRECTION OF THE MAGNETIC FIELD INDUCED BY THE WIRE.

A MAGNETIC FIELD WITH A CIRCULAR PATTERN IS GENERATED WHEN CURRENT FLOWS.

Ampère's law

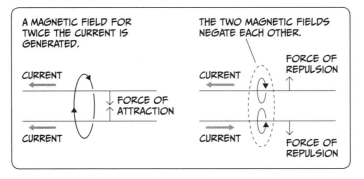

A MAGNETIC FIELD FOR TWICE THE CURRENT IS GENERATED.

THE TWO MAGNETIC FIELDS NEGATE EACH OTHER.

FORCE OF REPULSION

CURRENT

CURRENT

↓ FORCE OF
↑ ATTRACTION

CURRENT

CURRENT

FORCE OF REPULSION

Forces generated when current flows in two conductors

The additive property of magnetic fields also holds true for more than two wires (for example, a coil). In this way, a large magnetic field can be generated.

FLEMING'S LEFT-HAND RULE AND MOTORS

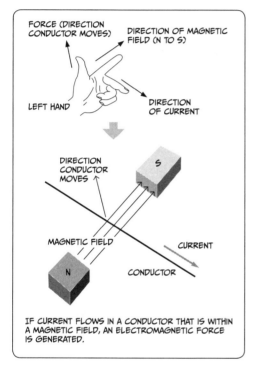

If current flows in a conductor that is within a magnetic field, an *electromagnetic force* is generated on the conductor. *Fleming's left-hand rule* indicates an easy-to-remember relationship among the directions of the magnetic field, the current, and the movement of the conductor. This rule says that when you extend the thumb, index finger, and middle finger of your left hand so they are mutually perpendicular, the index finger points in the direction of the magnetic field, the middle finger points in the direction of the current, and the thumb points in the direction that the conductor moves (the direction of the electromagnetic force). The name of this rule comes from the name of the English electrical engineer John Ambrose Fleming who defined it. You can determine the direction of rotation of a motor by using Fleming's left-hand rule.

FORCE (DIRECTION CONDUCTOR MOVES)
DIRECTION OF MAGNETIC FIELD (N TO S)
LEFT HAND
DIRECTION OF CURRENT
DIRECTION CONDUCTOR MOVES
S
MAGNETIC FIELD
CURRENT
N
CONDUCTOR
IF CURRENT FLOWS IN A CONDUCTOR THAT IS WITHIN A MAGNETIC FIELD, AN ELECTROMAGNETIC FORCE IS GENERATED.

Fleming's left-hand rule

DIRECTION OF ROTATION
N
CURRENT
MAGNETIC FIELD
FORCE
S
CURRENT
FORCE
CURRENT
A MOTOR ROTATES DUE TO THE ELECTROMAGNETIC FORCE GENERATED ON THE CONDUCTOR BY CURRENT FLOWING IN THE CONDUCTOR WITHIN A MAGNETIC FIELD.

Motor rotation

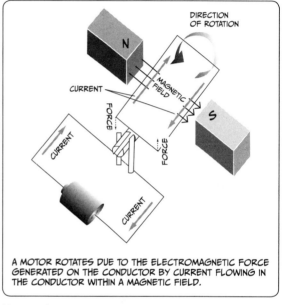

FLEMING'S RIGHT-HAND RULE AND ELECTRIC GENERATORS

You can determine the direction of the electromotive force created by an electric generator by using *Fleming's right-hand rule*. When a conductor moves between the poles of a magnet, the conductor crosses a magnetic field facing from the north (N) pole to the south (S) pole of the magnet; electromotive force is thus generated in the conductor, and current flows. *Fleming's right-hand rule* indicates an easy-to-remember relationship among the directions of the magnetic field, the movement of the conductor, and the current. When you extend the thumb, index finger, and middle finger of your right hand so they are mutually perpendicular, the index finger points in the direction of the magnetic field, the thumb points in the direction that the conductor moves, and the middle finger points in the direction of the current.

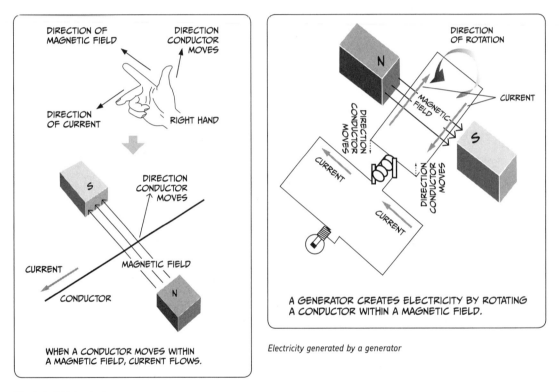

DIRECTION OF MAGNETIC FIELD
DIRECTION CONDUCTOR MOVES
DIRECTION OF CURRENT
RIGHT HAND
S
DIRECTION CONDUCTOR MOVES
CURRENT
MAGNETIC FIELD
CONDUCTOR
N
WHEN A CONDUCTOR MOVES WITHIN A MAGNETIC FIELD, CURRENT FLOWS.

Fleming's right-hand rule

N
DIRECTION OF ROTATION
CURRENT
MAGNETIC FIELD
DIRECTION CONDUCTOR MOVES
S
CURRENT
DIRECTION CONDUCTOR MOVES
CURRENT
A GENERATOR CREATES ELECTRICITY BY ROTATING A CONDUCTOR WITHIN A MAGNETIC FIELD.

Electricity generated by a generator

You must apply a force to keep the loop spinning within the magnetic field. This could be the force from falling water, like in a hydroelectric generator, or the force of pressurized steam, like in a coal power plant.

But why do Fleming's hand laws work? We can better understand why generators and motors work by understanding how magnetism and electricity are related.

ELECTRICITY AND COILS

An electric wire wound in loops is called a *coil*. If current flows in a coil, a magnetic field is generated that goes through the inside of the coil. If an iron core is inserted in the coil, the magnetic field is concentrated in the iron, and it becomes a strong *electromagnet*. The strength of an electromagnet is proportional to the product of the current and the number of loops in the coil; if the direction of the current is reversed, the polarity of the electromagnet is also reversed. If the current is stopped, the magnetic force of the electromagnet disappears.

CURRENT
MAGNETIC FIELD
CURRENT
CURRENT
CURRENT
MAGNETIC FIELD

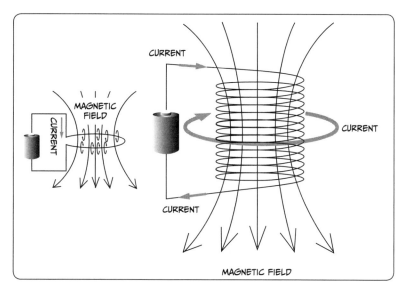

Magnetic field created by a coil of electric wire

You can use your right hand to find the orientation of the magnetic field induced by a coil. Just curl your fingers in the direction the current flows in the coil, and your thumb will point towards the N pole of the induced magnetic field.

COILS AND ELECTROMAGNETIC INDUCTION

When a bar magnet moves within a coil, current flows in that coil, which creates a magnetic field in order to oppose the change in magnetism. If the direction of the bar magnet's movement changes, the direction of the current in the wire also changes. This phenomenon is called *electromagnetic induction*, and the electricity that is generated during this process is called *induced electromotive force*. The current generated is called *induced current*.

Lenz's law, discovered by the Russian physicist Heinrich Friedrich Emil Lenz, states that the current due to electromagnetic induction flows in a direction such that the magnetic field produced by that current obstructs the motion of the magnet.

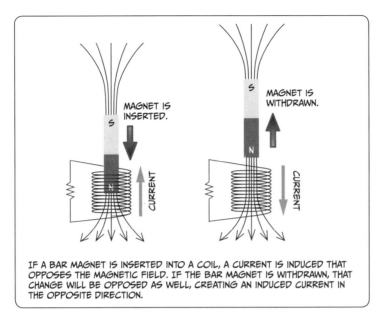

MAGNET IS
INSERTED.
S
N
CURRENT
S
MAGNET IS
WITHDRAWN.
N
CURRENT
IF A BAR MAGNET IS INSERTED INTO A COIL, A CURRENT IS INDUCED THAT OPPOSES THE MAGNETIC FIELD. IF THE BAR MAGNET IS WITHDRAWN, THAT CHANGE WILL BE OPPOSED AS WELL, CREATING AN INDUCED CURRENT IN THE OPPOSITE DIRECTION.

Electromagnetic induction

COILS AND INDUCTANCE

If a coil is connected to a battery and current begins flowing, the magnetic field generated becomes larger, and the coil becomes an electromagnet. At this time, an induced electromotive force is generated on the coil itself due to the varying magnetic fields. This is called *self-induction* or simply *inductance*.

When the current to the coil is cut off, the magnetic field begins to disappear, and an induced electromotive force is generated in the direction that obstructs the flow of current in the coil. This is called a *counter-electromotive force*. The counter-electromotive force can be easily verified. When a battery is connected to a coil and current flows, a magnetic field is generated. When the current is constant, no counter-electromotive force is generated, but when the battery is detached and the current is cut off, the magnetic field that was being generated becomes smaller. At this time, voltage due to the counter-electromotive force appears at both ends of the coil.

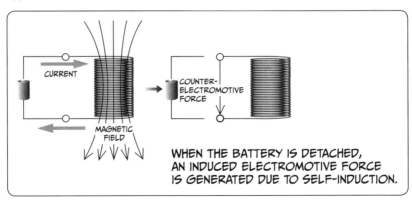

CURRENT
MAGNETIC
FIELD
COUNTER-
ELECTROMOTIVE
FORCE
WHEN THE BATTERY IS DETACHED,
AN INDUCED ELECTROMOTIVE FORCE
IS GENERATED DUE TO SELF-INDUCTION.

Self-induction of a coil

COILS AND ALTERNATING CURRENT

The magnitude of alternating current is always changing. If alternating current flows in a coil, an induced electromotive force is generated in the coil in the direction that obstructs the flow of the current, and current flows so that it lags behind the power supply voltage variation by one-fourth of a cycle. This is called the *lagging current*, and it flows in an electrical device such as a motor with a coil. This temporal lag is called a *phase difference*. The coil acts like a resistance to the current as described above. This is called *inductive reactance*, and its magnitude is proportional to the frequency of the alternating current.

Electric power consumption is represented by the product of voltage and current, and when the voltage and current waves match with respect to time, 100 percent work is done—in other words, "the power factor is 100 percent." If the current lags, the power factor will be less than 100 percent, and the circuit is said to have a "low power factor."

If the power factor is low, the electric power that is input from the power supply will not do 100 percent work, so a power supply having a correspondingly larger capacity is required. The ratio of the consumed power to the input power is the *power factor.*

$$\text{Power factor} = \frac{\text{Consumed power}}{\text{Input power}}$$

A low power factor means that some current returns to the power supply without doing work.

CURRENT I
AC POWER SUPPLY E
COIL
LAGGING CURRENT
POWER SUPPLY VOLTAGE E
O
THE CURRENT FLOW LAGS BEHIND THE VOLTAGE VARIATION BY ONE-FOURTH OF A CYCLE.

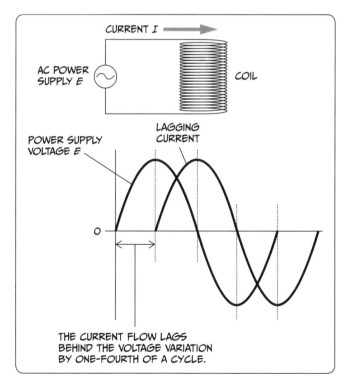

Lagging current flowing in a coil

COILS AND TRANSFORMERS

If an AC power supply is connected to coil 1, a magnetic field is generated. When this magnetic field varies within coil 2, an induced electromotive force is generated in coil 2. This phenomenon is called *mutual induction*. A *transformer* is an electrical device that uses this phenomenon to change voltage.

If two coils are wrapped around an iron core, and an AC power supply is connected to coil 1, a magnetic field is generated and passes through the inside of the iron core. Since coil 2 has been wrapped around the same iron core, the magnetic field varies inside coil 2, and an induced electromotive force is generated in coil 2.

The power supply side of a transformer is called the *primary side*, and the load side is called the *secondary side*. The voltage generated on the secondary side is determined by the ratio of the number of turns (n_1) of the primary coil and the number of turns (n_2) of the secondary coil. For example, if the number of turns of the secondary coil is twice that of the primary coil, twice the voltage is generated on the secondary side. The current that flows

MAGNETIC FIELD
COIL 1
AC POWER SUPPLY
COIL 2
AN INDUCED ELECTROMOTIVE FORCE IS GENERATED
THE MAGNETIC FIELD THAT IS GENERATED IN COIL 1 PASSES THROUGH THE MIDDLE OF COIL 2, AND AN INDUCED ELECTROMOTIVE FORCE IS GENERATED DUE TO MUTUAL INDUCTION.
TRANSFORMER
IRON CORE
PRIMARY SIDE
SECONDARY SIDE
I_1
I_2
AC POWER SUPPLY
V_1
V_2
LOAD
NUMBER OF TURNS OF N_2
MAGNETIC FIELD
TRANSFORMATION RATIO $N = \frac{V_1}{V_2}$
$V_1 \times I_1 = V_2 \times I_2$

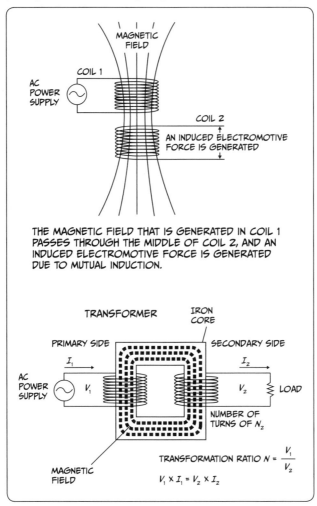

Mutual induction in a transformer

in the secondary coil at this time will be half the current that flows in the primary coil. The equation that describes this relationship is:

$$V_1 I_1 = V_2 I_2$$

The ratio of the primary voltage (V_1) and secondary voltage (V_2) is called the *transformation ratio*, and the product of the primary voltage and the current will be equal to the product of the secondary voltage and the current. In other words, a transformer only changes voltage—it does not change the magnitude of the electric power.

CAPACITORS

When an insulator is sandwiched between two metal plates and a battery is connected, electrons move from the negative pole of the battery to the bottom metal plate to charge it. Since the electrons in the top metal plate move to the positive pole of the battery, the top metal plate is positively charged. At this time, charge is stored on the metal plates. An object that stores charge in this way is called a *capacitor*.

Current flows from the instant the battery is connected, but eventually the electrons will stop moving, as charge builds on the capacitor. In other words, if a DC power supply is connected to a capacitor, current flows only at first and then stops because of the gap in the circuit. If the battery is detached in this state, the charge remains stored on the capacitor. If the battery is then connected in the reverse direction, the charge that had been stored discharges, and the capacitor is charged in the opposite direction.

The ability of the capacitor to store charge in this way is called *capacitance*; its magnitude is directly proportional to the area of the metal plate and inversely proportional to the distance between the metal plates. Capacitance is measured in *farads (F)*.

THE GREATER THE PLATE'S AREA, THE MORE CHARGE IS STORED.
+
-
TWO METAL PLATES
THE SMALLER THE SEPARATION BETWEEN THE PLATES, THE MORE CHARGE IS STORED.
CAPACITOR

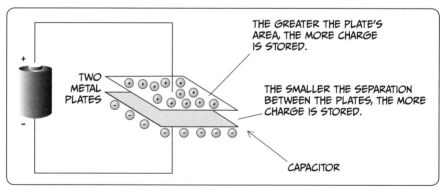

Charge stored on a capacitor

CAPACITORS AND ALTERNATING CURRENT

If AC voltage is applied to a capacitor, a charged current flows until the power supply voltage reaches its maximum (starting from 0V). The current is zero at the power supply's peak voltage. When the power supply voltage decreases from its peak voltage, discharging begins, and the discharge current reaches its maximum when the power supply voltage is 0V.

At this time, the polarity of the power supply voltage changes, and a current flows again. Charging stops when the power supply voltage reaches its peak voltage for the opposite polarity, and then discharging occurs again.

If a capacitor is connected to an AC power supply, the variation of the current is one-fourth of a cycle ahead of the variation of the power supply voltage; this current is called *leading current*.

A capacitor works like resistance to alternating current. This is called *capacitive reactance*, and its magnitude is inversely proportional to the frequency.

If an AC circuit has a coil, the current lags, and the power factor decreases. If a capacitor is connected to that circuit, the current leads, and the power factor increases.

In an AC circuit, capacitors and coils act like resistance, and are called *impedance*.

CURRENT I
CHARGING
DISCHARGING
AC POWER SUPPLY E
CAPACITOR
LEADING CURRENT I
POWER SUPPLY VOLTAGE E
DISCHARGE CURRENT
+V
0V
TIME
-V
CHARGE CURRENT
CHARGE CURRENT
DISCHARGE CURRENT
ONE-FOURTH OF A CYCLE
ONE CYCLE

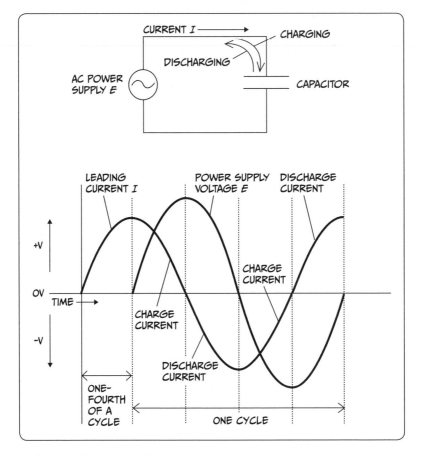

Leading current flowing in a capacitor

4

HOW DO YOU CREATE ELECTRICITY?

GENERATORS

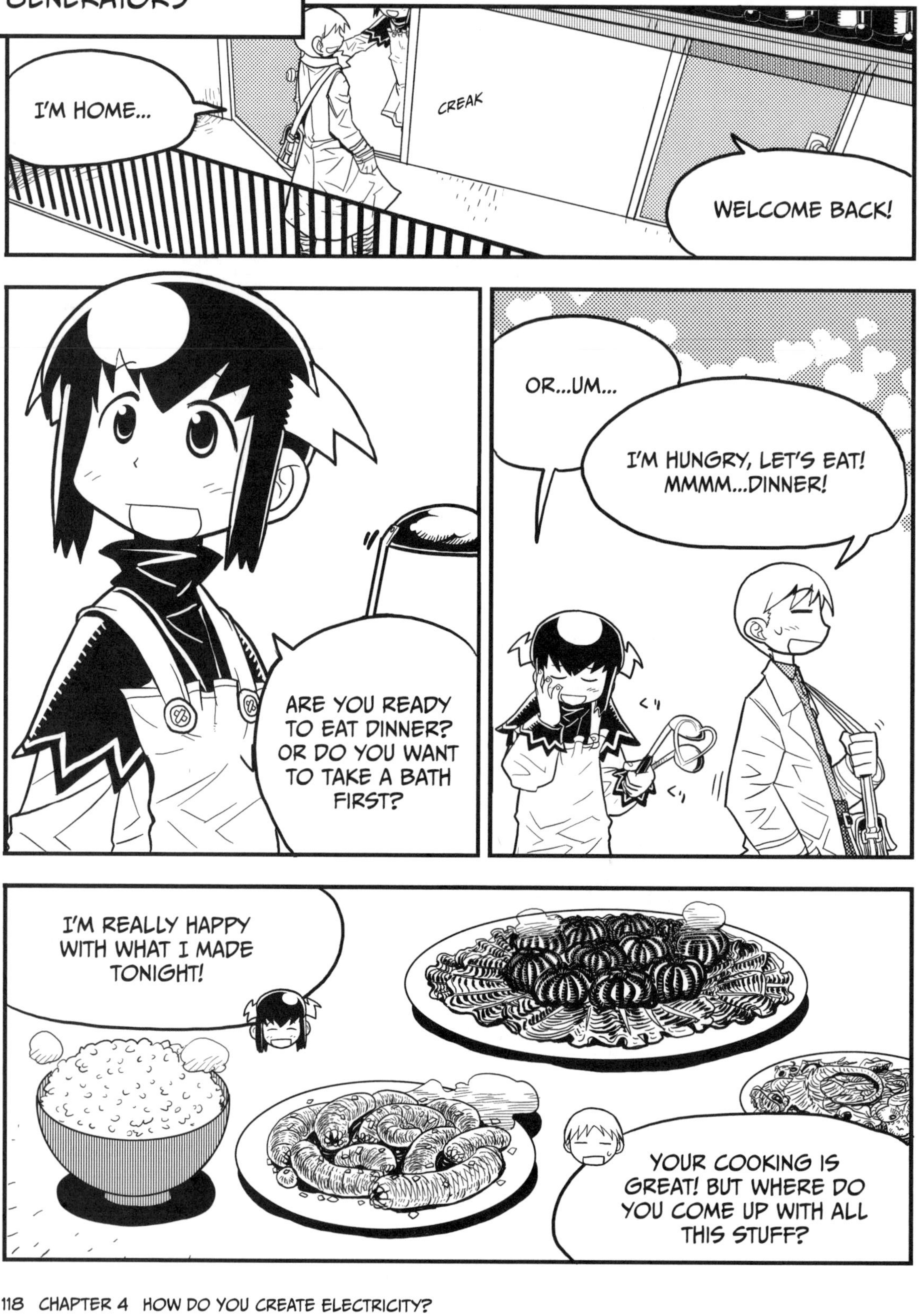

I'M HOME...
CREAK
WELCOME BACK!
ARE YOU READY TO EAT DINNER? OR DO YOU WANT TO TAKE A BATH FIRST?
OR...UM...
I'M HUNGRY, LET'S EAT! MMMM...DINNER!
I'M REALLY HAPPY WITH WHAT I MADE TONIGHT!
YOUR COOKING IS GREAT! BUT WHERE DO YOU COME UP WITH ALL THIS STUFF?

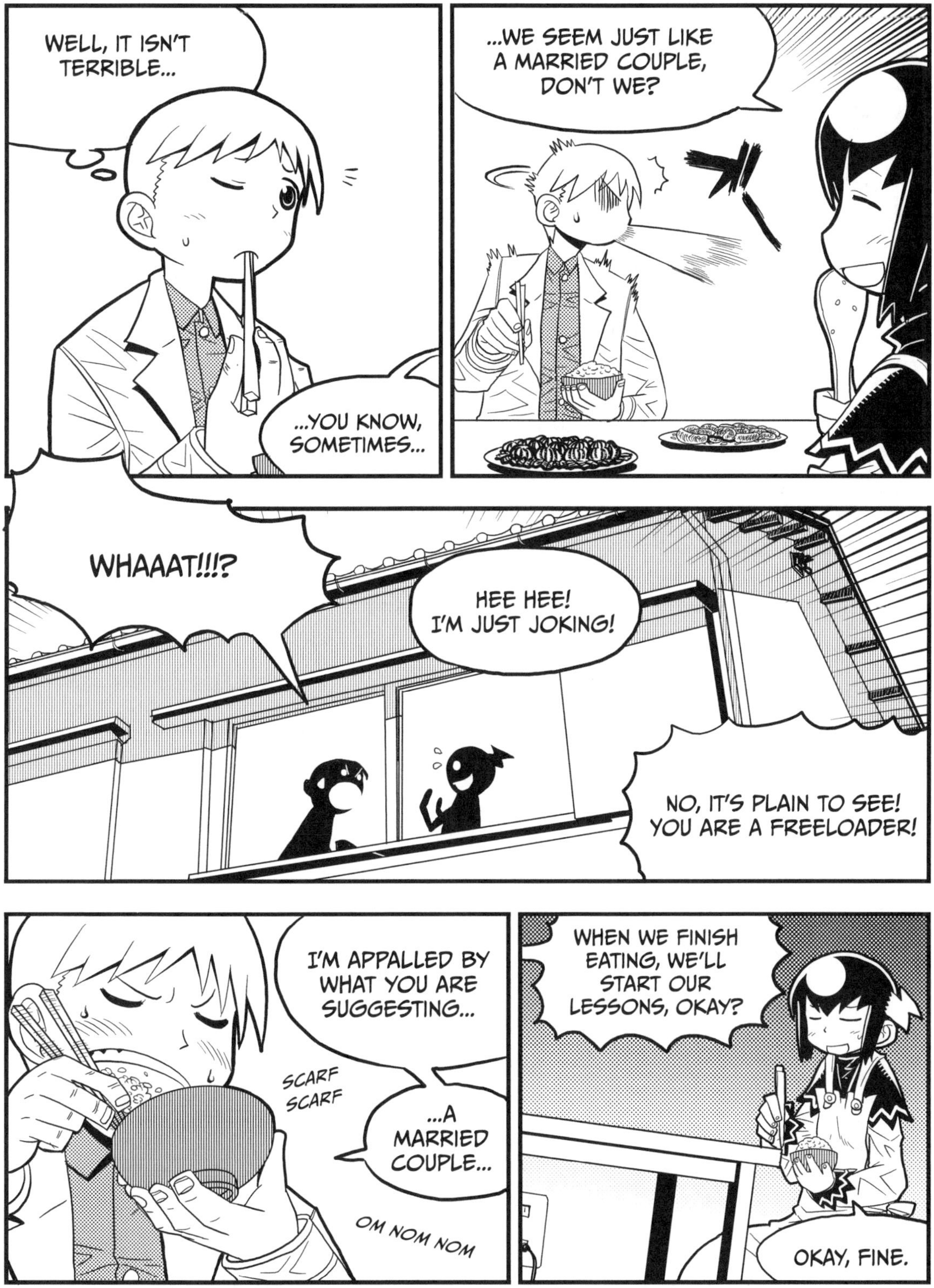

WELL, IT ISN'T TERRIBLE...
...YOU KNOW, SOMETIMES...
...WE SEEM JUST LIKE A MARRIED COUPLE, DON'T WE?
WHAAAT!!!?
HEE HEE! I'M JUST JOKING!
NO, IT'S PLAIN TO SEE! YOU ARE A FREELOADER!
I'M APPALLED BY WHAT YOU ARE SUGGESTING...
SCARF SCARF
...A MARRIED COUPLE...
OM NOM NOM
WHEN WE FINISH EATING, WE'LL START OUR LESSONS, OKAY?
OKAY, FINE.

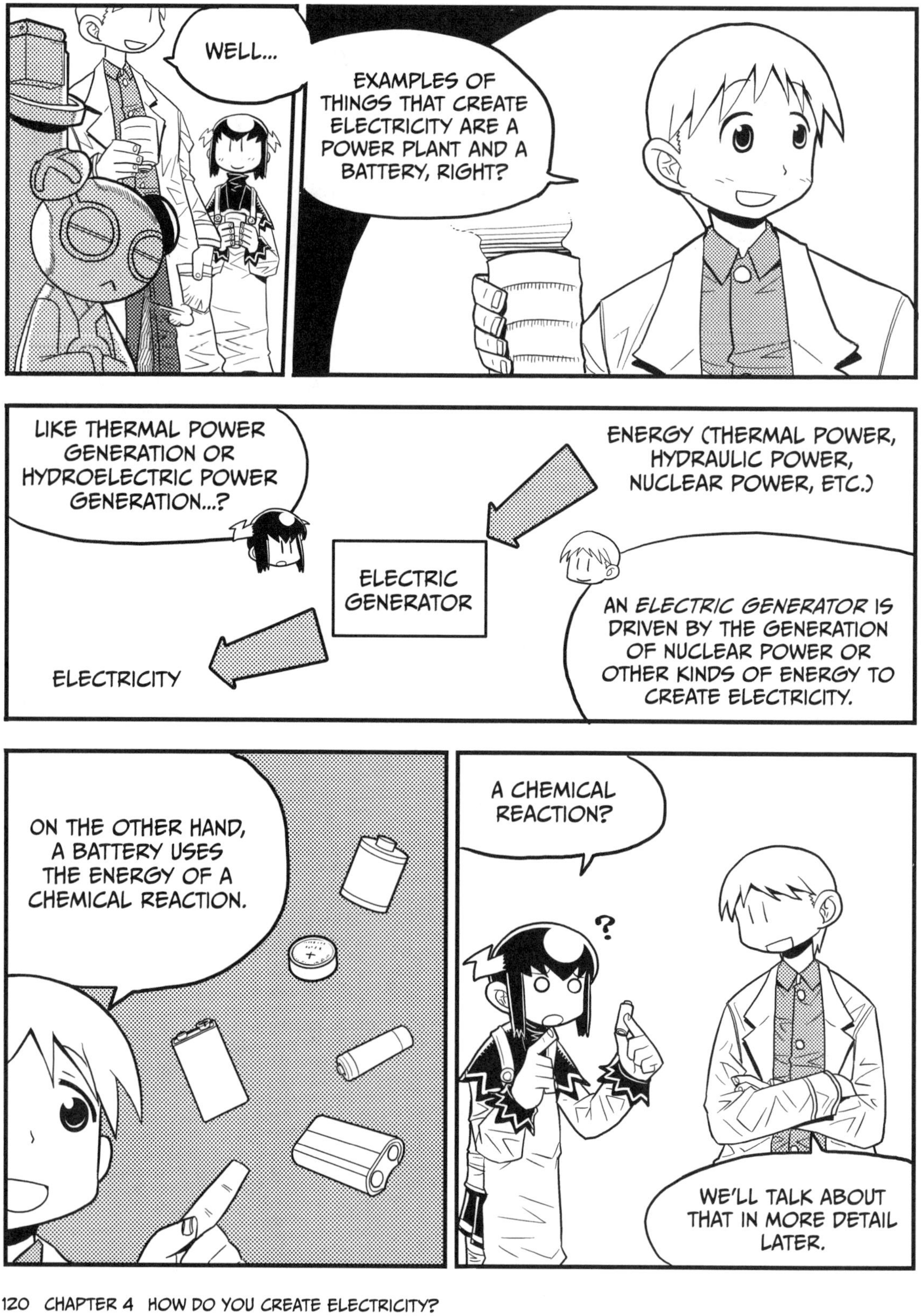

WELL...
EXAMPLES OF THINGS THAT CREATE ELECTRICITY ARE A POWER PLANT AND A BATTERY, RIGHT?
LIKE THERMAL POWER GENERATION OR HYDROELECTRIC POWER GENERATION...?
ENERGY (THERMAL POWER, HYDRAULIC POWER, NUCLEAR POWER, ETC.)
ELECTRIC GENERATOR
ELECTRICITY
AN *ELECTRIC GENERATOR* IS DRIVEN BY THE GENERATION OF NUCLEAR POWER OR OTHER KINDS OF ENERGY TO CREATE ELECTRICITY.
ON THE OTHER HAND, A BATTERY USES THE ENERGY OF A CHEMICAL REACTION.
A CHEMICAL REACTION?
?
WE'LL TALK ABOUT THAT IN MORE DETAIL LATER.

HOW DOES A POWER GENERATOR CREATE ELECTRICITY?

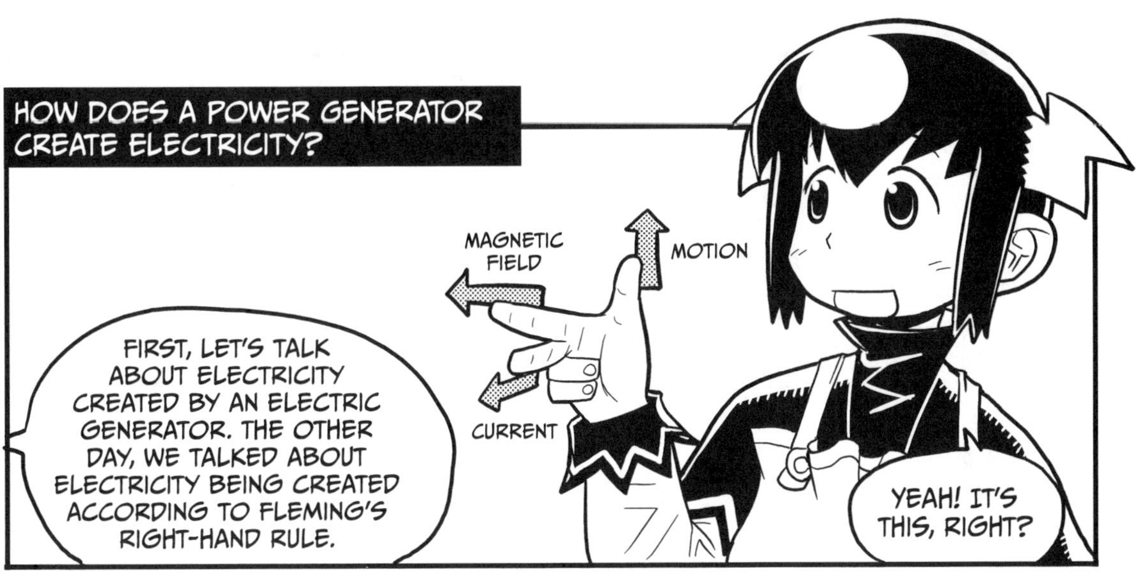

MAGNETIC FIELD
MOTION
CURRENT
FIRST, LET'S TALK ABOUT ELECTRICITY CREATED BY AN ELECTRIC GENERATOR. THE OTHER DAY, WE TALKED ABOUT ELECTRICITY BEING CREATED ACCORDING TO FLEMING'S RIGHT-HAND RULE.
YEAH! IT'S THIS, RIGHT?

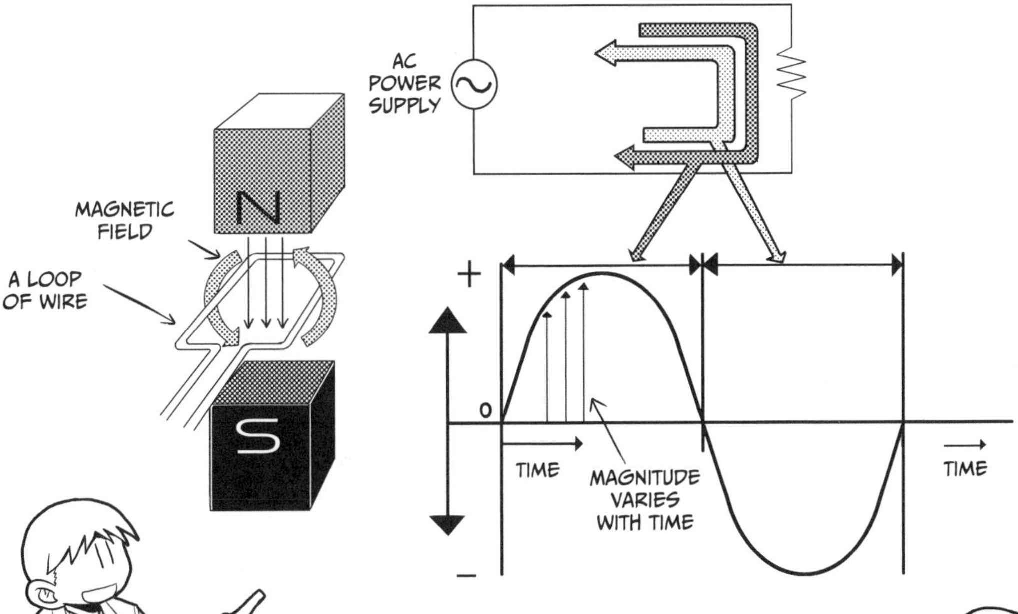

AC POWER SUPPLY
MAGNETIC FIELD
A LOOP OF WIRE
N
S
+
0
–
TIME
MAGNITUDE VARIES WITH TIME
TIME

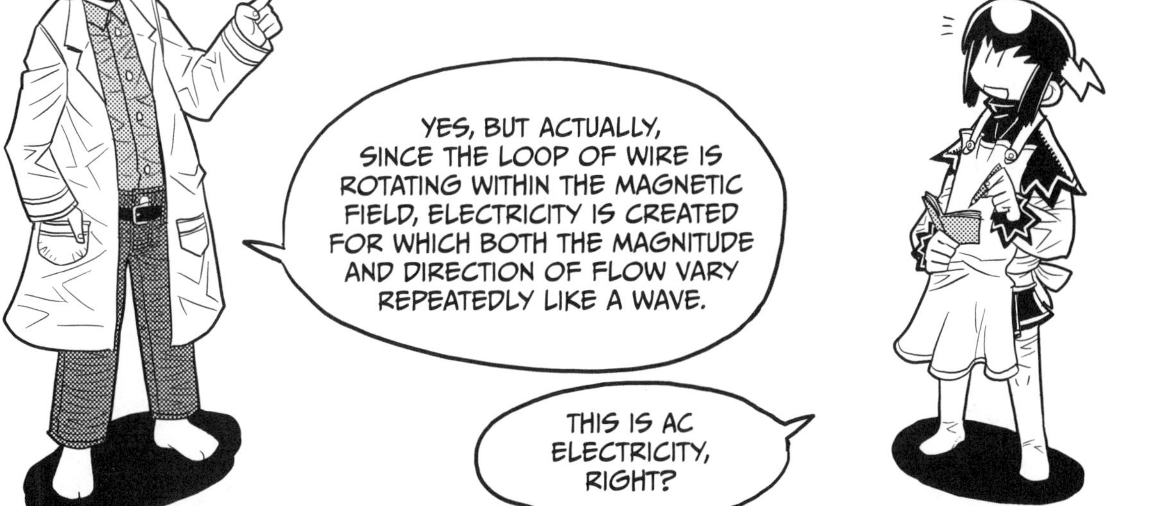

YES, BUT ACTUALLY, SINCE THE LOOP OF WIRE IS ROTATING WITHIN THE MAGNETIC FIELD, ELECTRICITY IS CREATED FOR WHICH BOTH THE MAGNITUDE AND DIRECTION OF FLOW VARY REPEATEDLY LIKE A WAVE.
THIS IS AC ELECTRICITY, RIGHT?

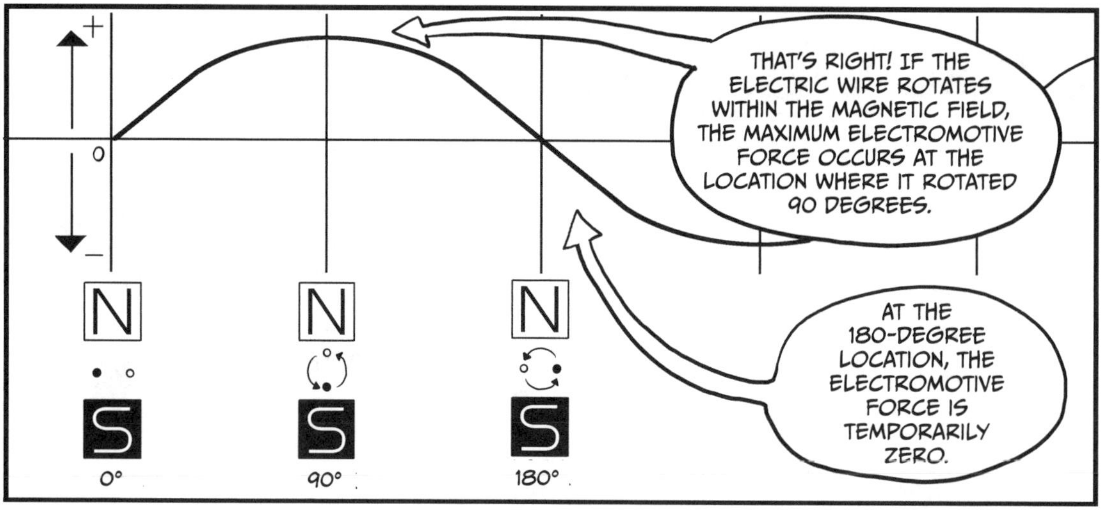
+
0
–
THAT'S RIGHT! IF THE ELECTRIC WIRE ROTATES WITHIN THE MAGNETIC FIELD, THE MAXIMUM ELECTROMOTIVE FORCE OCCURS AT THE LOCATION WHERE IT ROTATED 90 DEGREES.
N
N
N
S
S
S
0°
90°
180°
AT THE 180-DEGREE LOCATION, THE ELECTROMOTIVE FORCE IS TEMPORARILY ZERO.
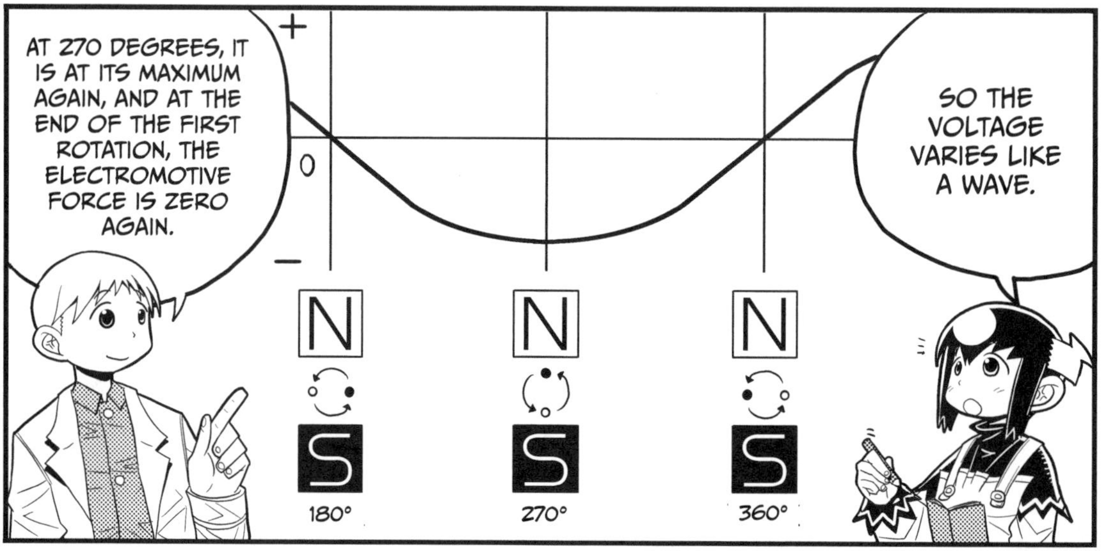
AT 270 DEGREES, IT IS AT ITS MAXIMUM AGAIN, AND AT THE END OF THE FIRST ROTATION, THE ELECTROMOTIVE FORCE IS ZERO AGAIN.
+
0
–
N
N
N
S
S
S
180°
270°
360°
SO THE VOLTAGE VARIES LIKE A WAVE.
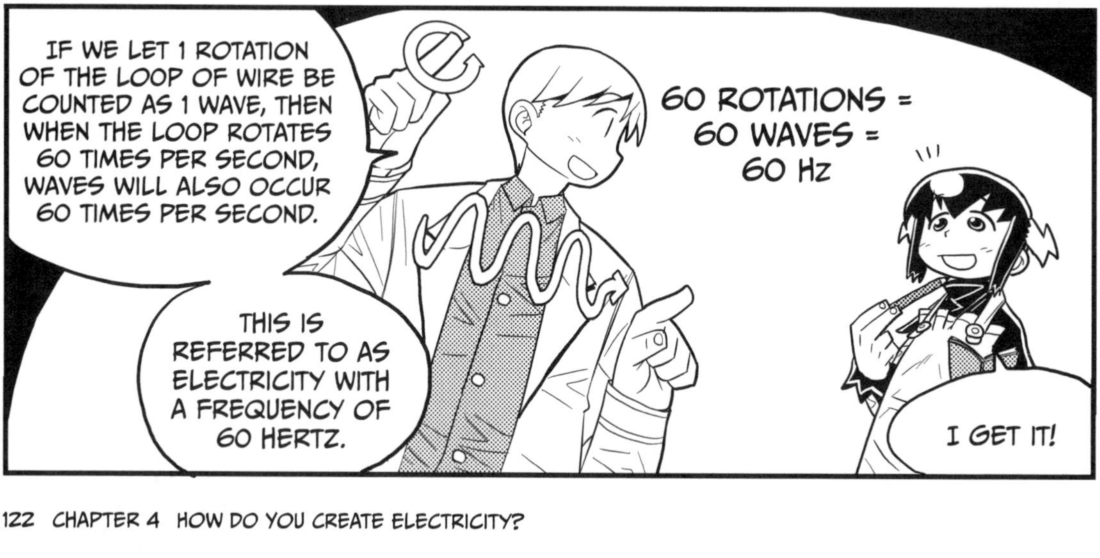
IF WE LET 1 ROTATION OF THE LOOP OF WIRE BE COUNTED AS 1 WAVE, THEN WHEN THE LOOP ROTATES 60 TIMES PER SECOND, WAVES WILL ALSO OCCUR 60 TIMES PER SECOND.
60 ROTATIONS = 60 WAVES = 60 Hz
THIS IS REFERRED TO AS ELECTRICITY WITH A FREQUENCY OF 60 HERTZ.
I GET IT!

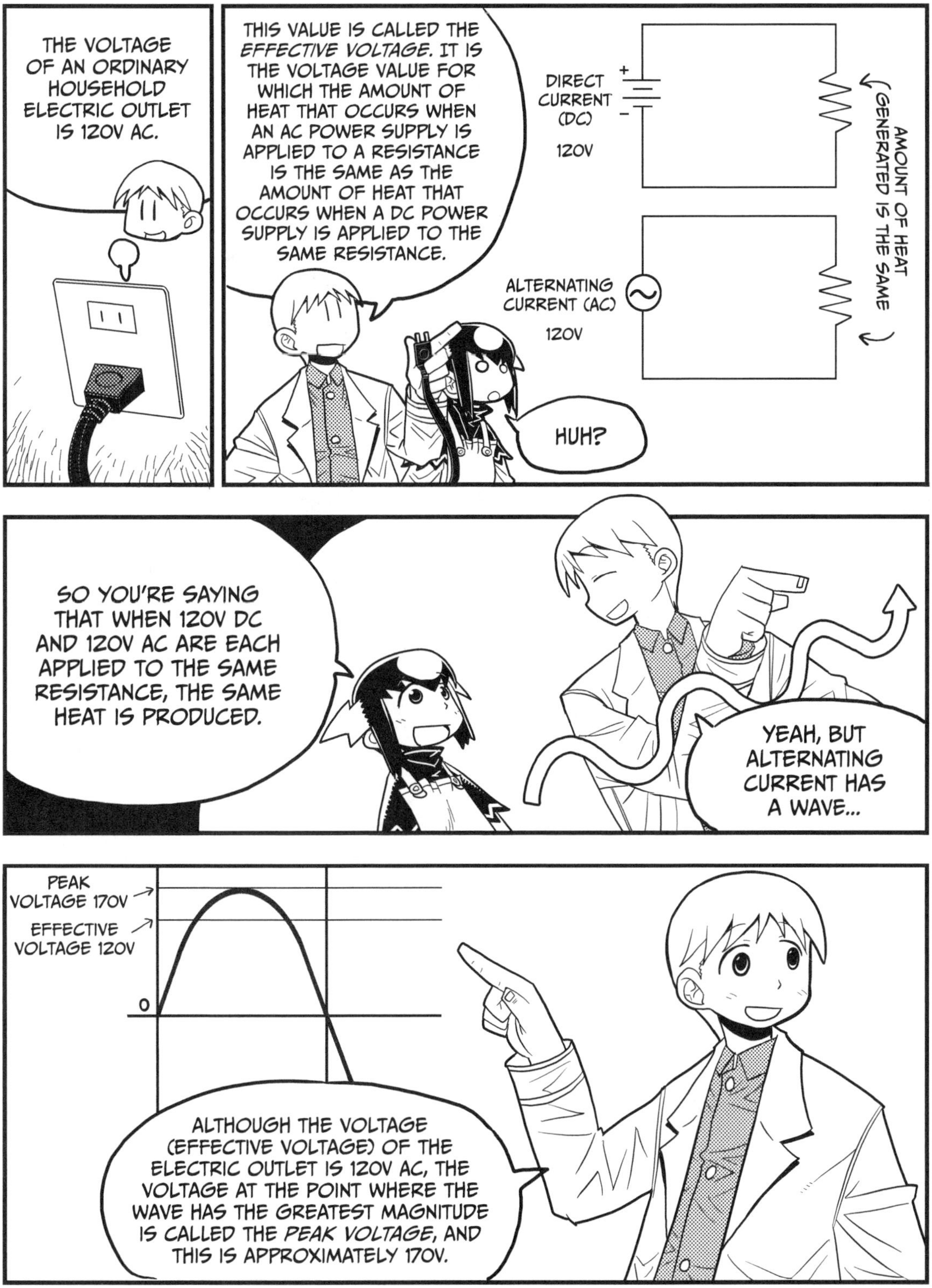

THE VOLTAGE OF AN ORDINARY HOUSEHOLD ELECTRIC OUTLET IS 120V AC.
THIS VALUE IS CALLED THE *EFFECTIVE VOLTAGE*. IT IS THE VOLTAGE VALUE FOR WHICH THE AMOUNT OF HEAT THAT OCCURS WHEN AN AC POWER SUPPLY IS APPLIED TO A RESISTANCE IS THE SAME AS THE AMOUNT OF HEAT THAT OCCURS WHEN A DC POWER SUPPLY IS APPLIED TO THE SAME RESISTANCE.
DIRECT CURRENT (DC)
120V
ALTERNATING CURRENT (AC)
120V
AMOUNT OF HEAT GENERATED IS THE SAME
HUH?
SO YOU'RE SAYING THAT WHEN 120V DC AND 120V AC ARE EACH APPLIED TO THE SAME RESISTANCE, THE SAME HEAT IS PRODUCED.
YEAH, BUT ALTERNATING CURRENT HAS A WAVE...
PEAK VOLTAGE 170V
EFFECTIVE VOLTAGE 120V
0
ALTHOUGH THE VOLTAGE (EFFECTIVE VOLTAGE) OF THE ELECTRIC OUTLET IS 120V AC, THE VOLTAGE AT THE POINT WHERE THE WAVE HAS THE GREATEST MAGNITUDE IS CALLED THE *PEAK VOLTAGE*, AND THIS IS APPROXIMATELY 170V.

BATTERIES AND OTHER SOURCES OF ELECTRICITY

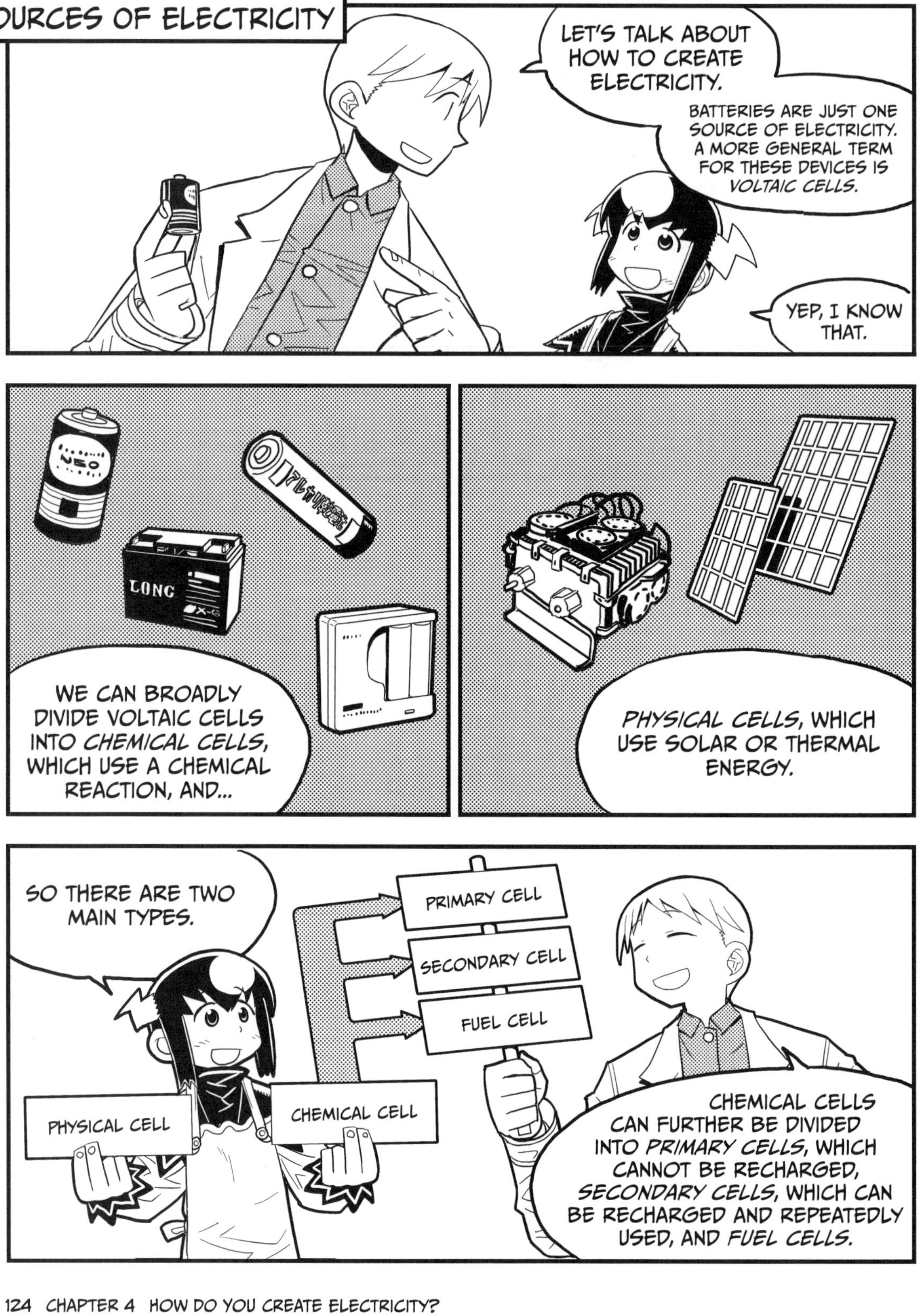
LET'S TALK ABOUT HOW TO CREATE ELECTRICITY.
BATTERIES ARE JUST ONE SOURCE OF ELECTRICITY. A MORE GENERAL TERM FOR THESE DEVICES IS *VOLTAIC CELLS*.
YEP, I KNOW THAT.
LONG
WE CAN BROADLY DIVIDE VOLTAIC CELLS INTO *CHEMICAL CELLS*, WHICH USE A CHEMICAL REACTION, AND...
PHYSICAL CELLS, WHICH USE SOLAR OR THERMAL ENERGY.
SO THERE ARE TWO MAIN TYPES.
PHYSICAL CELL
CHEMICAL CELL
PRIMARY CELL
SECONDARY CELL
FUEL CELL
CHEMICAL CELLS CAN FURTHER BE DIVIDED INTO *PRIMARY CELLS*, WHICH CANNOT BE RECHARGED, *SECONDARY CELLS*, WHICH CAN BE RECHARGED AND REPEATEDLY USED, AND *FUEL CELLS*.

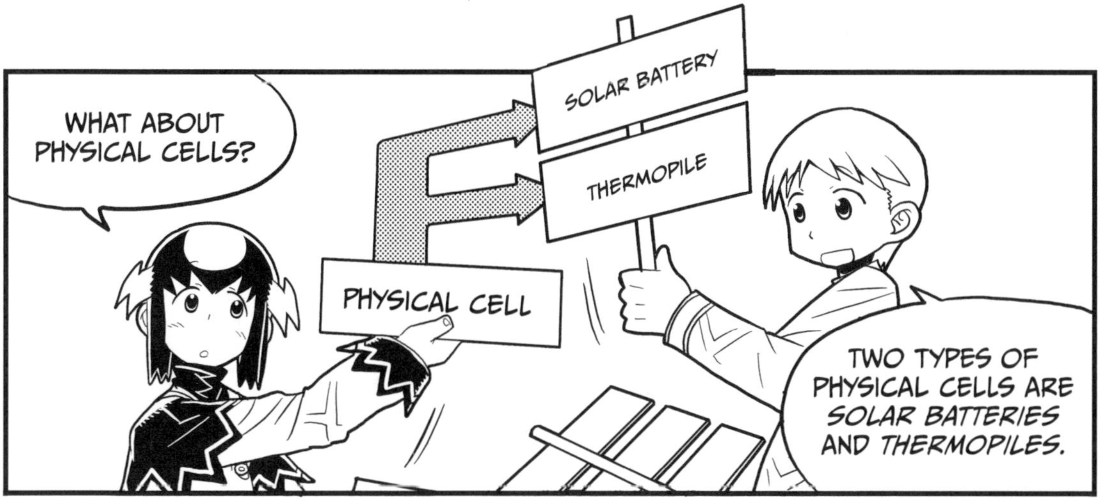
WHAT ABOUT PHYSICAL CELLS?
SOLAR BATTERY
THERMOPILE
PHYSICAL CELL
TWO TYPES OF PHYSICAL CELLS ARE *SOLAR BATTERIES* AND *THERMOPILES.*

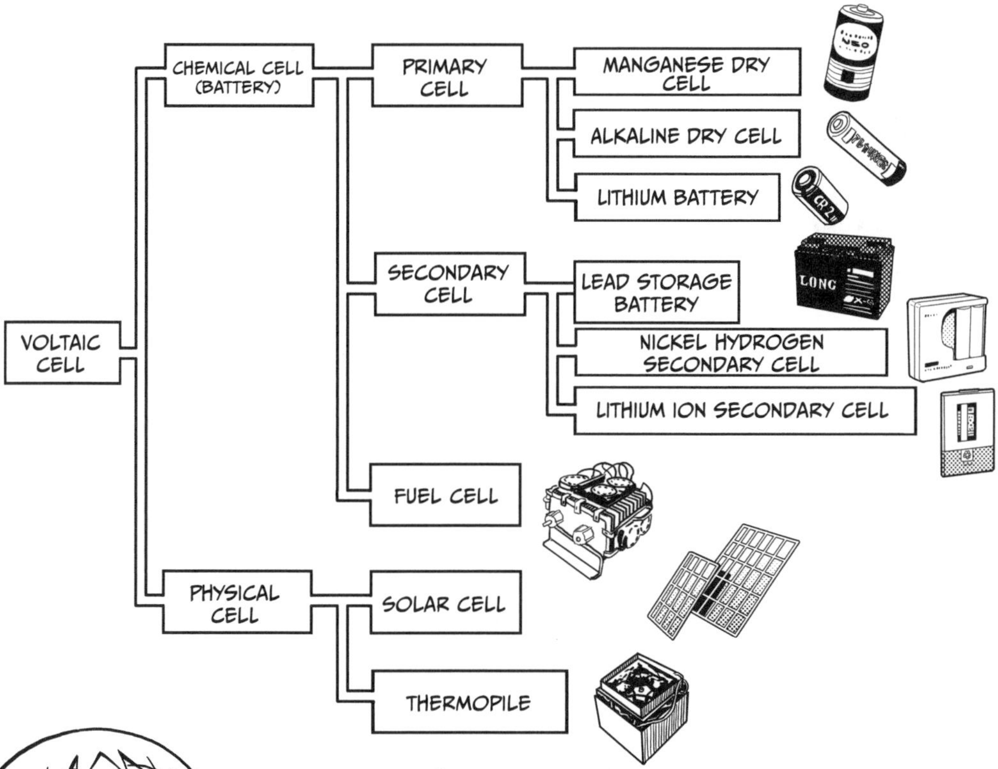
VOLTAIC CELL
CHEMICAL CELL (BATTERY)
PRIMARY CELL
MANGANESE DRY CELL
ALKALINE DRY CELL
LITHIUM BATTERY
SECONDARY CELL
LEAD STORAGE BATTERY
NICKEL HYDROGEN SECONDARY CELL
LITHIUM ION SECONDARY CELL
FUEL CELL
PHYSICAL CELL
SOLAR CELL
THERMOPILE

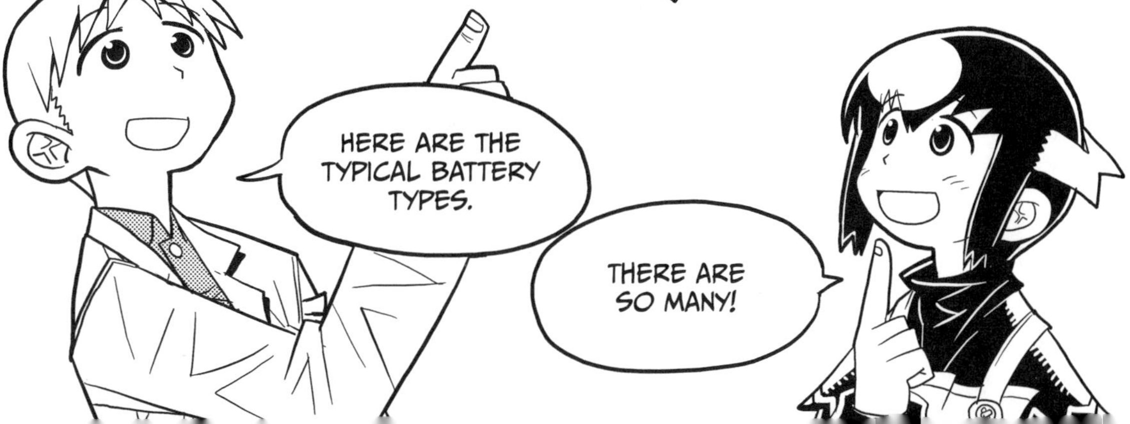
HERE ARE THE TYPICAL BATTERY TYPES.
THERE ARE SO MANY!

CHEMICAL CELLS

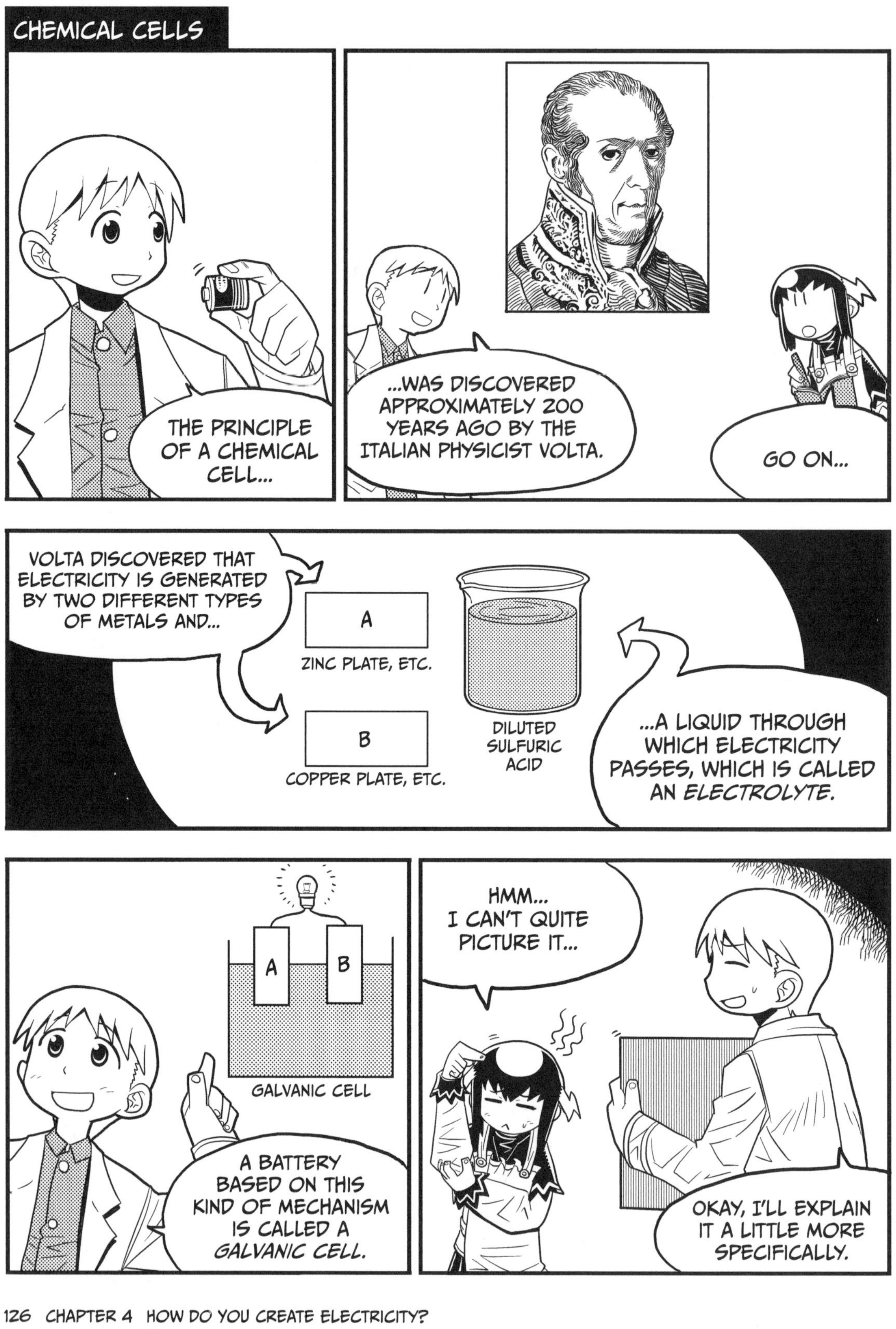
THE PRINCIPLE OF A CHEMICAL CELL...
...WAS DISCOVERED APPROXIMATELY 200 YEARS AGO BY THE ITALIAN PHYSICIST VOLTA.
GO ON...
VOLTA DISCOVERED THAT ELECTRICITY IS GENERATED BY TWO DIFFERENT TYPES OF METALS AND...
A
ZINC PLATE, ETC.
B
COPPER PLATE, ETC.
DILUTED SULFURIC ACID
...A LIQUID THROUGH WHICH ELECTRICITY PASSES, WHICH IS CALLED AN *ELECTROLYTE*.
A
B
GALVANIC CELL
A BATTERY BASED ON THIS KIND OF MECHANISM IS CALLED A *GALVANIC CELL*.
HMM... I CAN'T QUITE PICTURE IT...
OKAY, I'LL EXPLAIN IT A LITTLE MORE SPECIFICALLY.

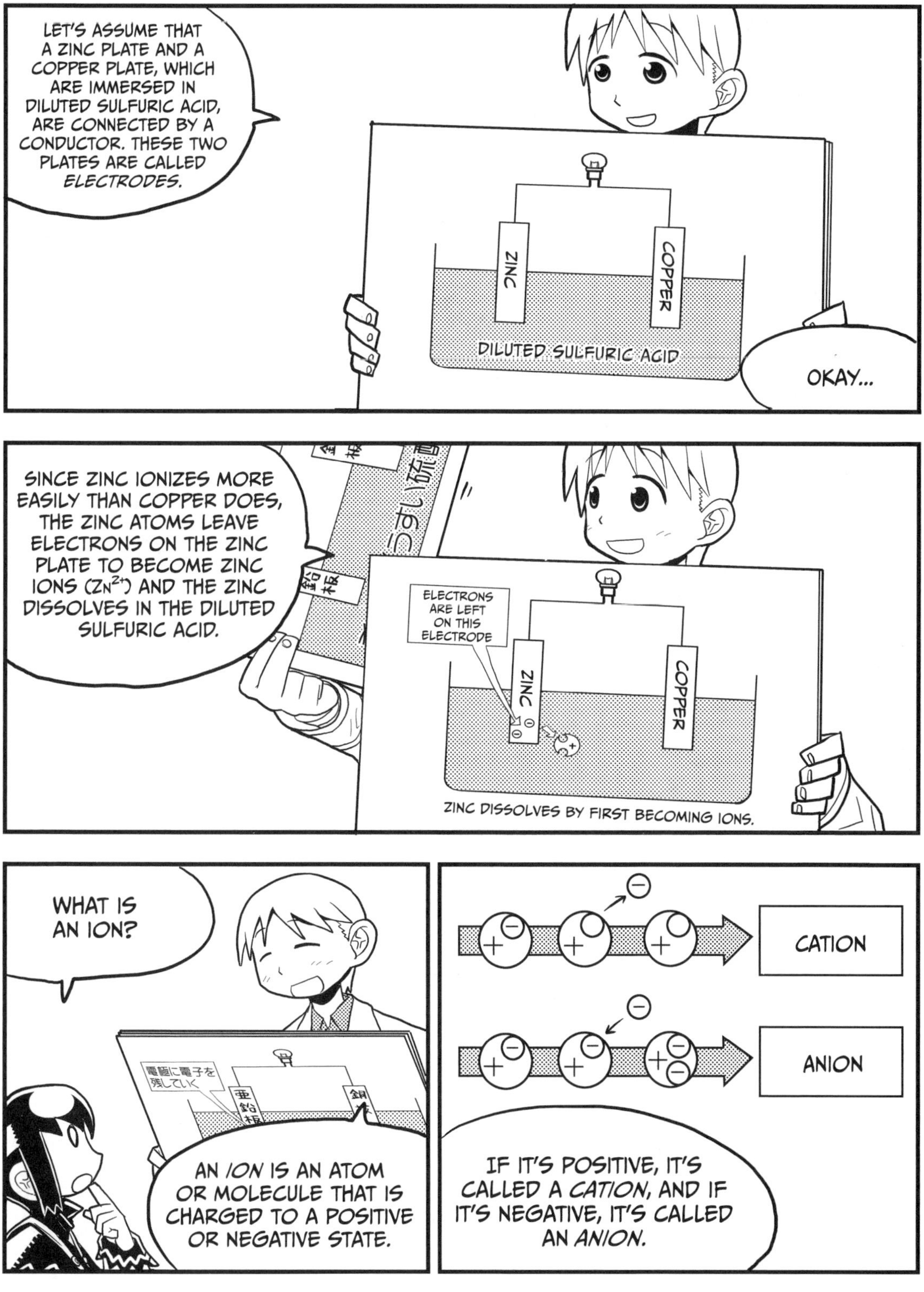
LET'S ASSUME THAT A ZINC PLATE AND A COPPER PLATE, WHICH ARE IMMERSED IN DILUTED SULFURIC ACID, ARE CONNECTED BY A CONDUCTOR. THESE TWO PLATES ARE CALLED *ELECTRODES*.
ZINC
COPPER
DILUTED SULFURIC ACID
OKAY...
SINCE ZINC IONIZES MORE EASILY THAN COPPER DOES, THE ZINC ATOMS LEAVE ELECTRONS ON THE ZINC PLATE TO BECOME ZINC IONS (Zn^{2+}) AND THE ZINC DISSOLVES IN THE DILUTED SULFURIC ACID.
ELECTRONS ARE LEFT ON THIS ELECTRODE
ZINC
COPPER
ZINC DISSOLVES BY FIRST BECOMING IONS.
WHAT IS AN ION?
電極に電子を残していく
亜鉛板
AN *ION* IS AN ATOM OR MOLECULE THAT IS CHARGED TO A POSITIVE OR NEGATIVE STATE.
CATION
ANION
IF IT'S POSITIVE, IT'S CALLED A *CATION*, AND IF IT'S NEGATIVE, IT'S CALLED AN *ANION*.

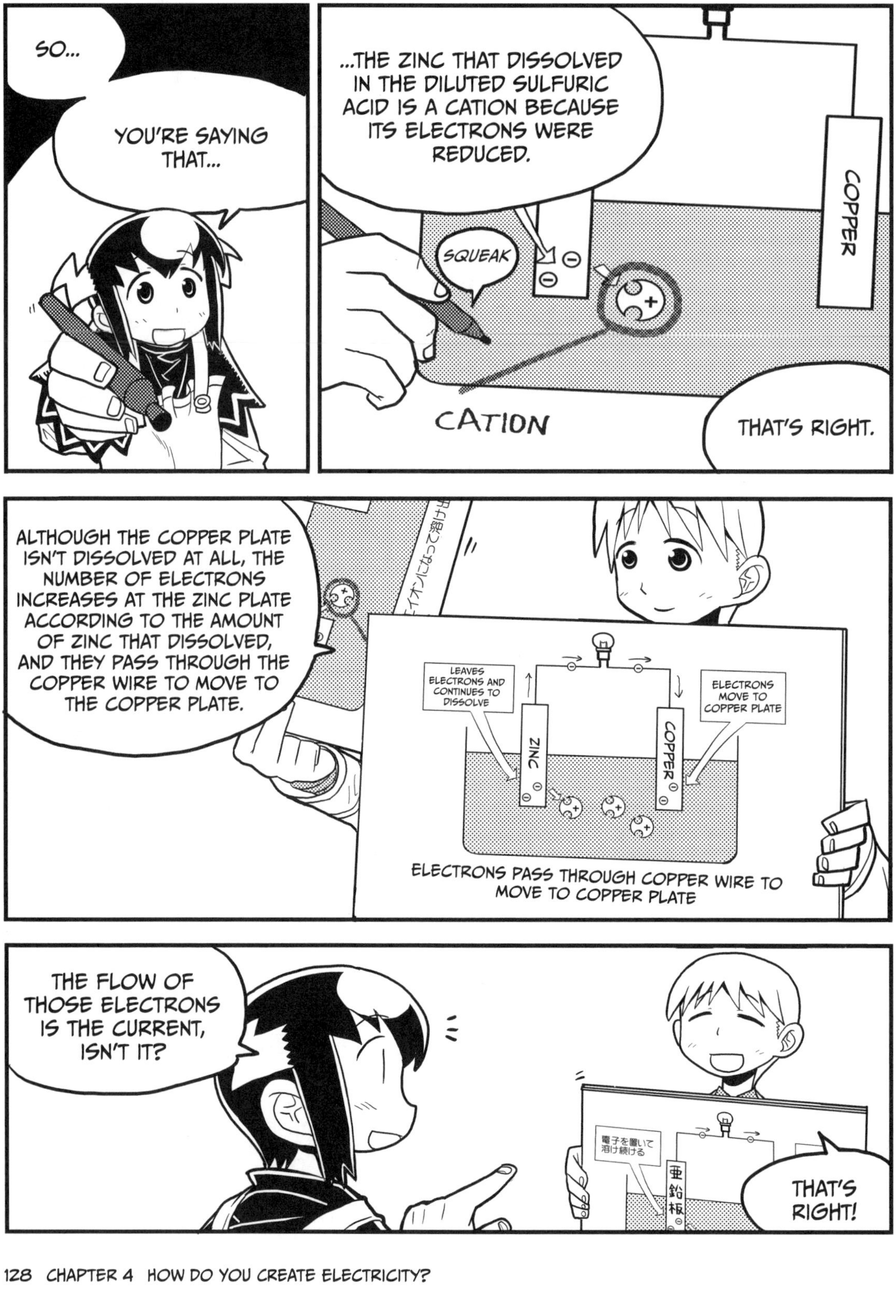
SO...
YOU'RE SAYING THAT...
...THE ZINC THAT DISSOLVED IN THE DILUTED SULFURIC ACID IS A CATION BECAUSE ITS ELECTRONS WERE REDUCED.
SQUEAK
COPPER
CATION
THAT'S RIGHT.
ALTHOUGH THE COPPER PLATE ISN'T DISSOLVED AT ALL, THE NUMBER OF ELECTRONS INCREASES AT THE ZINC PLATE ACCORDING TO THE AMOUNT OF ZINC THAT DISSOLVED, AND THEY PASS THROUGH THE COPPER WIRE TO MOVE TO THE COPPER PLATE.
LEAVES ELECTRONS AND CONTINUES TO DISSOLVE
ELECTRONS MOVE TO COPPER PLATE
ZINC
COPPER
ELECTRONS PASS THROUGH COPPER WIRE TO MOVE TO COPPER PLATE
THE FLOW OF THOSE ELECTRONS IS THE CURRENT, ISN'T IT?
亜鉛板
THAT'S RIGHT!

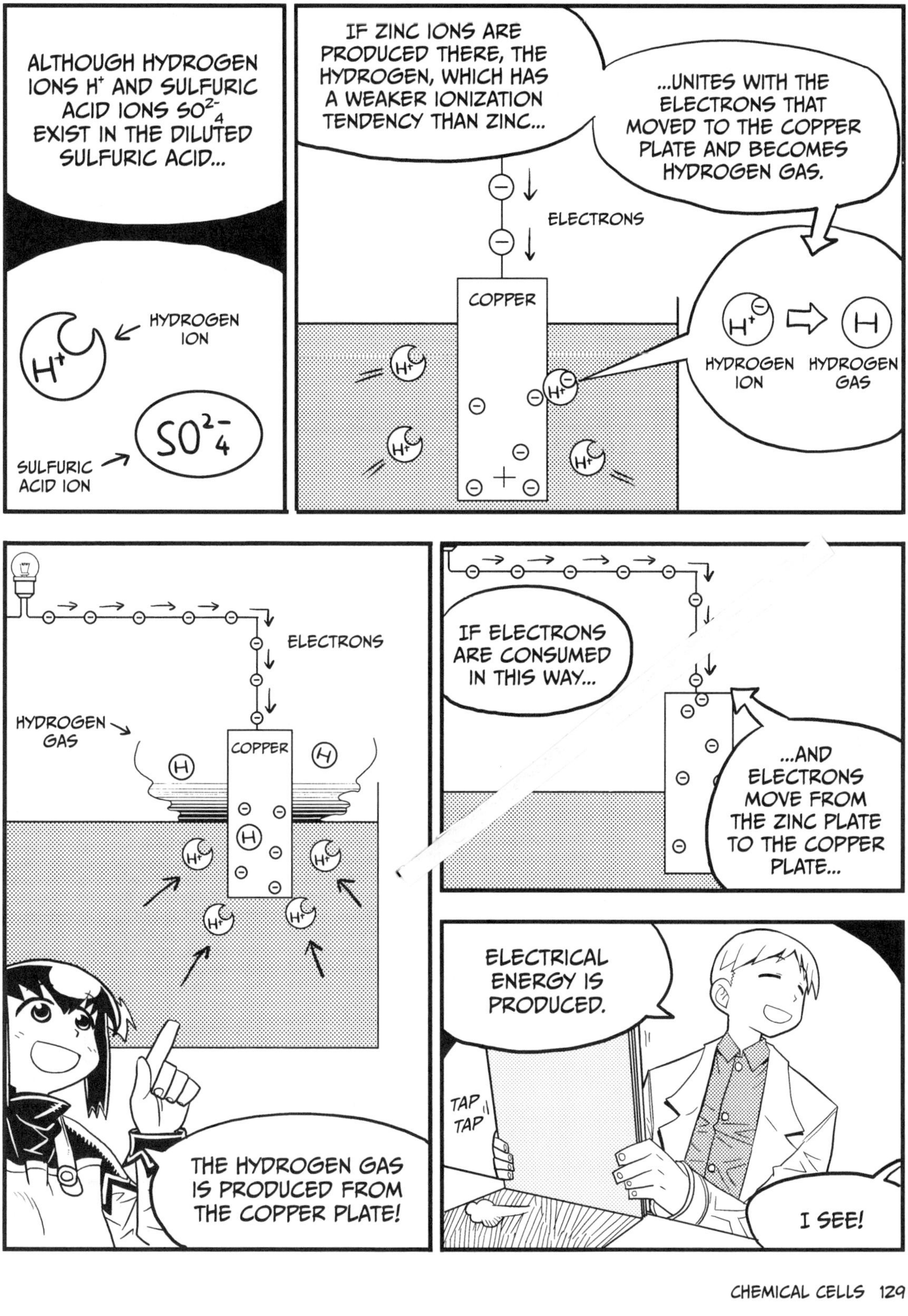
ALTHOUGH HYDROGEN IONS H+ AND SULFURIC ACID IONS SO2-4 EXIST IN THE DILUTED SULFURIC ACID...
HYDROGEN ION
SULFURIC ACID ION
IF ZINC IONS ARE PRODUCED THERE, THE HYDROGEN, WHICH HAS A WEAKER IONIZATION TENDENCY THAN ZINC...
...UNITES WITH THE ELECTRONS THAT MOVED TO THE COPPER PLATE AND BECOMES HYDROGEN GAS.
ELECTRONS
COPPER
HYDROGEN ION
HYDROGEN GAS
ELECTRONS
HYDROGEN GAS
COPPER
THE HYDROGEN GAS IS PRODUCED FROM THE COPPER PLATE!
IF ELECTRONS ARE CONSUMED IN THIS WAY...
...AND ELECTRONS MOVE FROM THE ZINC PLATE TO THE COPPER PLATE...
ELECTRICAL ENERGY IS PRODUCED.
TAP TAP
I SEE!

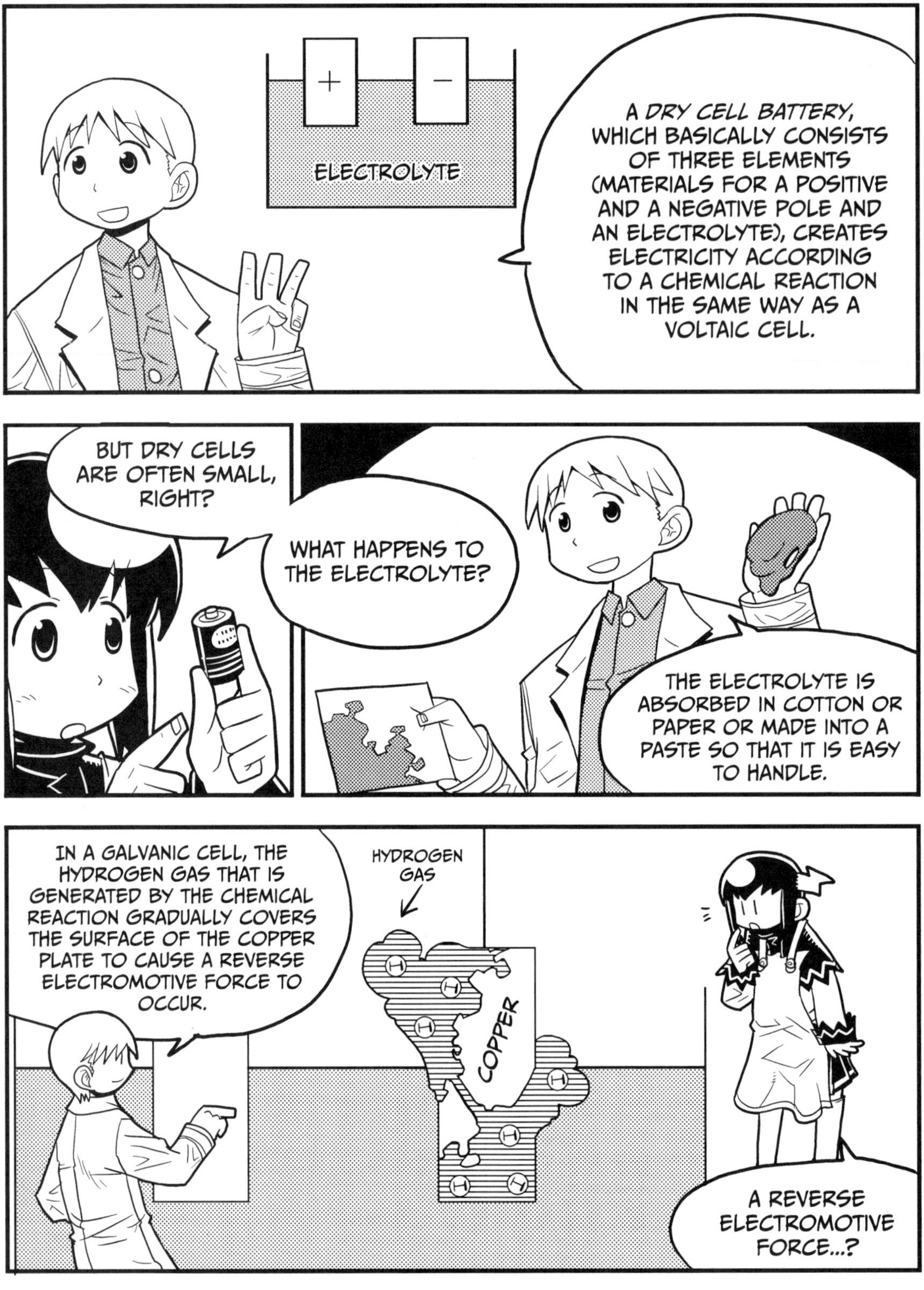
ELECTROLYTE
A *DRY CELL BATTERY*, WHICH BASICALLY CONSISTS OF THREE ELEMENTS (MATERIALS FOR A POSITIVE AND A NEGATIVE POLE AND AN ELECTROLYTE), CREATES ELECTRICITY ACCORDING TO A CHEMICAL REACTION IN THE SAME WAY AS A VOLTAIC CELL.
BUT DRY CELLS ARE OFTEN SMALL, RIGHT?
WHAT HAPPENS TO THE ELECTROLYTE?
THE ELECTROLYTE IS ABSORBED IN COTTON OR PAPER OR MADE INTO A PASTE SO THAT IT IS EASY TO HANDLE.
IN A GALVANIC CELL, THE HYDROGEN GAS THAT IS GENERATED BY THE CHEMICAL REACTION GRADUALLY COVERS THE SURFACE OF THE COPPER PLATE TO CAUSE A REVERSE ELECTROMOTIVE FORCE TO OCCUR.
HYDROGEN GAS
COPPER
A REVERSE ELECTROMOTIVE FORCE...?

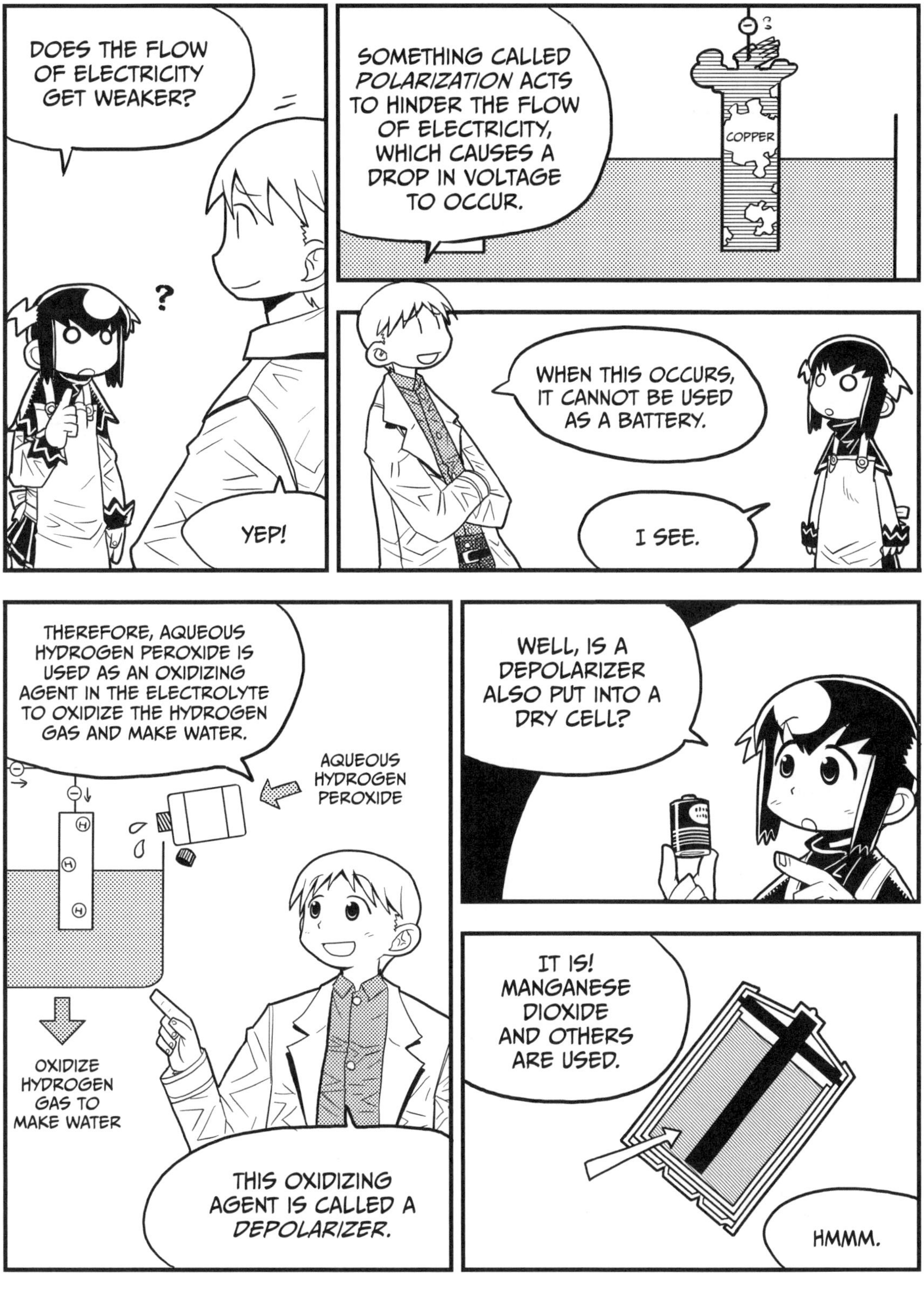
DOES THE FLOW OF ELECTRICITY GET WEAKER?
YEP!
SOMETHING CALLED *POLARIZATION* ACTS TO HINDER THE FLOW OF ELECTRICITY, WHICH CAUSES A DROP IN VOLTAGE TO OCCUR.
COPPER
WHEN THIS OCCURS, IT CANNOT BE USED AS A BATTERY.
I SEE.
THEREFORE, AQUEOUS HYDROGEN PEROXIDE IS USED AS AN OXIDIZING AGENT IN THE ELECTROLYTE TO OXIDIZE THE HYDROGEN GAS AND MAKE WATER.
AQUEOUS HYDROGEN PEROXIDE
OXIDIZE HYDROGEN GAS TO MAKE WATER
THIS OXIDIZING AGENT IS CALLED A *DEPOLARIZER*.
WELL, IS A DEPOLARIZER ALSO PUT INTO A DRY CELL?
IT IS! MANGANESE DIOXIDE AND OTHERS ARE USED.
HMMM.

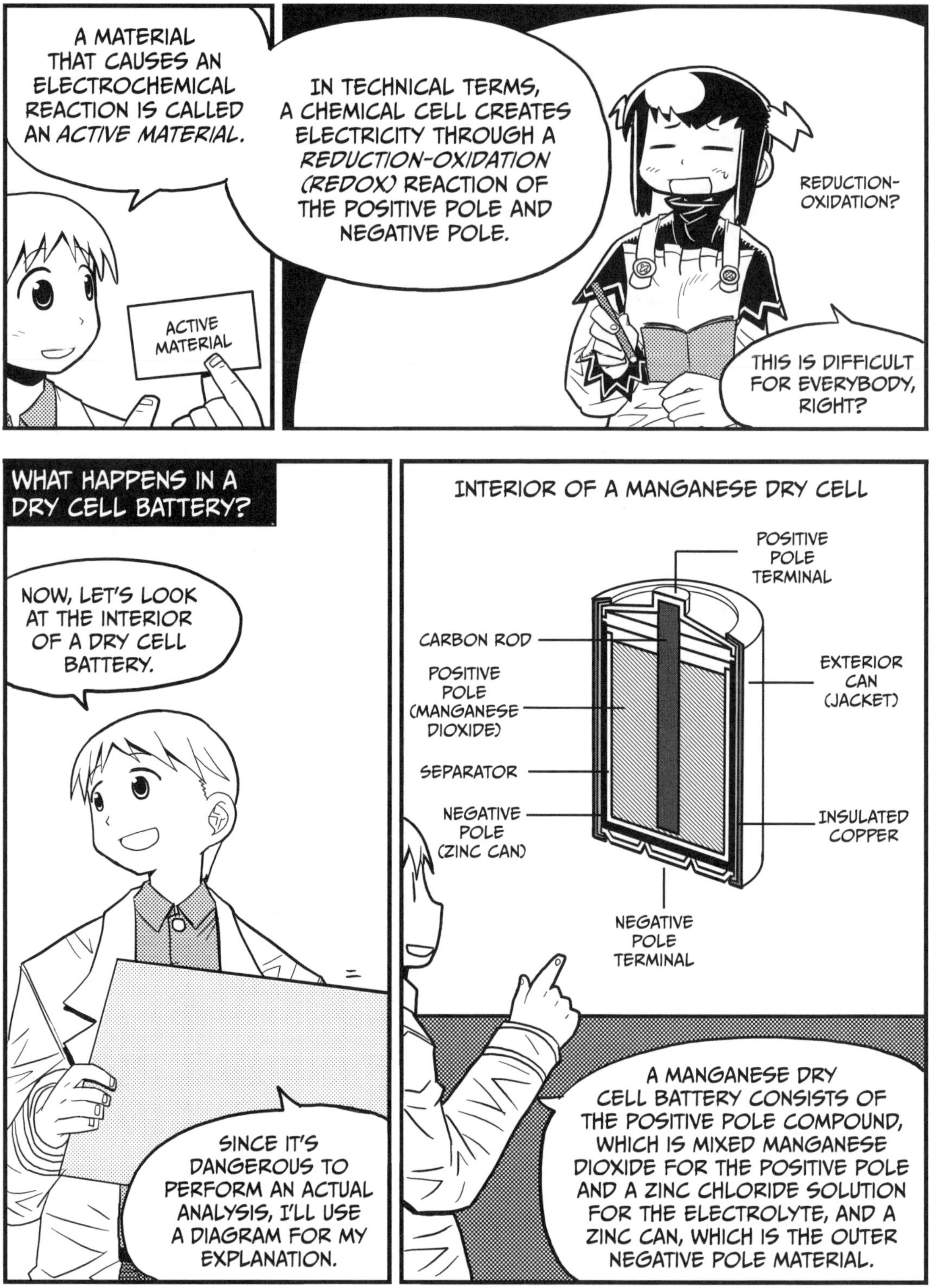
A MATERIAL THAT CAUSES AN ELECTROCHEMICAL REACTION IS CALLED AN *ACTIVE MATERIAL*.
ACTIVE MATERIAL
IN TECHNICAL TERMS, A CHEMICAL CELL CREATES ELECTRICITY THROUGH A *REDUCTION-OXIDATION (REDOX)* REACTION OF THE POSITIVE POLE AND NEGATIVE POLE.
REDUCTION-OXIDATION?
THIS IS DIFFICULT FOR EVERYBODY, RIGHT?
WHAT HAPPENS IN A DRY CELL BATTERY?
NOW, LET'S LOOK AT THE INTERIOR OF A DRY CELL BATTERY.
SINCE IT'S DANGEROUS TO PERFORM AN ACTUAL ANALYSIS, I'LL USE A DIAGRAM FOR MY EXPLANATION.
INTERIOR OF A MANGANESE DRY CELL
POSITIVE POLE TERMINAL
CARBON ROD
POSITIVE POLE (MANGANESE DIOXIDE)
SEPARATOR
NEGATIVE POLE (ZINC CAN)
EXTERIOR CAN (JACKET)
INSULATED COPPER
NEGATIVE POLE TERMINAL
A MANGANESE DRY CELL BATTERY CONSISTS OF THE POSITIVE POLE COMPOUND, WHICH IS MIXED MANGANESE DIOXIDE FOR THE POSITIVE POLE AND A ZINC CHLORIDE SOLUTION FOR THE ELECTROLYTE, AND A ZINC CAN, WHICH IS THE OUTER NEGATIVE POLE MATERIAL.

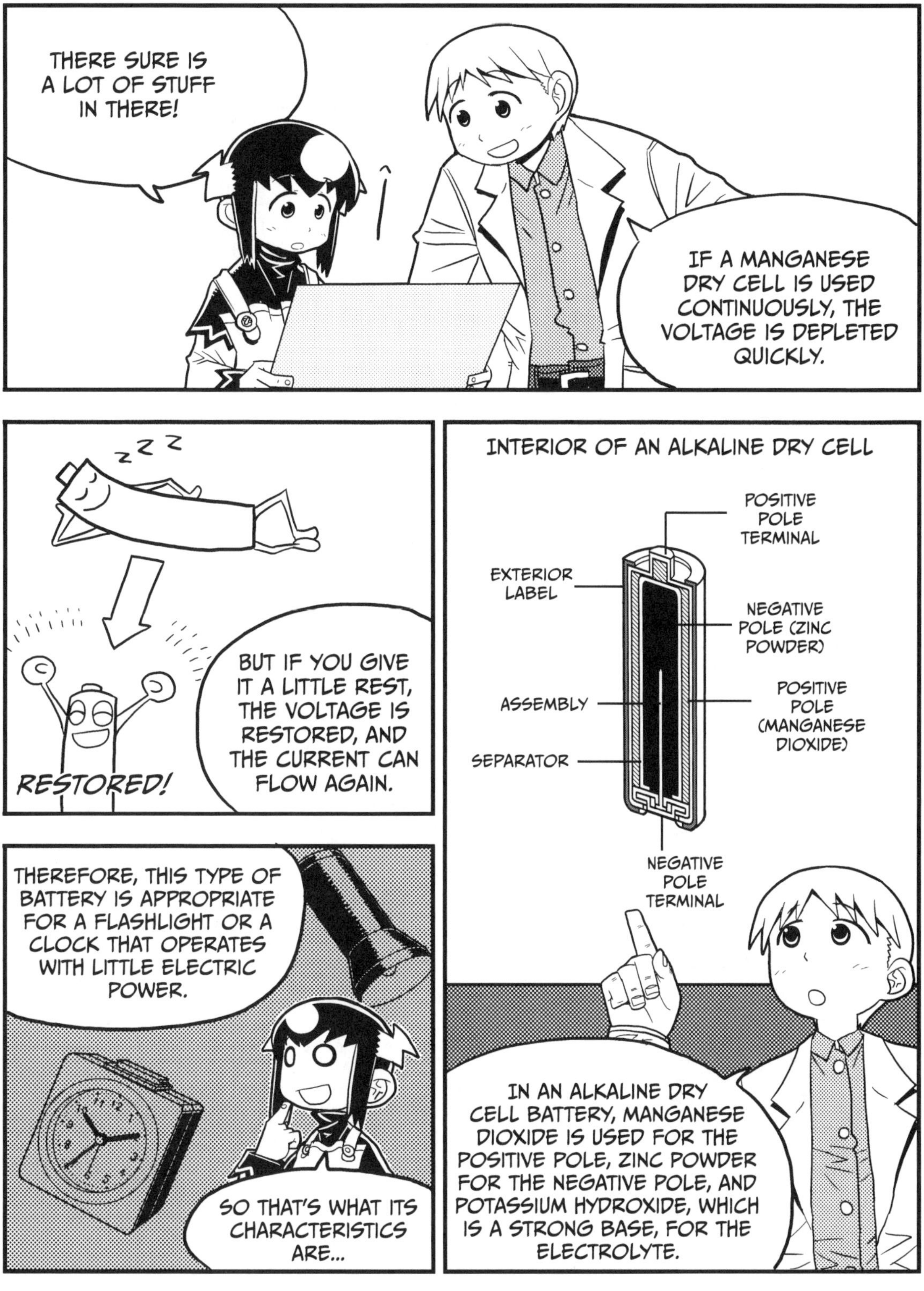

THERE SURE IS A LOT OF STUFF IN THERE!
IF A MANGANESE DRY CELL IS USED CONTINUOUSLY, THE VOLTAGE IS DEPLETED QUICKLY.
Z Z Z
RESTORED!
BUT IF YOU GIVE IT A LITTLE REST, THE VOLTAGE IS RESTORED, AND THE CURRENT CAN FLOW AGAIN.
THEREFORE, THIS TYPE OF BATTERY IS APPROPRIATE FOR A FLASHLIGHT OR A CLOCK THAT OPERATES WITH LITTLE ELECTRIC POWER.
SO THAT'S WHAT ITS CHARACTERISTICS ARE...
INTERIOR OF AN ALKALINE DRY CELL
POSITIVE POLE TERMINAL
EXTERIOR LABEL
NEGATIVE POLE (ZINC POWDER)
ASSEMBLY
POSITIVE POLE (MANGANESE DIOXIDE)
SEPARATOR
NEGATIVE POLE TERMINAL
IN AN ALKALINE DRY CELL BATTERY, MANGANESE DIOXIDE IS USED FOR THE POSITIVE POLE, ZINC POWDER FOR THE NEGATIVE POLE, AND POTASSIUM HYDROXIDE, WHICH IS A STRONG BASE, FOR THE ELECTROLYTE.

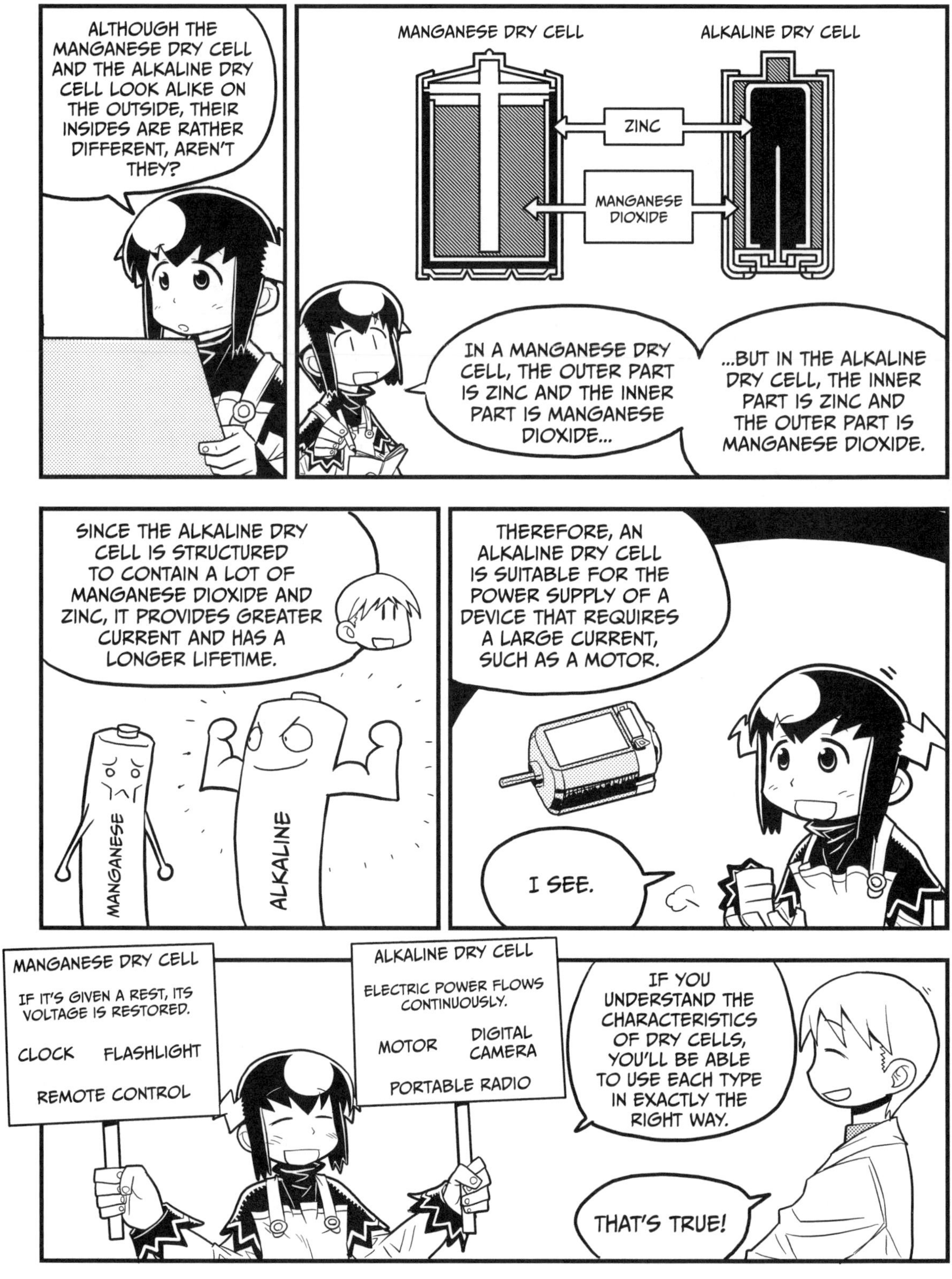
ALTHOUGH THE MANGANESE DRY CELL AND THE ALKALINE DRY CELL LOOK ALIKE ON THE OUTSIDE, THEIR INSIDES ARE RATHER DIFFERENT, AREN'T THEY?
MANGANESE DRY CELL
ALKALINE DRY CELL
ZINC
MANGANESE DIOXIDE
IN A MANGANESE DRY CELL, THE OUTER PART IS ZINC AND THE INNER PART IS MANGANESE DIOXIDE...
...BUT IN THE ALKALINE DRY CELL, THE INNER PART IS ZINC AND THE OUTER PART IS MANGANESE DIOXIDE.
SINCE THE ALKALINE DRY CELL IS STRUCTURED TO CONTAIN A LOT OF MANGANESE DIOXIDE AND ZINC, IT PROVIDES GREATER CURRENT AND HAS A LONGER LIFETIME.
MANGANESE
ALKALINE
THEREFORE, AN ALKALINE DRY CELL IS SUITABLE FOR THE POWER SUPPLY OF A DEVICE THAT REQUIRES A LARGE CURRENT, SUCH AS A MOTOR.
I SEE.
MANGANESE DRY CELL
IF IT'S GIVEN A REST, ITS VOLTAGE IS RESTORED.
CLOCK
FLASHLIGHT
REMOTE CONTROL
ALKALINE DRY CELL
ELECTRIC POWER FLOWS CONTINUOUSLY.
MOTOR
DIGITAL CAMERA
PORTABLE RADIO
IF YOU UNDERSTAND THE CHARACTERISTICS OF DRY CELLS, YOU'LL BE ABLE TO USE EACH TYPE IN EXACTLY THE RIGHT WAY.
THAT'S TRUE!

WATER AND FUEL CELLS

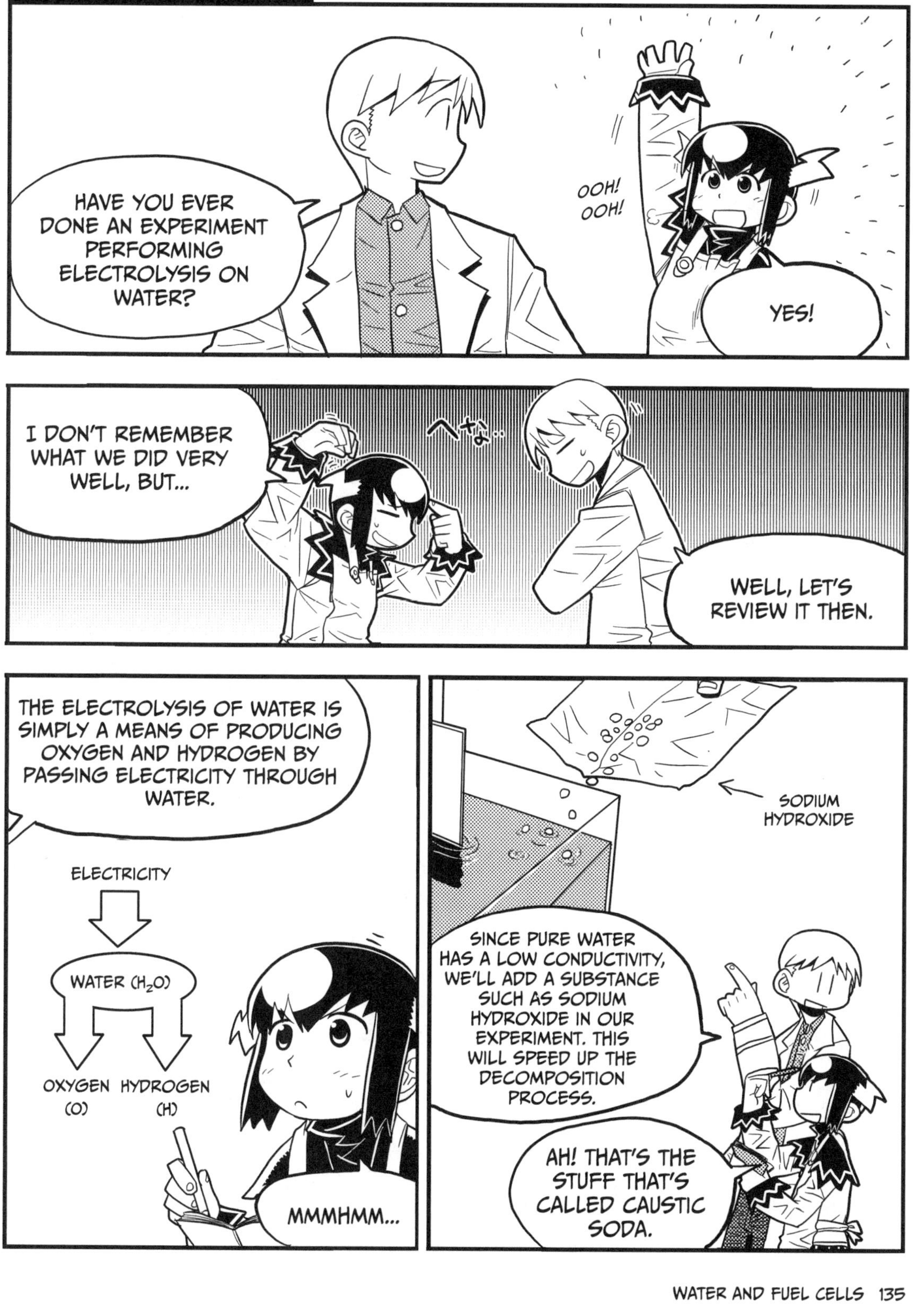
HAVE YOU EVER DONE AN EXPERIMENT PERFORMING ELECTROLYSIS ON WATER?
OOH! OOH!
YES!
I DON'T REMEMBER WHAT WE DID VERY WELL, BUT...
WELL, LET'S REVIEW IT THEN.
THE ELECTROLYSIS OF WATER IS SIMPLY A MEANS OF PRODUCING OXYGEN AND HYDROGEN BY PASSING ELECTRICITY THROUGH WATER.
ELECTRICITY
WATER (H_2O)
OXYGEN (O)
HYDROGEN (H)
MMMHMM...
SODIUM HYDROXIDE
SINCE PURE WATER HAS A LOW CONDUCTIVITY, WE'LL ADD A SUBSTANCE SUCH AS SODIUM HYDROXIDE IN OUR EXPERIMENT. THIS WILL SPEED UP THE DECOMPOSITION PROCESS.
AH! THAT'S THE STUFF THAT'S CALLED CAUSTIC SODA.

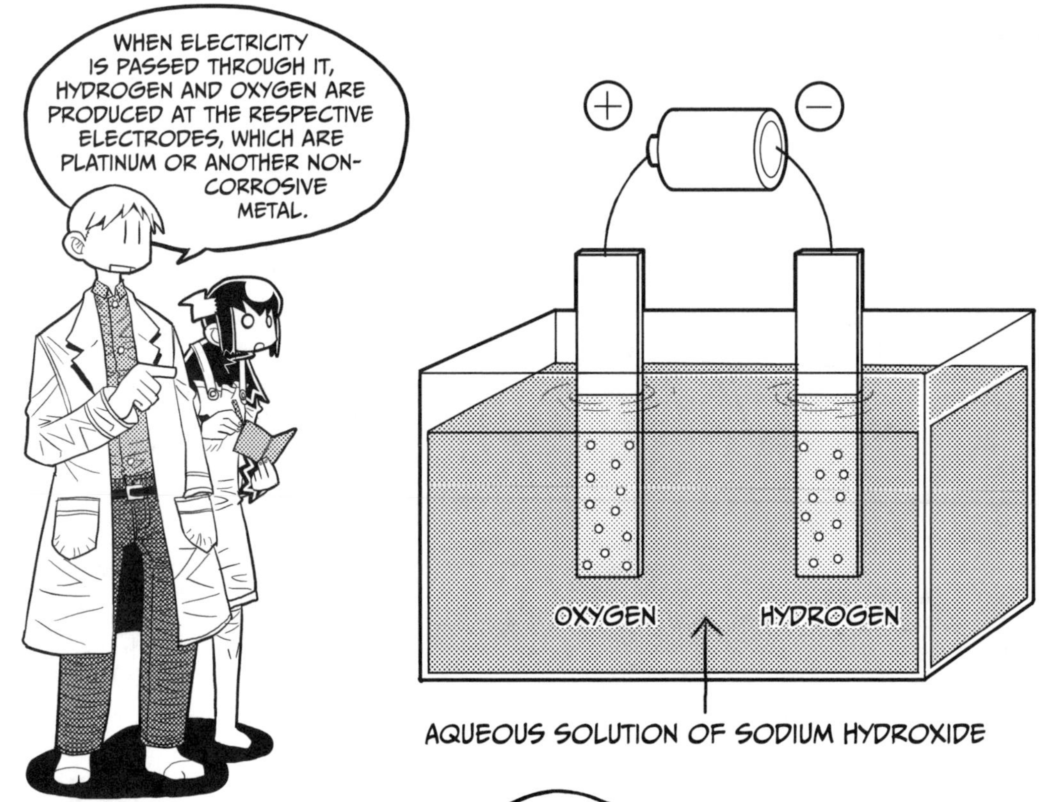
WHEN ELECTRICITY IS PASSED THROUGH IT, HYDROGEN AND OXYGEN ARE PRODUCED AT THE RESPECTIVE ELECTRODES, WHICH ARE PLATINUM OR ANOTHER NON-CORROSIVE METAL.
+
−
OXYGEN
HYDROGEN
AQUEOUS SOLUTION OF SODIUM HYDROXIDE

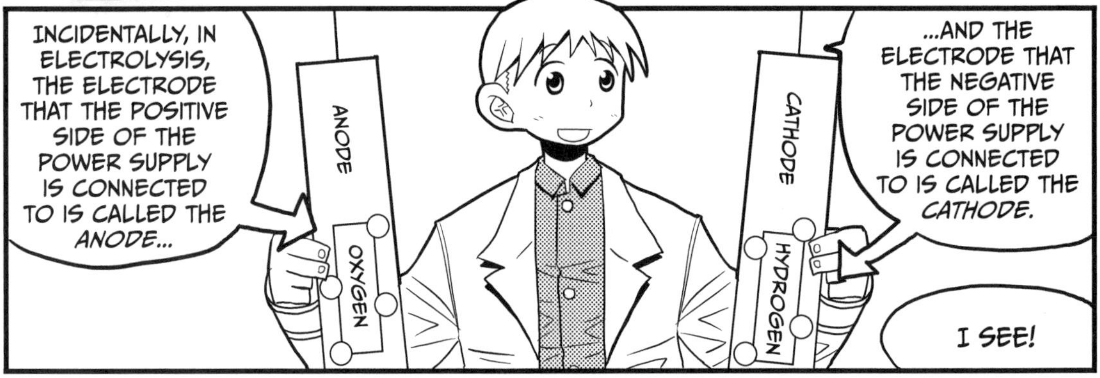
INCIDENTALLY, IN ELECTROLYSIS, THE ELECTRODE THAT THE POSITIVE SIDE OF THE POWER SUPPLY IS CONNECTED TO IS CALLED THE *ANODE*...
ANODE
OXYGEN
CATHODE
HYDROGEN
...AND THE ELECTRODE THAT THE NEGATIVE SIDE OF THE POWER SUPPLY IS CONNECTED TO IS CALLED THE *CATHODE*.
I SEE!

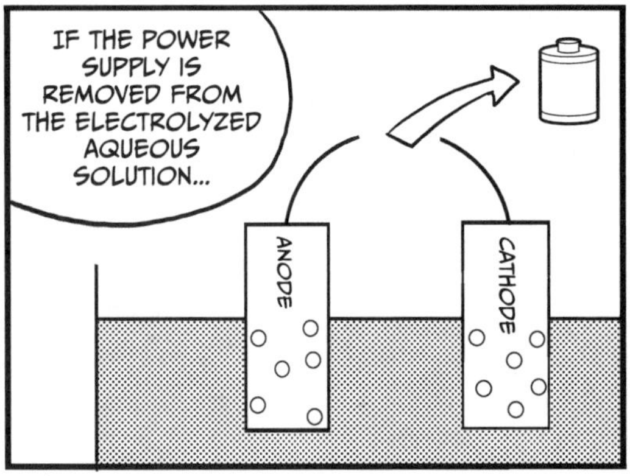
IF THE POWER SUPPLY IS REMOVED FROM THE ELECTROLYZED AQUEOUS SOLUTION...
ANODE
CATHODE

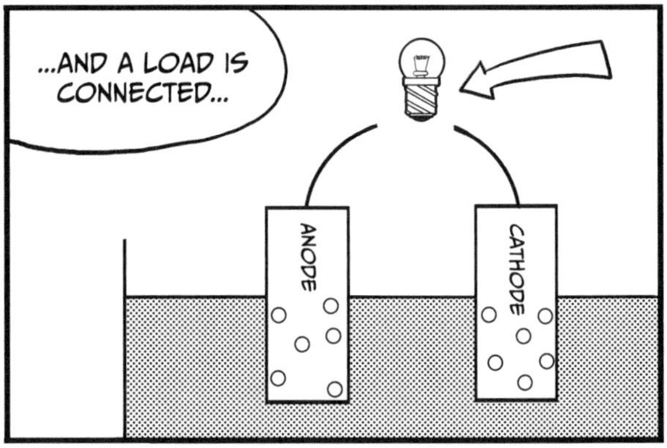
...AND A LOAD IS CONNECTED...
ANODE
CATHODE

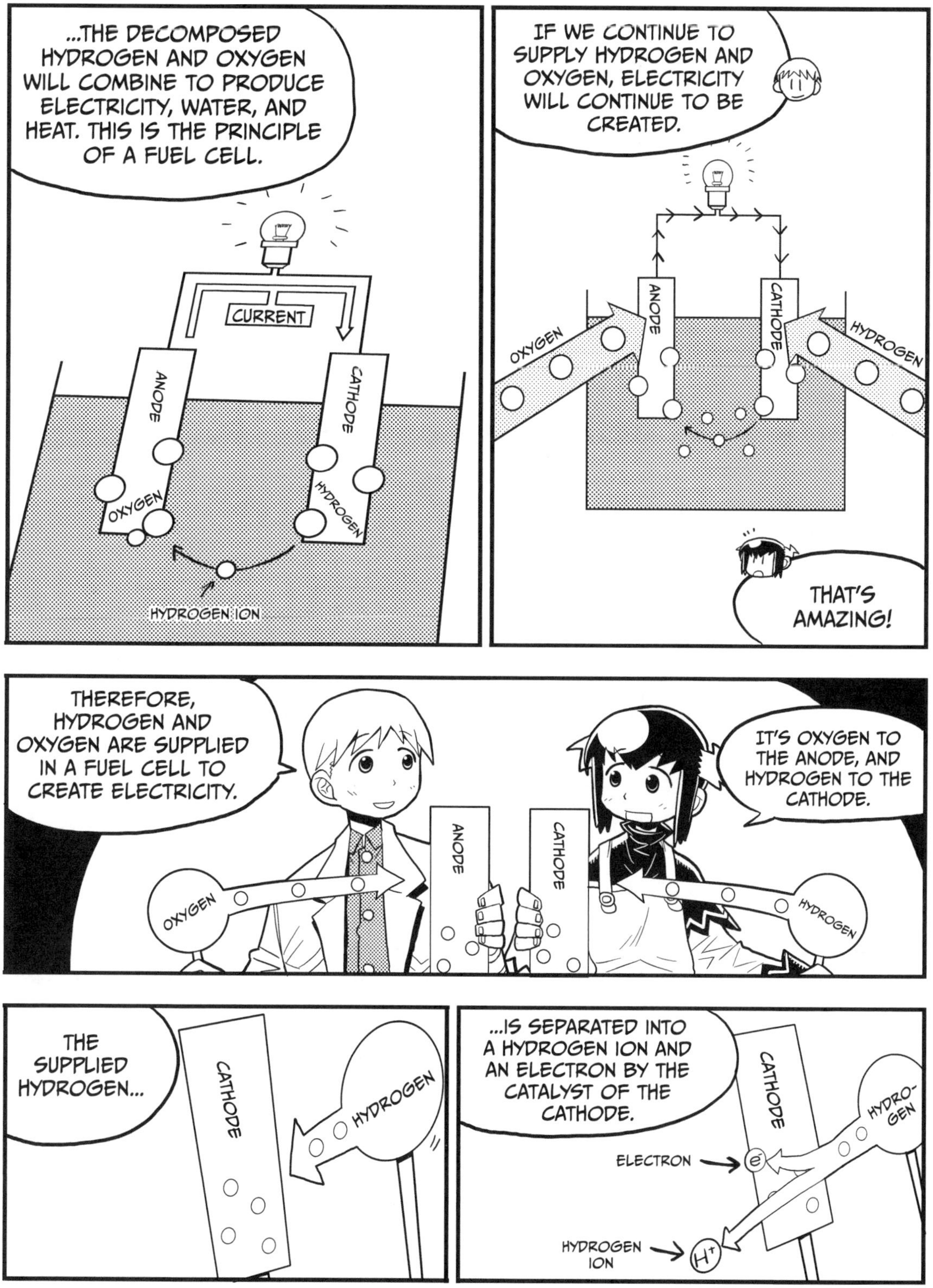
...THE DECOMPOSED HYDROGEN AND OXYGEN WILL COMBINE TO PRODUCE ELECTRICITY, WATER, AND HEAT. THIS IS THE PRINCIPLE OF A FUEL CELL.
CURRENT
ANODE
CATHODE
OXYGEN
HYDROGEN
HYDROGEN ION
IF WE CONTINUE TO SUPPLY HYDROGEN AND OXYGEN, ELECTRICITY WILL CONTINUE TO BE CREATED.
ANODE
CATHODE
OXYGEN
HYDROGEN
THAT'S AMAZING!
THEREFORE, HYDROGEN AND OXYGEN ARE SUPPLIED IN A FUEL CELL TO CREATE ELECTRICITY.
IT'S OXYGEN TO THE ANODE, AND HYDROGEN TO THE CATHODE.
ANODE
CATHODE
OXYGEN
HYDROGEN
THE SUPPLIED HYDROGEN...
CATHODE
HYDROGEN
...IS SEPARATED INTO A HYDROGEN ION AND AN ELECTRON BY THE CATALYST OF THE CATHODE.
CATHODE
HYDRO-GEN
ELECTRON
HYDROGEN ION

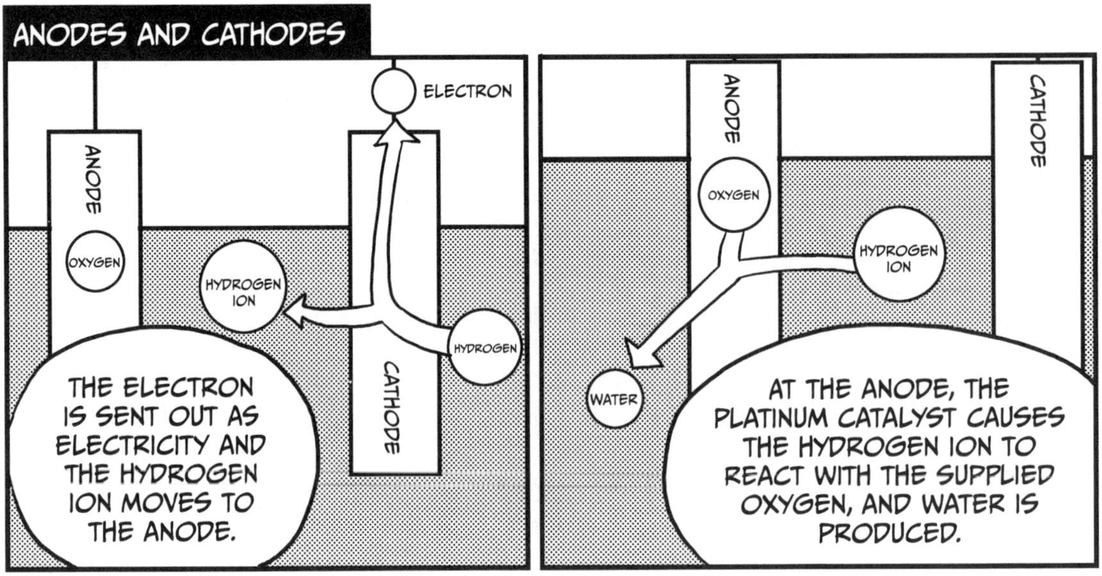
ANODES AND CATHODES
ELECTRON
ANODE
OXYGEN
HYDROGEN ION
HYDROGEN
CATHODE
THE ELECTRON IS SENT OUT AS ELECTRICITY AND THE HYDROGEN ION MOVES TO THE ANODE.
ANODE
CATHODE
OXYGEN
HYDROGEN ION
WATER
AT THE ANODE, THE PLATINUM CATALYST CAUSES THE HYDROGEN ION TO REACT WITH THE SUPPLIED OXYGEN, AND WATER IS PRODUCED.

MECHANISM OF A FUEL CELL

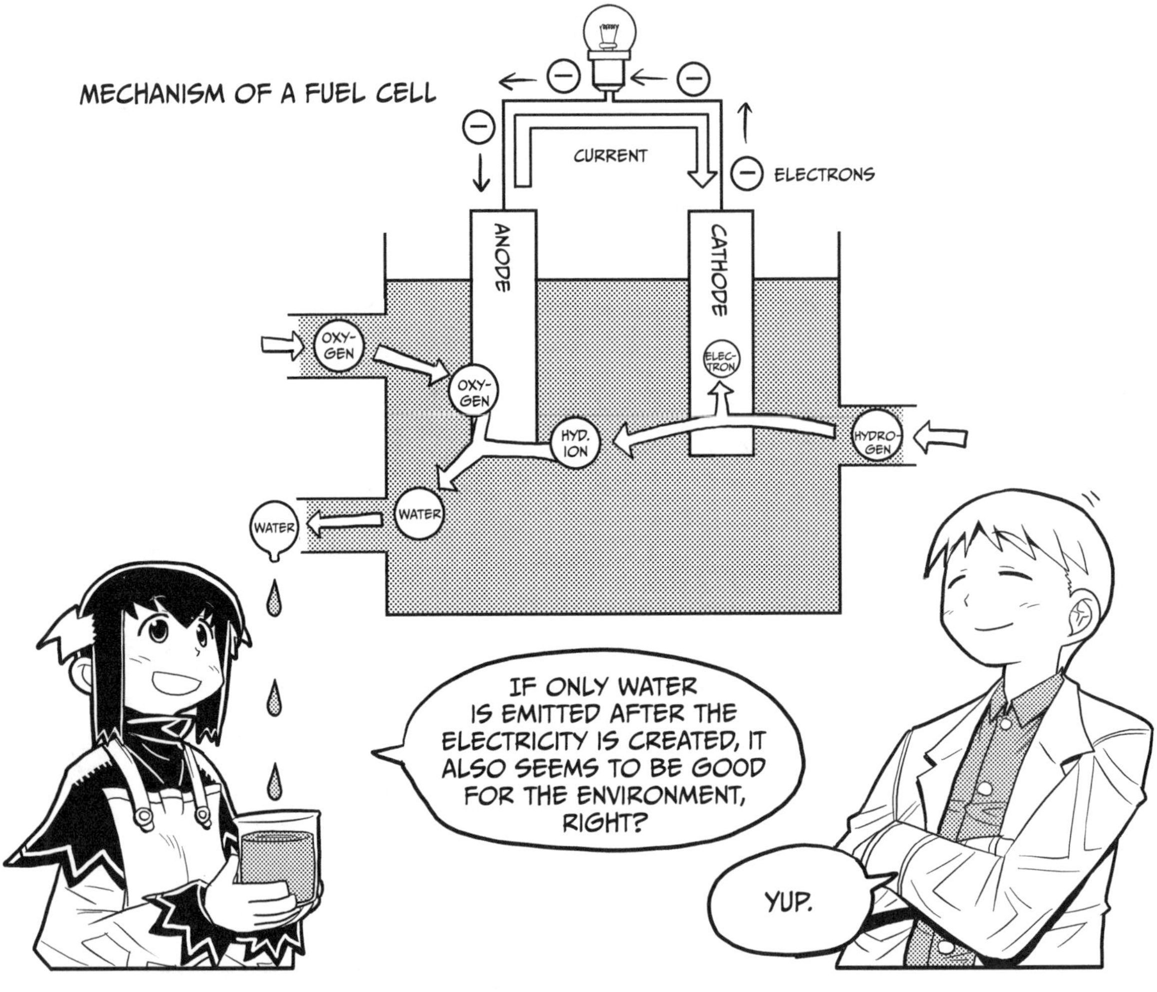
CURRENT
ELECTRONS
ANODE
CATHODE
OXY-GEN
OXY-GEN
ELEC-TRON
HYD. ION
HYDRO-GEN
WATER
WATER
IF ONLY WATER IS EMITTED AFTER THE ELECTRICITY IS CREATED, IT ALSO SEEMS TO BE GOOD FOR THE ENVIRONMENT, RIGHT?
YUP.

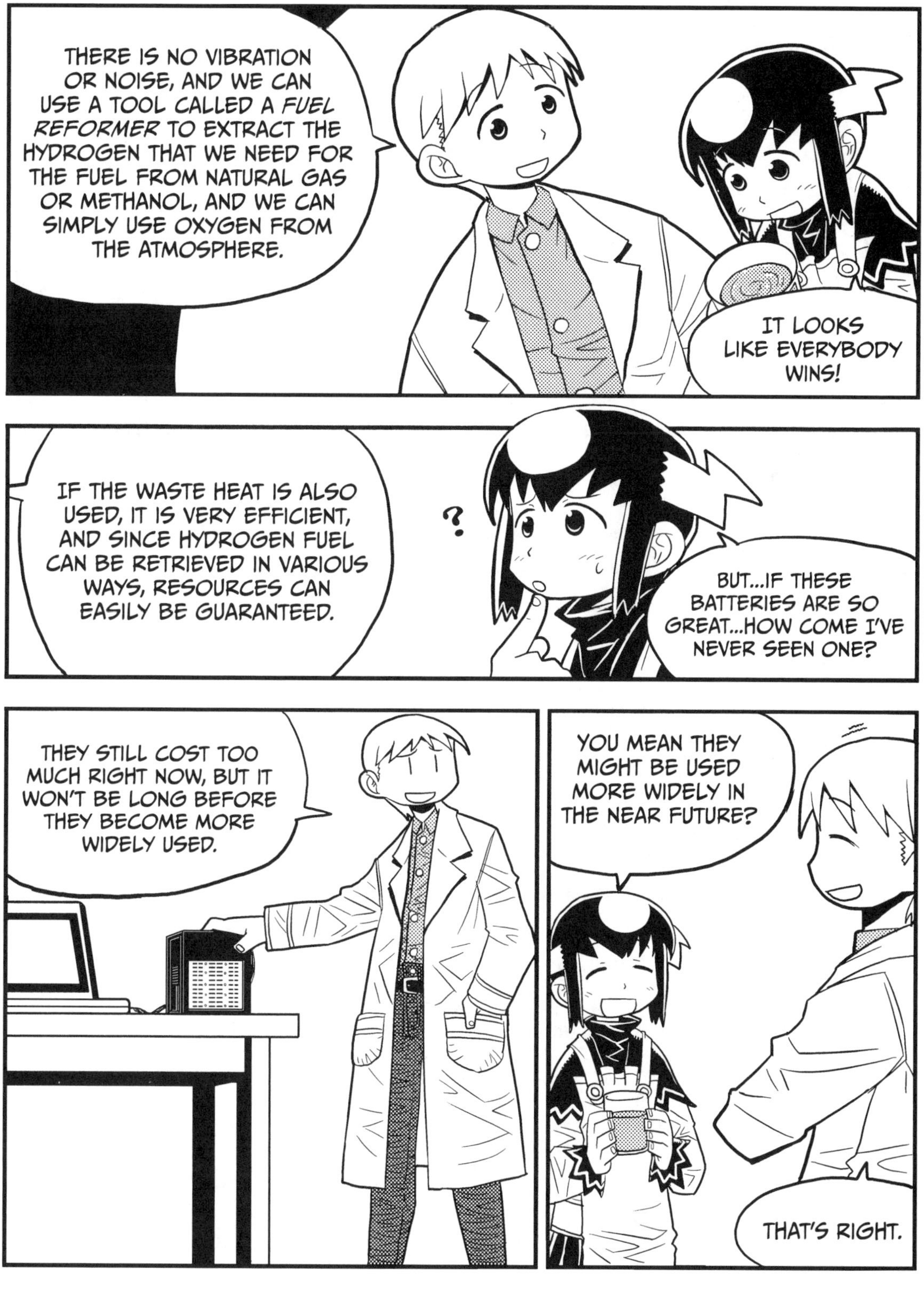
THERE IS NO VIBRATION OR NOISE, AND WE CAN USE A TOOL CALLED A *FUEL REFORMER* TO EXTRACT THE HYDROGEN THAT WE NEED FOR THE FUEL FROM NATURAL GAS OR METHANOL, AND WE CAN SIMPLY USE OXYGEN FROM THE ATMOSPHERE.
IT LOOKS LIKE EVERYBODY WINS!
IF THE WASTE HEAT IS ALSO USED, IT IS VERY EFFICIENT, AND SINCE HYDROGEN FUEL CAN BE RETRIEVED IN VARIOUS WAYS, RESOURCES CAN EASILY BE GUARANTEED.
?
BUT...IF THESE BATTERIES ARE SO GREAT...HOW COME I'VE NEVER SEEN ONE?
THEY STILL COST TOO MUCH RIGHT NOW, BUT IT WON'T BE LONG BEFORE THEY BECOME MORE WIDELY USED.
YOU MEAN THEY MIGHT BE USED MORE WIDELY IN THE NEAR FUTURE?
THAT'S RIGHT.

CREATING YOUR OWN COIN BATTERY

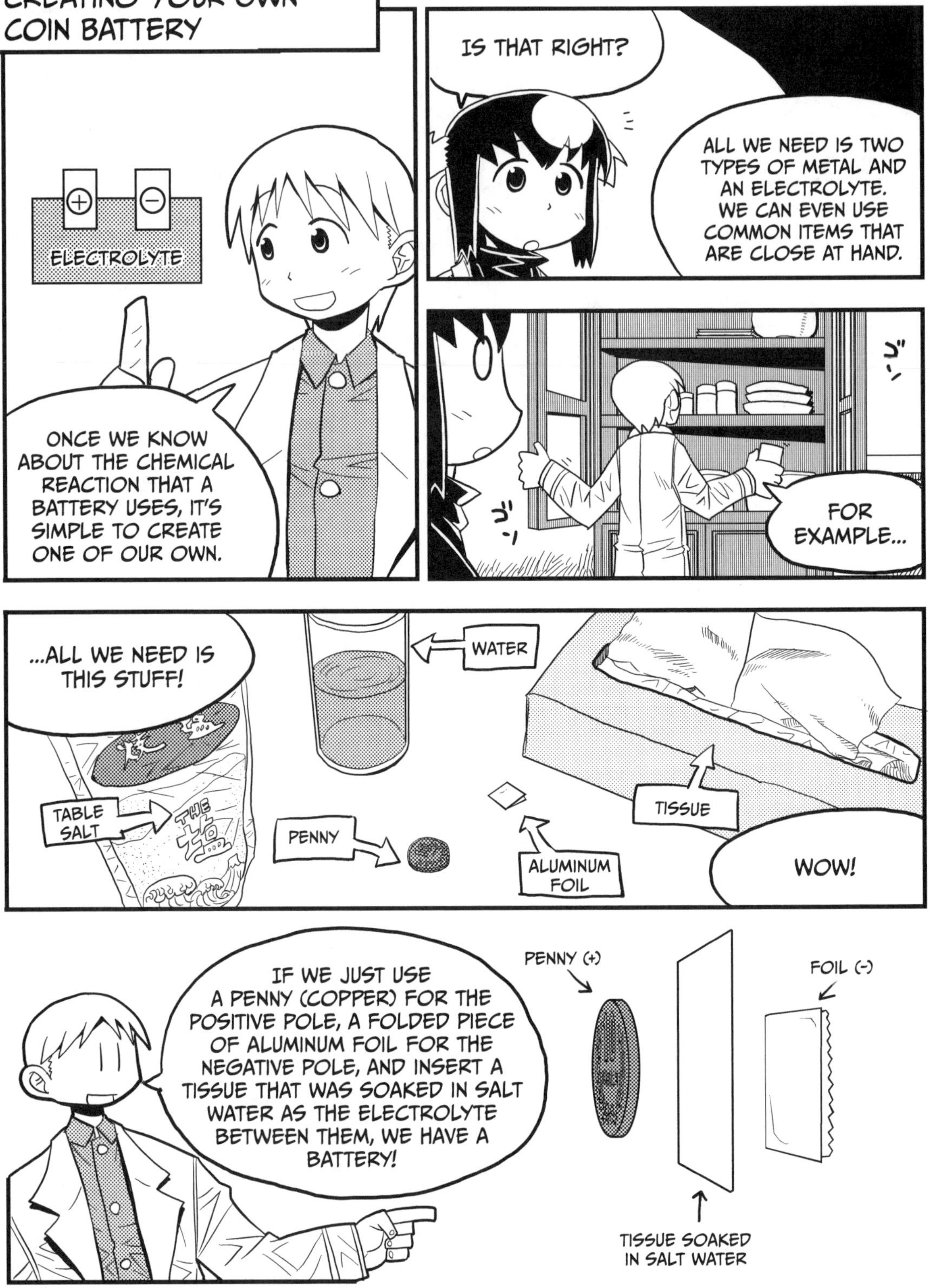

ELECTROLYTE
ONCE WE KNOW ABOUT THE CHEMICAL REACTION THAT A BATTERY USES, IT'S SIMPLE TO CREATE ONE OF OUR OWN.
IS THAT RIGHT?
ALL WE NEED IS TWO TYPES OF METAL AND AN ELECTROLYTE. WE CAN EVEN USE COMMON ITEMS THAT ARE CLOSE AT HAND.
FOR EXAMPLE...
...ALL WE NEED IS THIS STUFF!
WATER
TABLE SALT
PENNY
ALUMINUM FOIL
TISSUE
WOW!
IF WE JUST USE A PENNY (COPPER) FOR THE POSITIVE POLE, A FOLDED PIECE OF ALUMINUM FOIL FOR THE NEGATIVE POLE, AND INSERT A TISSUE THAT WAS SOAKED IN SALT WATER AS THE ELECTROLYTE BETWEEN THEM, WE HAVE A BATTERY!
PENNY (+)
FOIL (-)
TISSUE SOAKED IN SALT WATER

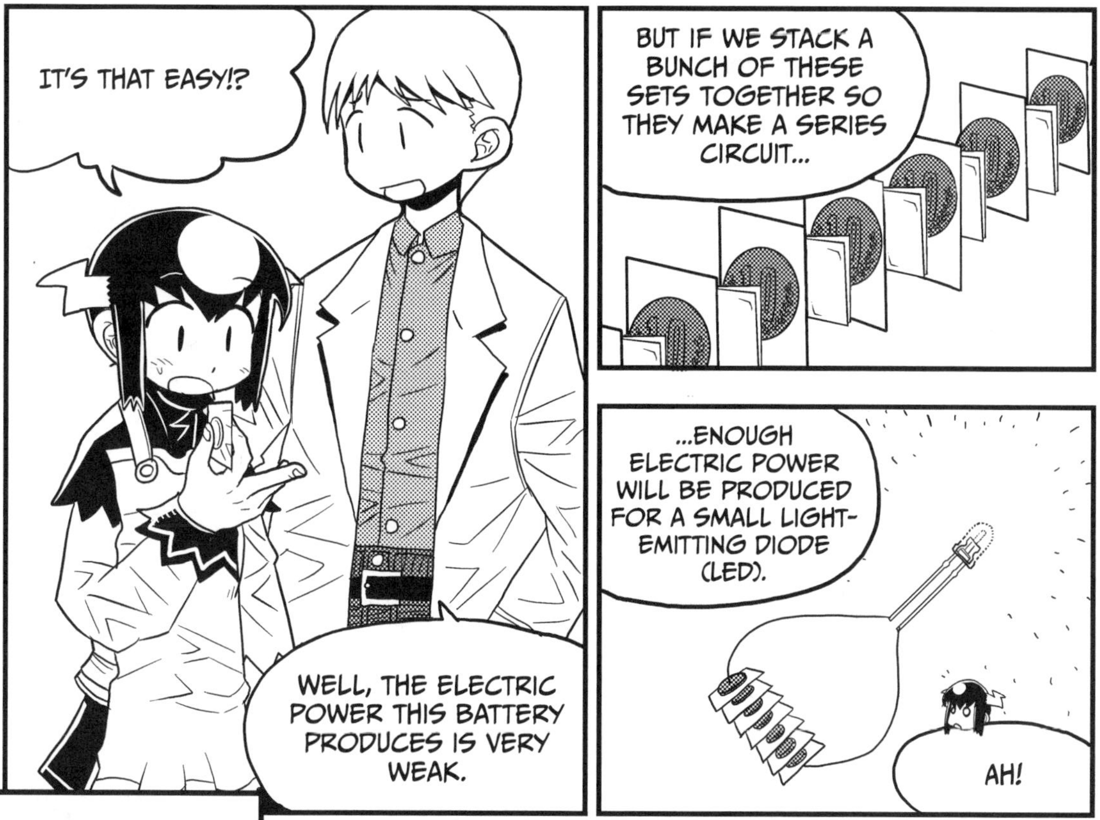
IT'S THAT EASY!?
WELL, THE ELECTRIC POWER THIS BATTERY PRODUCES IS VERY WEAK.
BUT IF WE STACK A BUNCH OF THESE SETS TOGETHER SO THEY MAKE A SERIES CIRCUIT...
...ENOUGH ELECTRIC POWER WILL BE PRODUCED FOR A SMALL LIGHT-EMITTING DIODE (LED).
AH!

THERMOPILES

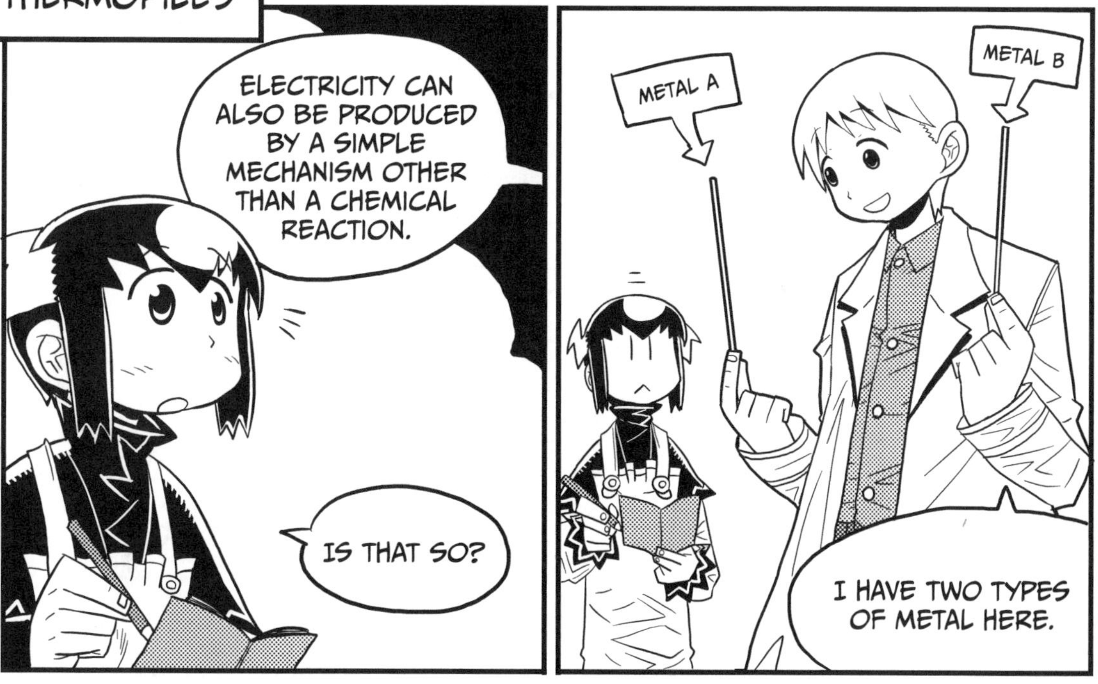
ELECTRICITY CAN ALSO BE PRODUCED BY A SIMPLE MECHANISM OTHER THAN A CHEMICAL REACTION.
IS THAT SO?
METAL A
METAL B
I HAVE TWO TYPES OF METAL HERE.

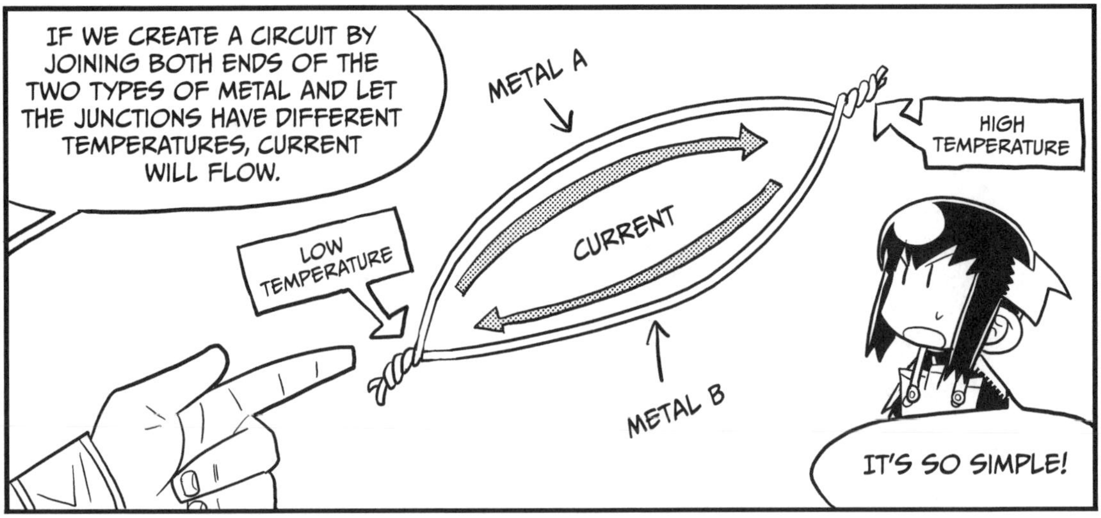
IF WE CREATE A CIRCUIT BY JOINING BOTH ENDS OF THE TWO TYPES OF METAL AND LET THE JUNCTIONS HAVE DIFFERENT TEMPERATURES, CURRENT WILL FLOW.
METAL A
HIGH TEMPERATURE
CURRENT
LOW TEMPERATURE
METAL B
IT'S SO SIMPLE!

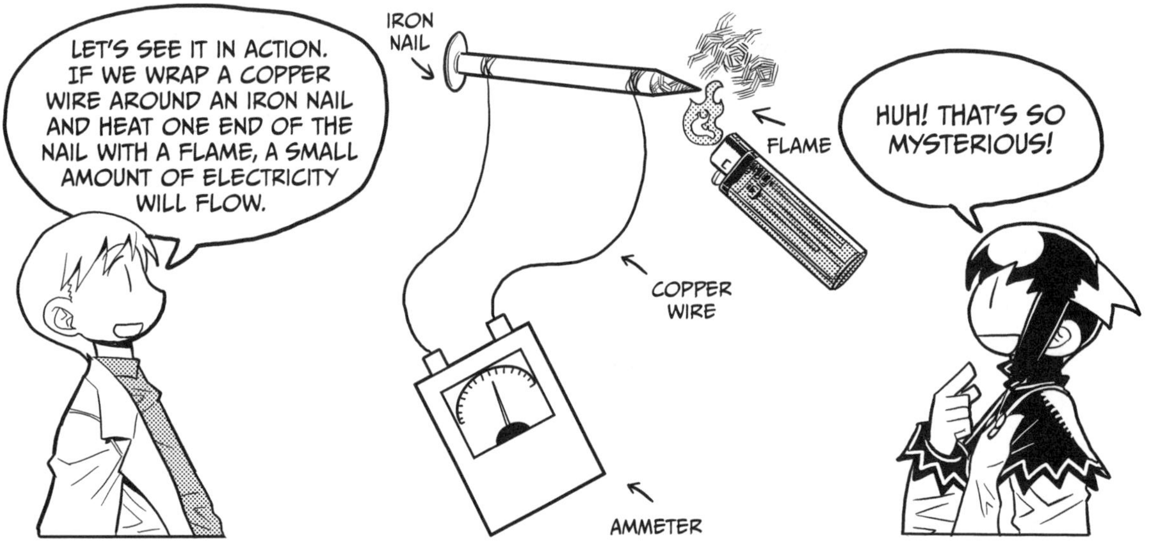
LET'S SEE IT IN ACTION. IF WE WRAP A COPPER WIRE AROUND AN IRON NAIL AND HEAT ONE END OF THE NAIL WITH A FLAME, A SMALL AMOUNT OF ELECTRICITY WILL FLOW.
IRON NAIL
FLAME
COPPER WIRE
AMMETER
HUH! THAT'S SO MYSTERIOUS!

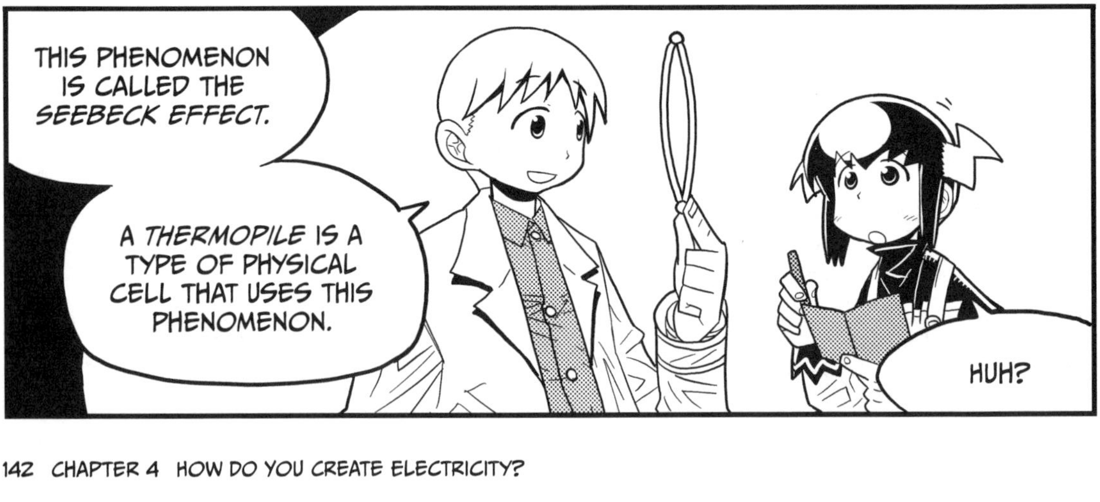
THIS PHENOMENON IS CALLED THE *SEEBECK EFFECT*.
A *THERMOPILE* IS A TYPE OF PHYSICAL CELL THAT USES THIS PHENOMENON.
HUH?

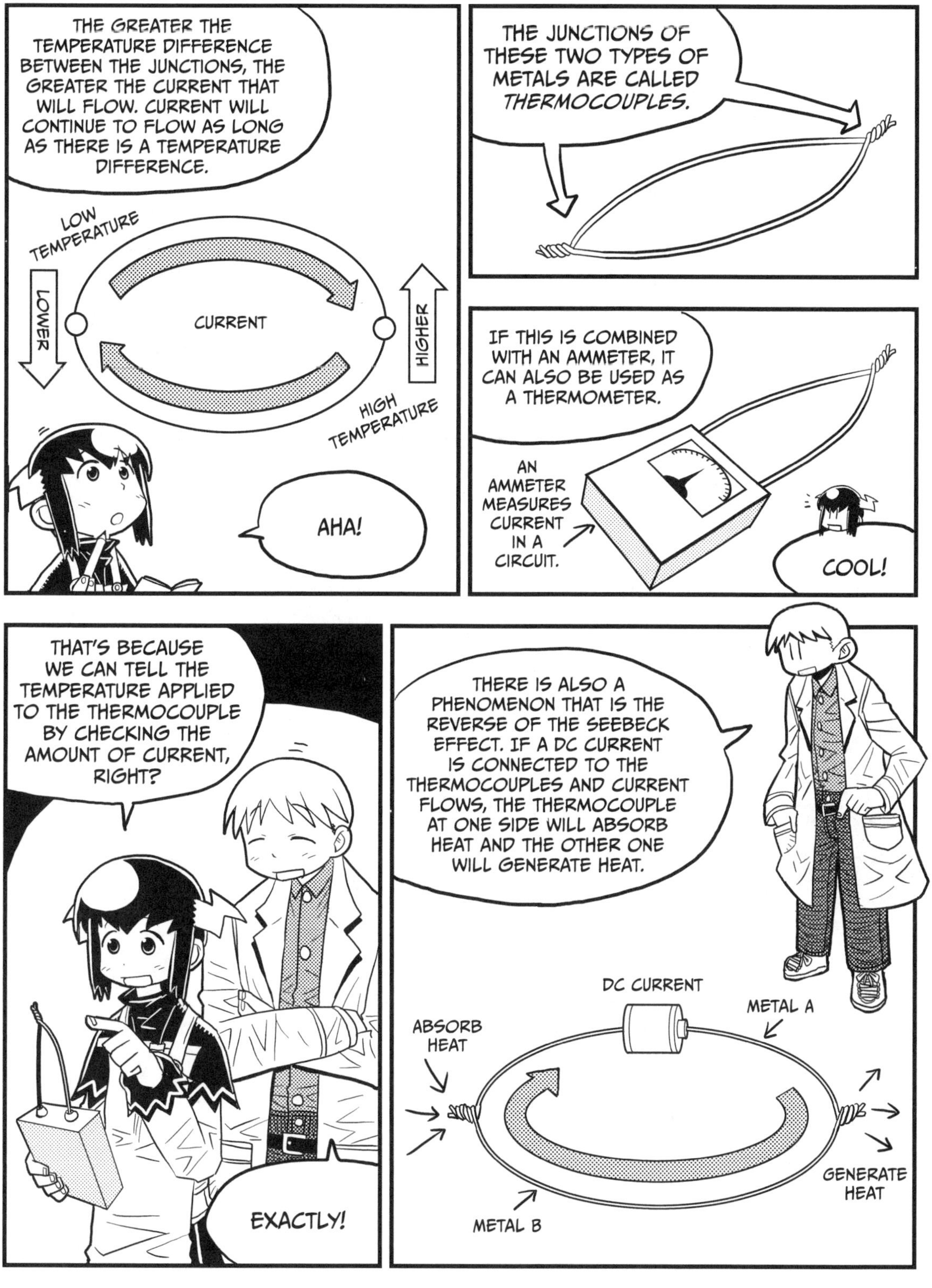
THE GREATER THE TEMPERATURE DIFFERENCE BETWEEN THE JUNCTIONS, THE GREATER THE CURRENT THAT WILL FLOW. CURRENT WILL CONTINUE TO FLOW AS LONG AS THERE IS A TEMPERATURE DIFFERENCE.
LOW TEMPERATURE
LOWER
CURRENT
HIGHER
HIGH TEMPERATURE
AHA!
THE JUNCTIONS OF THESE TWO TYPES OF METALS ARE CALLED *THERMOCOUPLES*.
IF THIS IS COMBINED WITH AN AMMETER, IT CAN ALSO BE USED AS A THERMOMETER.
AN AMMETER MEASURES CURRENT IN A CIRCUIT.
COOL!
THAT'S BECAUSE WE CAN TELL THE TEMPERATURE APPLIED TO THE THERMOCOUPLE BY CHECKING THE AMOUNT OF CURRENT, RIGHT?
EXACTLY!
THERE IS ALSO A PHENOMENON THAT IS THE REVERSE OF THE SEEBECK EFFECT. IF A DC CURRENT IS CONNECTED TO THE THERMOCOUPLES AND CURRENT FLOWS, THE THERMOCOUPLE AT ONE SIDE WILL ABSORB HEAT AND THE OTHER ONE WILL GENERATE HEAT.
DC CURRENT
METAL A
ABSORB HEAT
GENERATE HEAT
METAL B

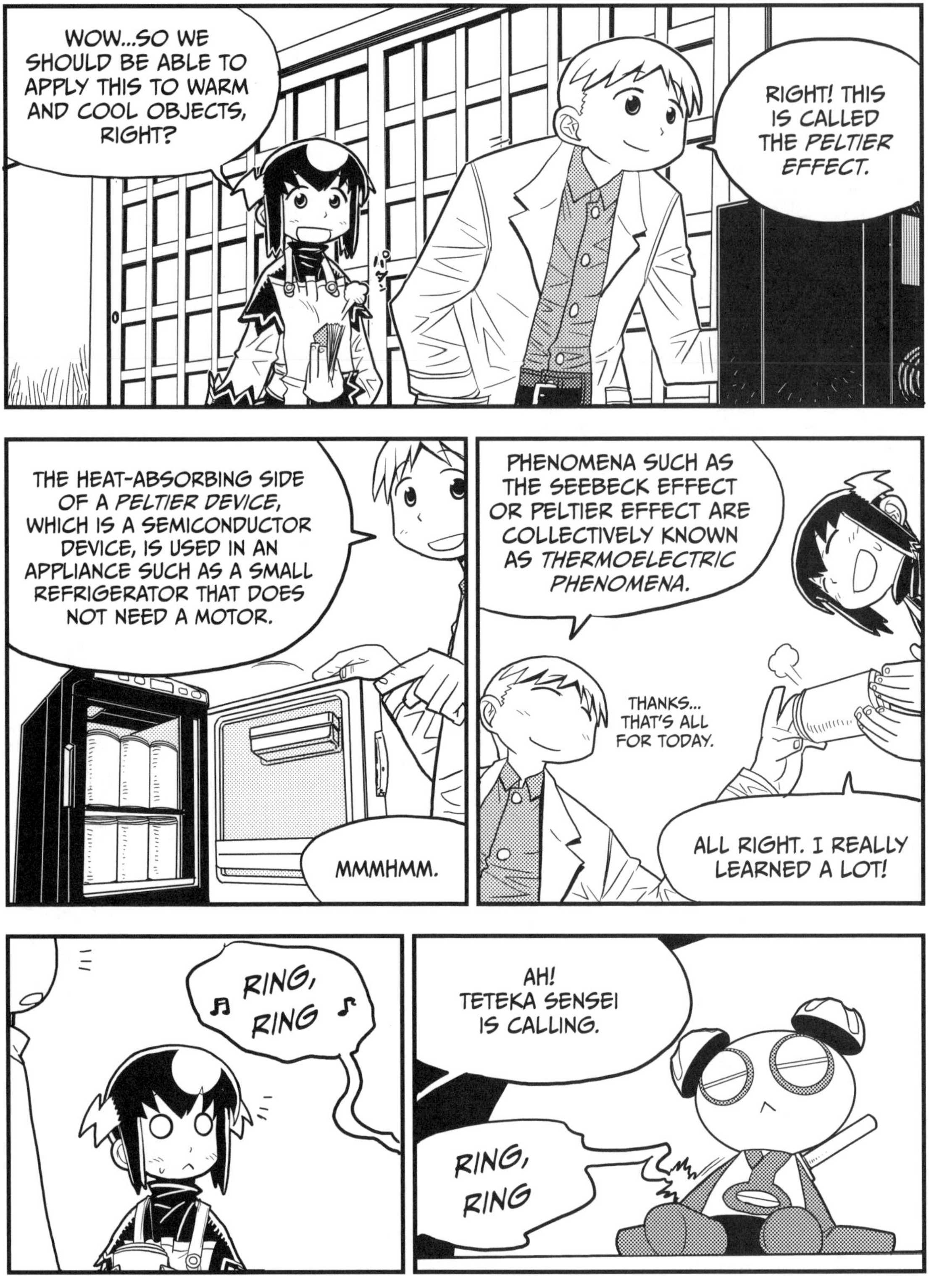
WOW...SO WE SHOULD BE ABLE TO APPLY THIS TO WARM AND COOL OBJECTS, RIGHT?
RIGHT! THIS IS CALLED THE *PELTIER EFFECT*.
THE HEAT-ABSORBING SIDE OF A *PELTIER DEVICE*, WHICH IS A SEMICONDUCTOR DEVICE, IS USED IN AN APPLIANCE SUCH AS A SMALL REFRIGERATOR THAT DOES NOT NEED A MOTOR.
MMMHMM.
PHENOMENA SUCH AS THE SEEBECK EFFECT OR PELTIER EFFECT ARE COLLECTIVELY KNOWN AS *THERMOELECTRIC PHENOMENA*.
THANKS... THAT'S ALL FOR TODAY.
ALL RIGHT. I REALLY LEARNED A LOT!
♫ RING, RING ♪
AH! TETEKA SENSEI IS CALLING.
RING, RING

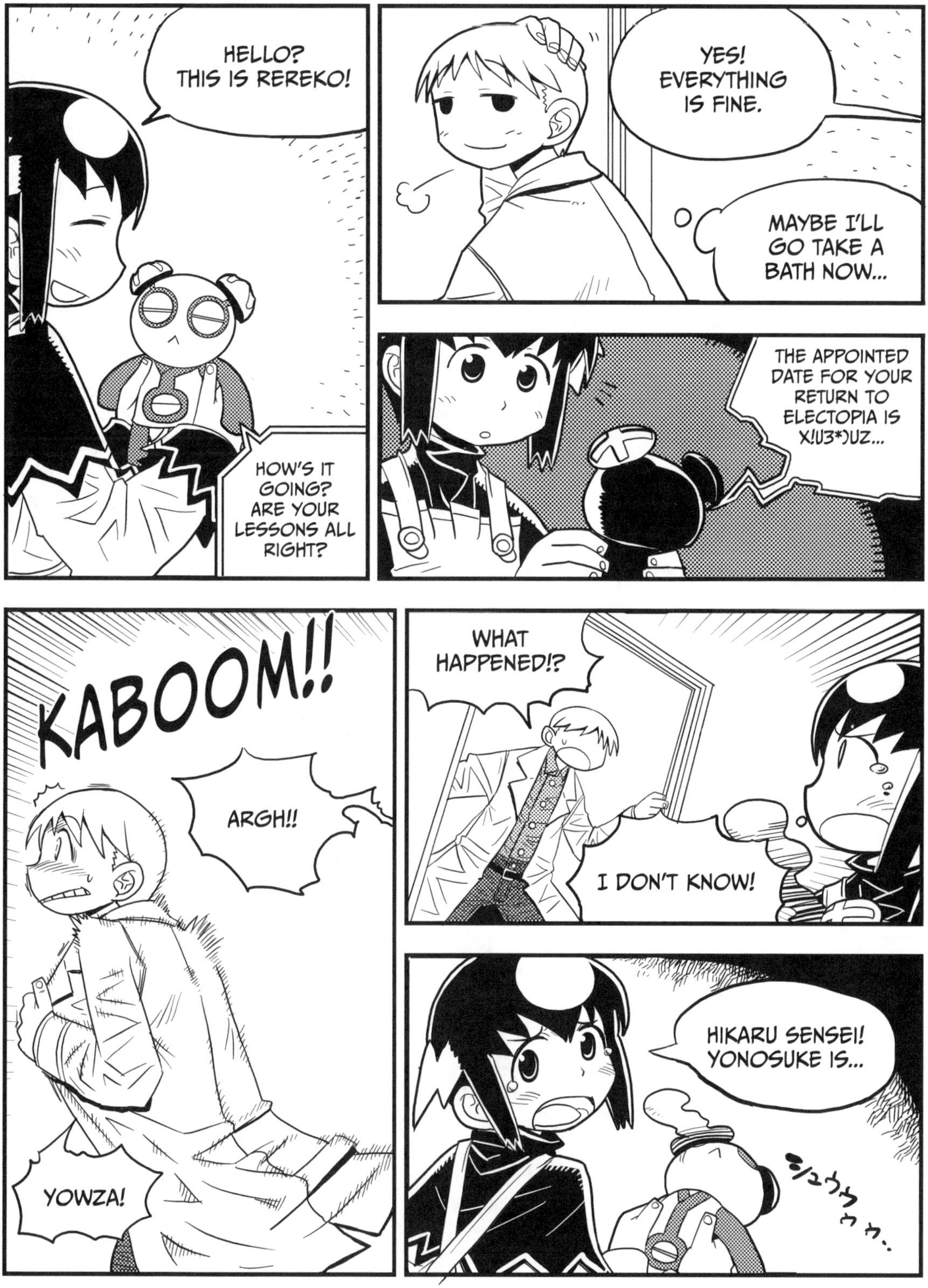
HELLO? THIS IS REREKO!
HOW'S IT GOING? ARE YOUR LESSONS ALL RIGHT?
YES! EVERYTHING IS FINE.
MAYBE I'LL GO TAKE A BATH NOW...
THE APPOINTED DATE FOR YOUR RETURN TO ELECTOPIA IS X!U3*)UZ...
KABOOM!!
ARGH!!
YOWZA!
WHAT HAPPENED!?
I DON'T KNOW!
HIKARU SENSEI! YONOSUKE IS...

[redacted]

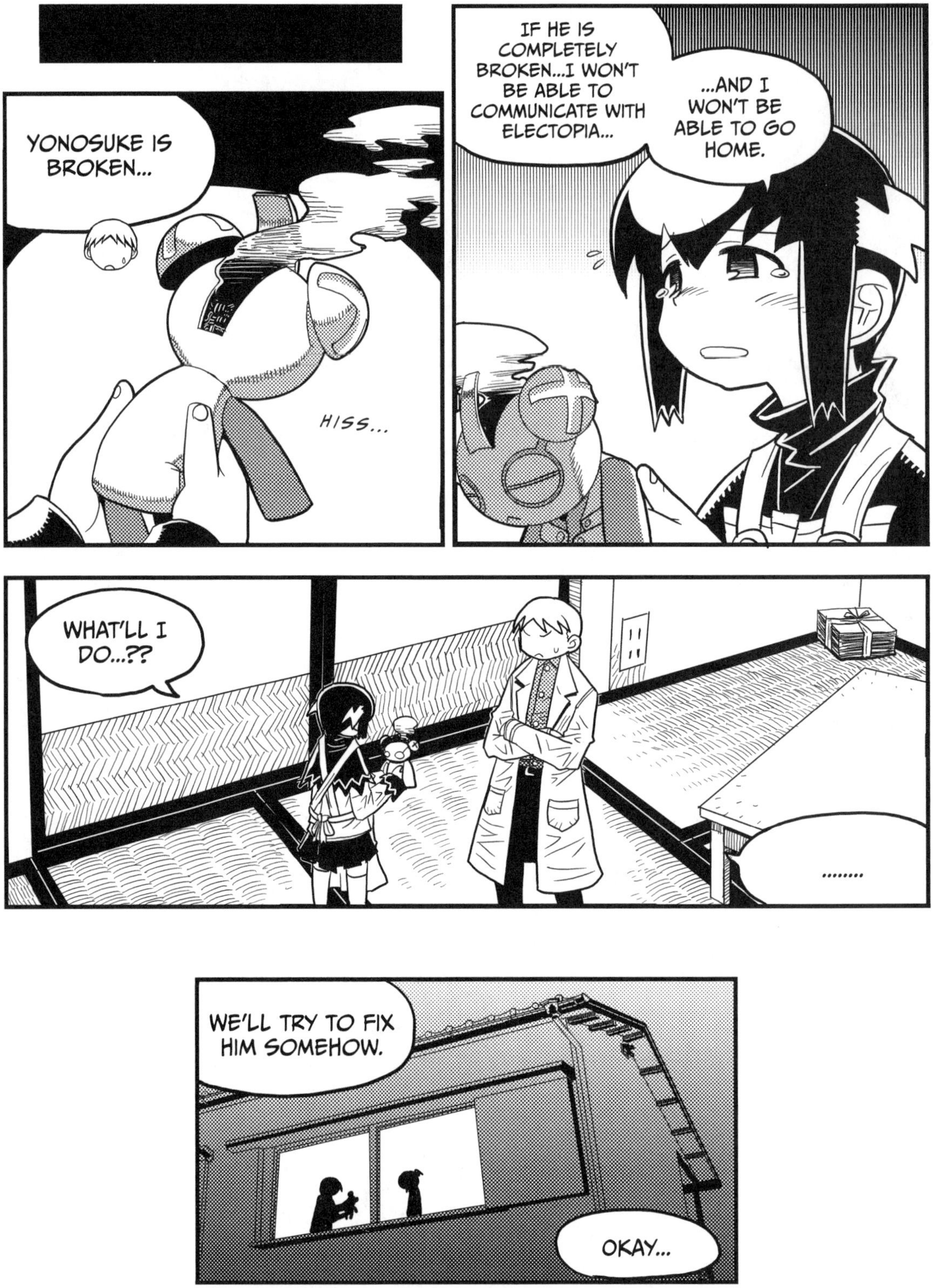

YONOSUKE IS BROKEN...
HISS...
IF HE IS COMPLETELY BROKEN...I WON'T BE ABLE TO COMMUNICATE WITH ELECTOPIA...
...AND I WON'T BE ABLE TO GO HOME.
WHAT'LL I DO...??
....
WE'LL TRY TO FIX HIM SOMEHOW.
OKAY...

ELECTRICITY CREATED BY A POWER PLANT

In power plants, no matter what the source of motion, a turbine spins, which generates electricity in a generator.

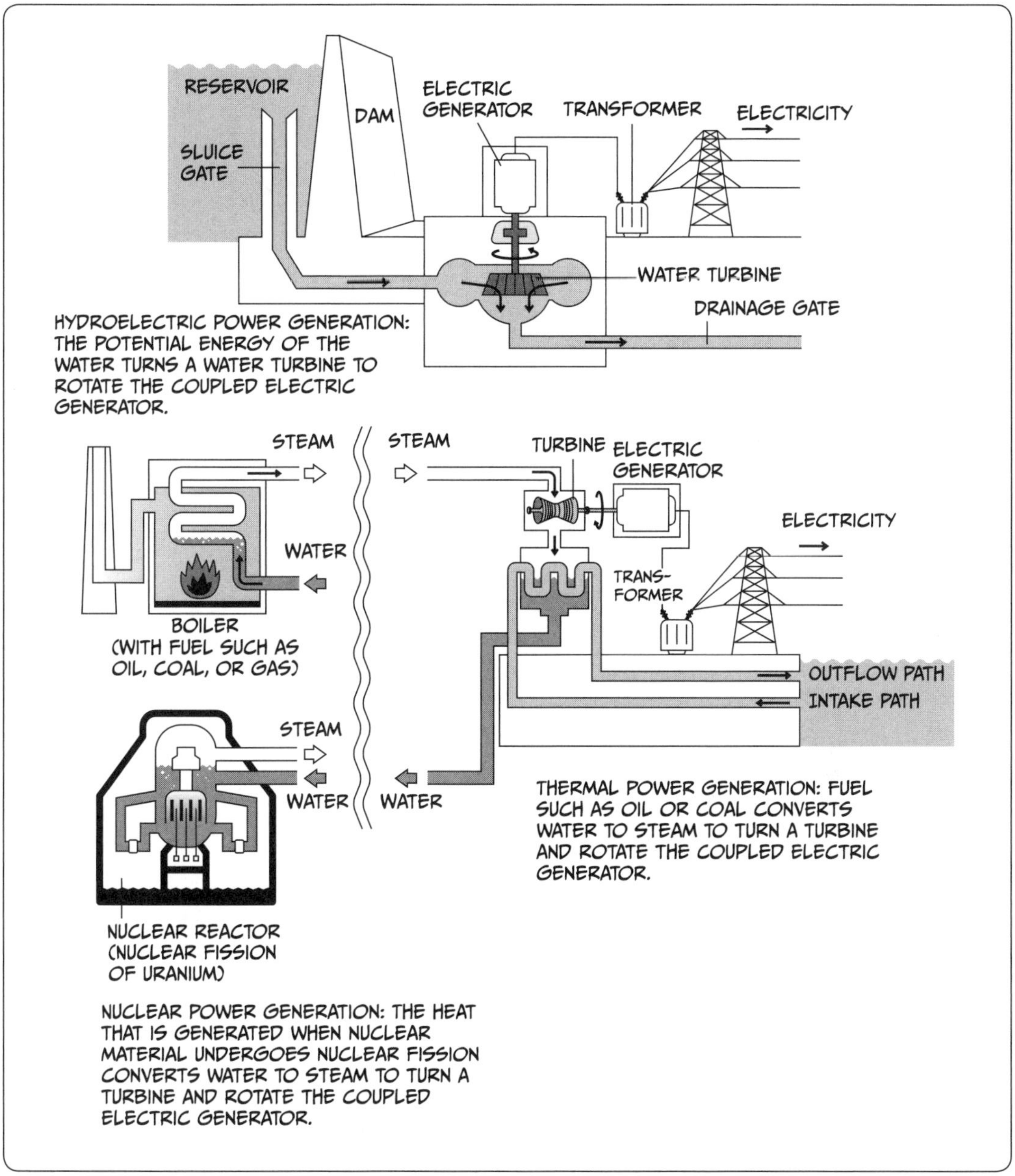

RESERVOIR
SLUICE GATE
DAM
ELECTRIC GENERATOR
TRANSFORMER
ELECTRICITY
WATER TURBINE
DRAINAGE GATE
HYDROELECTRIC POWER GENERATION: THE POTENTIAL ENERGY OF THE WATER TURNS A WATER TURBINE TO ROTATE THE COUPLED ELECTRIC GENERATOR.
STEAM
STEAM
TURBINE
ELECTRIC GENERATOR
ELECTRICITY
WATER
TRANS-FORMER
BOILER (WITH FUEL SUCH AS OIL, COAL, OR GAS)
OUTFLOW PATH
INTAKE PATH
STEAM
WATER
WATER
THERMAL POWER GENERATION: FUEL SUCH AS OIL OR COAL CONVERTS WATER TO STEAM TO TURN A TURBINE AND ROTATE THE COUPLED ELECTRIC GENERATOR.
NUCLEAR REACTOR (NUCLEAR FISSION OF URANIUM)
NUCLEAR POWER GENERATION: THE HEAT THAT IS GENERATED WHEN NUCLEAR MATERIAL UNDERGOES NUCLEAR FISSION CONVERTS WATER TO STEAM TO TURN A TURBINE AND ROTATE THE COUPLED ELECTRIC GENERATOR.

Electricity created by a power plant

An electric generator creates electricity according to Fleming's right-hand rule. And since the conductor is rotating within the magnetic field, both the magnitude and direction of the electricity's flow vary repeatedly like a wave. The maximum voltage is generated when the loop of wire cuts through the magnetic field at a right angle, and the voltage is zero when the direction of the magnetic field and the direction of the loop's movement are the same.

MAGNETIC FIELD
N
CONDUCTOR
CURRENT
S
N
CONDUCTOR
MAGNETIC FIELD
WHEN THE ELECTRIC WIRE MOVES IN THIS DIRECTION, NO CURRENT FLOWS
S
WHEN THE CURRENT MOVES PERPENDICULAR TO THE MAGNETIC FIELD, THE VOLTAGE THAT IS GENERATED IS THE MAXIMUM.

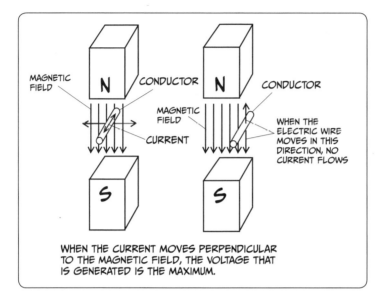

Electricity generated in a conductor

N
MAGNETIC FIELD
LOOP OF WIRE
S
WHEN A LOOP OF WIRE ROTATES WITHIN THE MAGNETIC FIELD, WAVEFORM VOLTAGE IS GENERATED.
+
VOLTAGE O
-
0°
90°
180°
270°
360°

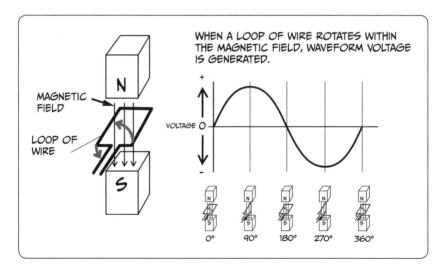

Electricity created by an electric generator

The electricity that is created by this process is called *alternating current (AC)*. This is the electricity that comes from a household electric outlet. One wave is created by one rotation of the conductor within the magnetic field. If the conductor rotates 60 times per second, 60 waves are produced per second. This would be electricity with a frequency of 60 hertz (Hz).

The voltage of an ordinary household electric outlet is 120V AC. The *peak voltage* of the wave of this electricity is approximately 170V. The value of 120V represents the *effective voltage*, which is the value for which direct current (DC) electricity does the same work—in other words, the amount of heat generated when 120V AC is applied to a resistance is the same as the amount of heat generated when 120V DC is applied to the same resistance.

THE VOLTAGE OF AN ORDINARY ELECTRIC OUTLET IN A HOUSE IS 120V AC.
PEAK VOLTAGE 170V
EFFECTIVE VOLTAGE 120V
0
-120V
MINIMUM VOLTAGE -170V
120V DC
RESISTANCE
120V AC
RESISTANCE
WHEN A 120V DC POWER SUPPLY IS CONNECTED TO A RESISTANCE, THE SAME AMOUNT OF HEAT IS GENERATED AS WHEN A 120V AC POWER SUPPLY IS CONNECTED TO THE SAME RESISTANCE.

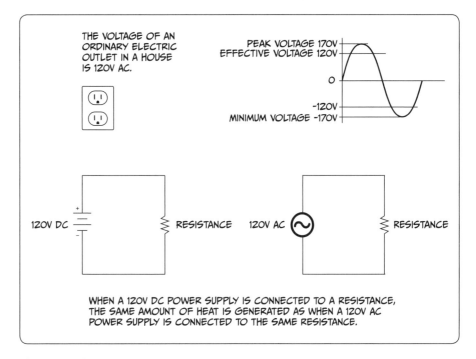

AC voltage and effective voltage

THERMAL POWER GENERATION

The types of thermal power generation that generate the most electrical power are steam generation, internal combustion generation, gas turbine generation, and combined cycle generation.

Steam generation burns fuel such as oil, coal, or liquefied natural gas (LNG) in a boiler to generate high-temperature, high-pressure steam. The force of that steam turns a turbine that is coupled with an electric generator to generate electricity.

The steam that was used to generate electricity is cooled in a device called a *condenser*; once it returns to liquid water, it is then sent to the boiler again.

TURBINE
ELECTRIC GENERATOR
MAIN SHAFT
STEAM
STEAM
ROTATES
STEAM
BLADE
CHIMNEY
STEAM
COMBUSTION EXHAUST GAS
TO TURBINE
WATER
FUEL
BOILER

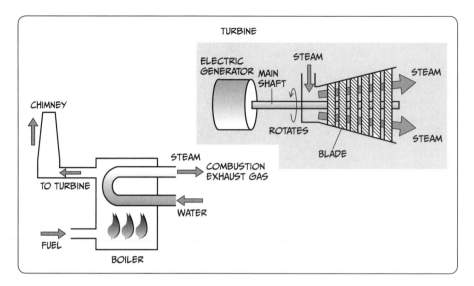

Turbine and steam power generation

Internal combustion generation uses an internal combustion engine (like a diesel engine) to generate electricity.

Gas turbine generation uses a combustion gas such as kerosene or diesel oil to turn a gas turbine and create electricity.

CHIMNEY
EXHAUST HEAT RECOVERY BOILER
WATER
EXHAUST GAS
STEAM
WATER
BOILER
COMBUSTION EXHAUST GAS
GENERATOR
COOLANT (SUCH AS OCEAN WATER)
STEAM TURBINE
GAS TURBINE
CONDENSER
COMBUSTION GAS

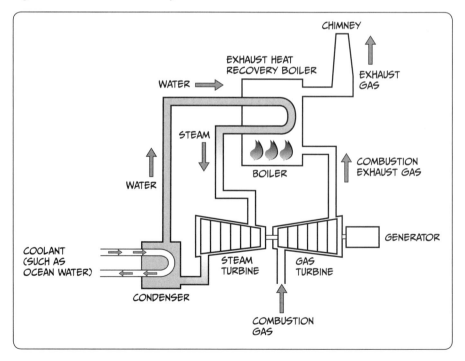

Combined cycle generation

Combined cycle generation combines steam generation and gas turbine generation. Electricity is generated by a gas turbine, and then the heat of the exhaust gas is used to create steam to turn a steam turbine and generate more electricity; this is an efficient method of generating electricity.

NUCLEAR POWER GENERATION

Nuclear power generation uses the heat generated when the nuclear fission of uranium occurs in a nuclear reactor to create high-temperature, high-pressure steam, which turns a turbine and creates electricity. When a neutron collides with uranium-235, it decays to thorium-231; several neutrons as well as heat are emitted. The neutrons successively collide with other uranium-235 nuclei, causing *nuclear fission* to occur and generating a great deal of thermal energy.

Nuclear power generation uses this heat to create steam, which turns a turbine and generates electricity in a manner similar to thermal power generation. *Control rods*, which absorb neutrons, and a *moderator*, which reduces the speed of the neutrons, are used in the nuclear reactor to control nuclear fission and regulate the reactor's output.

THORIUM-231
NEUTRON
NEUTRON
URANIUM-235
DECAYS TO
NEUTRON
NEUTRON
THERMAL ENERGY

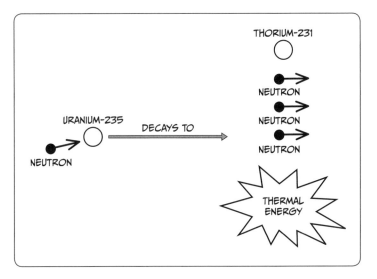

The nuclear fission of uranium-235 leads to more neutrons, which may strike other U-235 atoms, causing a chain reaction.

There are various types of nuclear reactors. Currently, the type that is used most often is called a *light water reactor*, which uses light water (ordinary water) as a moderator and coolant. Light water reactors include boiling water reactors and pressurized water reactors.

A *boiling water reactor* sends steam that was generated in the reactor pressure vessel directly to the turbine; after it turns the turbine, the steam turns back into liquid water in a device called a condenser, and the water is reused. A *condenser* uses ocean water to cool the steam so it turns back into liquid water and can be reused.

A *pressurized water reactor* passes boiling water that was created in the reactor pressure vessel to the steam generator, where water from a separate system is changed to steam, which turns the turbine.

STEAM
STEAM
TO TURBINE
REACTOR PRESSURE VESSEL
WATER
WATER
URANIUM FUEL
CONTROL RODS

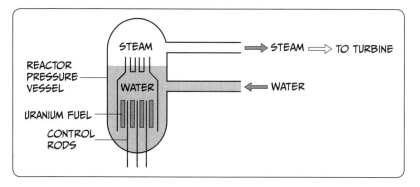

Boiling water reactor

STEAM GENERATOR
PRESSURIZER
STEAM
STEAM
TO TURBINE
WATER
CONTROL ROD
HIGH TEMP. WATER
REACTOR PRESSURE VESSEL
URANIUM FUEL
PUMP
WATER

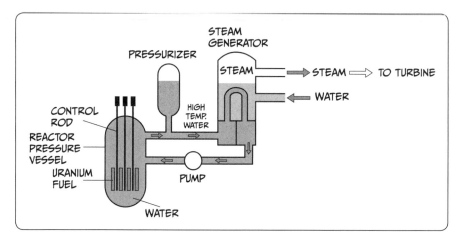

Pressurized water reactor

HYDROELECTRIC POWER GENERATION

Hydroelectric power generation uses the potential energy of water to generate electricity. *Dam-type power generation* stores water at a high location and lets the water drop from there to turn a water turbine coupled with an electric generator to generate electricity.

Since it is easier to start and stop power generation and to increase or decrease the amount of power generated via hydroelectric power generation than it is to do so for thermal or nuclear power generation, hydroelectric power can be generated corresponding to varying power demands. Also, during periods of low power demand, a lift pump can be used to draw water up to the higher location to store it as potential energy.

For hydroelectric power generation to use the energy of water efficiently, several types of water turbines are used for different purposes according to the *head (height difference)* or amount of water.

DAM
SLUICE GATE
POWER PLANT
GENERATOR

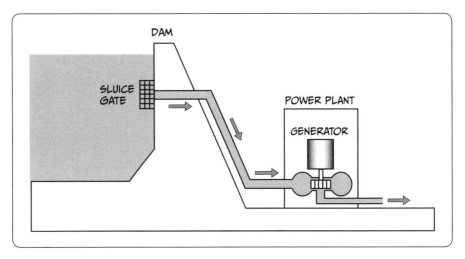

Dam-type power plant

A *Francis turbine* is used for a large amount of running water and a medium to high head. This type of water turbine is used for approximately 70 percent of the hydroelectric power generation in Japan. The water is directed perpendicular to the main shaft from all directions, the blades inside the turbine change the water's direction to the axial direction, and the turbine is rotated by hydraulic power when the water is discharged.

IMPELLER (RUNNER)
CASING

Francis turbine

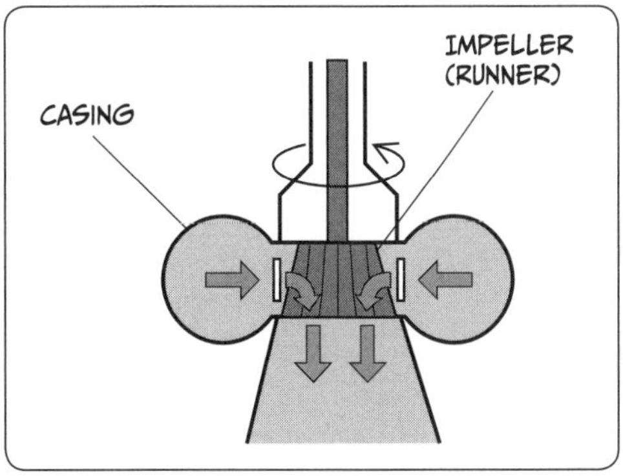

A *Pelton wheel* is a water turbine that rotates from the recoil that occurs when water sprayed from a nozzle hits a spoon-shaped bucket (blade). It is useful in locations that have a high head.

BUCKET
NOZZLE

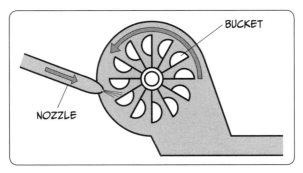

Pelton wheel

A *Kaplan turbine* is a water turbine that rotates because several propeller blades connected to a shaft adjust their angles according to variations in the amount of water flowing or the head. This kind of turbine is useful in locations having a low head. A type of Kaplan turbine that does not adjust the angle of the blades is called a *propeller turbine*.

Although hydroelectric power generation accounts for only 10 percent of Japan's power generation, it is a valuable method for a country with few resources.

SHAFT
WATER FLOW
BLADE

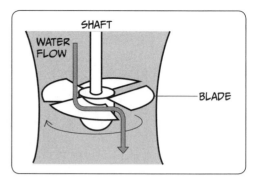

Kaplan turbine

WIND POWER GENERATION

Wind power generation uses wind power to turn a turbine; the turbine in turn rotates an electric generator to create electricity.

WIND
BLADE (PROPELLER)
GEARBOX
VARIABLE PITCH
GENERATOR

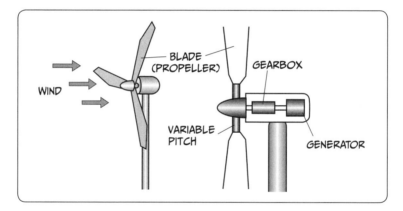

Mechanism of wind power generation (propeller wind turbine)

There are various types of wind turbines for wind power generation. A *propeller wind turbine*, which uses wind power energy very efficiently, is common. When wind hits the turbine's blades, it creates rotational motion, and the rotation speed is increased by a gearbox to turn an electrical generator. An *anemoscope (wind vane)* and an *anemometer (wind gauge)* constantly measure the wind conditions, and the direction of the propeller and angle of the blades are adjusted to the optimum state to use the wind power most effectively.

Although the power supplied by wind power generation is greatly affected by variations in the direction and speed of the wind, and the noise that is generated when the wind turbine rotates can be a problem, this is a clean, environmentally friendly power generation method that requires no fuel and produces no exhaust gases.

5
HOW CAN YOU CONVENIENTLY USE ELECTRICITY?

FORTUNATELY, HIKARU WAS ABLE TO REPAIR YONOSUKE'S BREAKDOWN WITH THE TECHNOLOGY AVAILABLE ON EARTH.

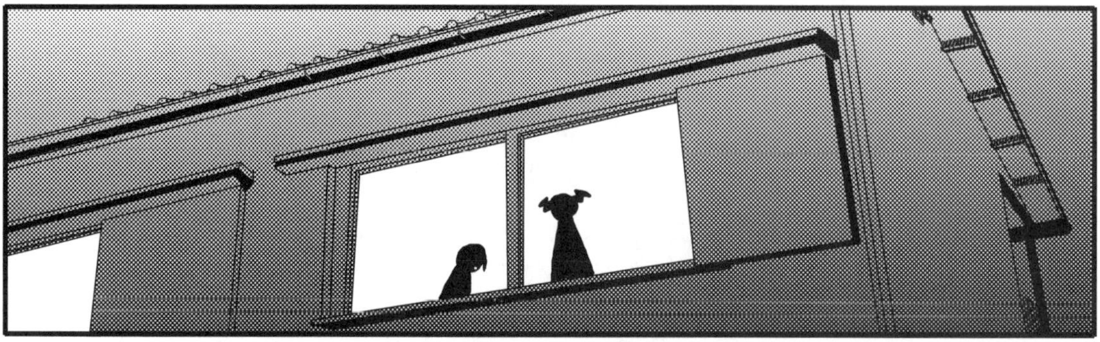

HE WORKED ON YONOSUKE EVERY NIGHT...

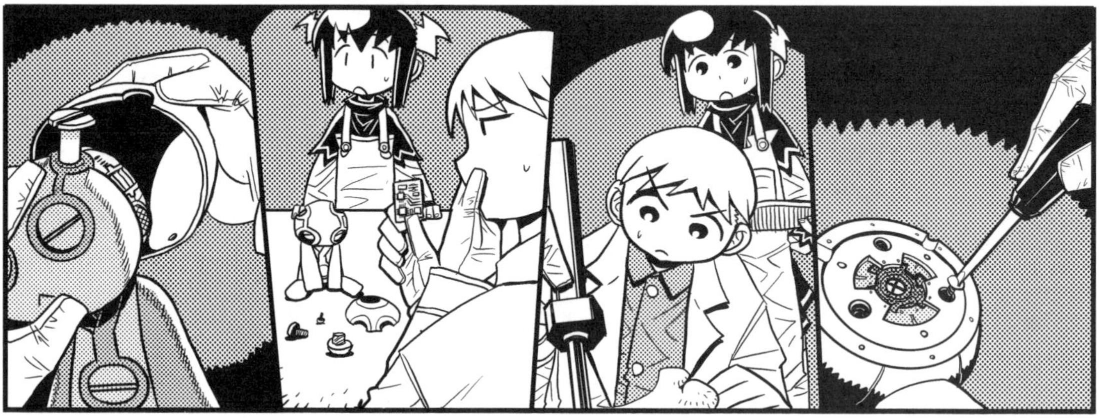

AND SEVERAL DAYS LATER...

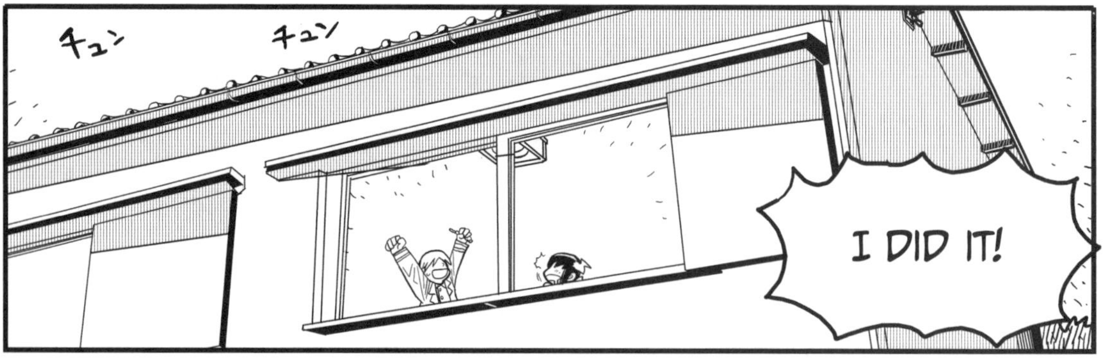
チュン
チュン
I DID IT!

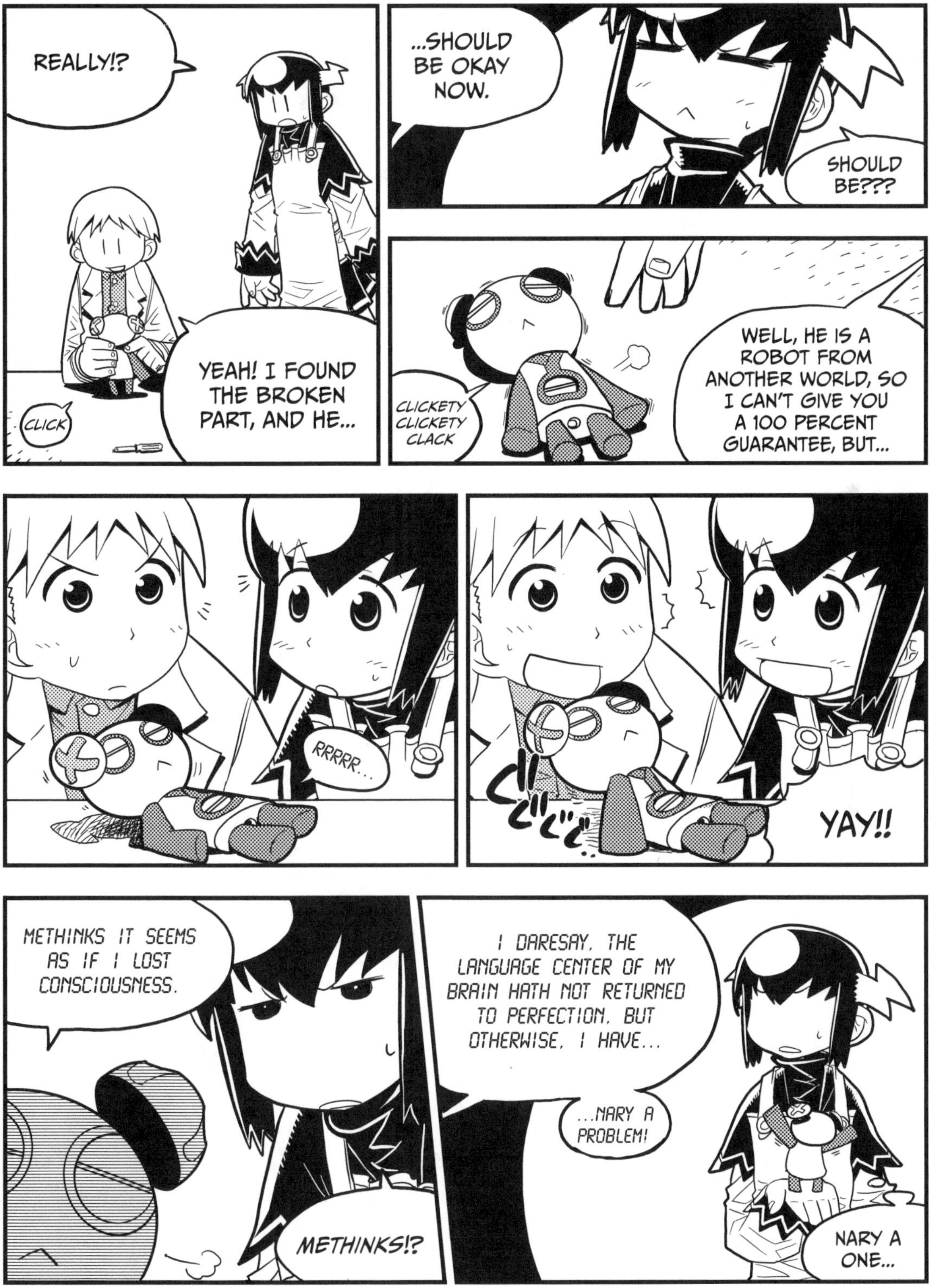
REALLY!?
YEAH! I FOUND THE BROKEN PART, AND HE...
CLICK
...SHOULD BE OKAY NOW.
SHOULD BE???
CLICKETY CLICKETY CLACK
WELL, HE IS A ROBOT FROM ANOTHER WORLD, SO I CAN'T GIVE YOU A 100 PERCENT GUARANTEE, BUT...
RRRRR...
YAY!!
METHINKS IT SEEMS AS IF I LOST CONSCIOUSNESS.
METHINKS!?
I DARESAY, THE LANGUAGE CENTER OF MY BRAIN HATH NOT RETURNED TO PERFECTION, BUT OTHERWISE, I HAVE...
...NARY A PROBLEM!
NARY A ONE...

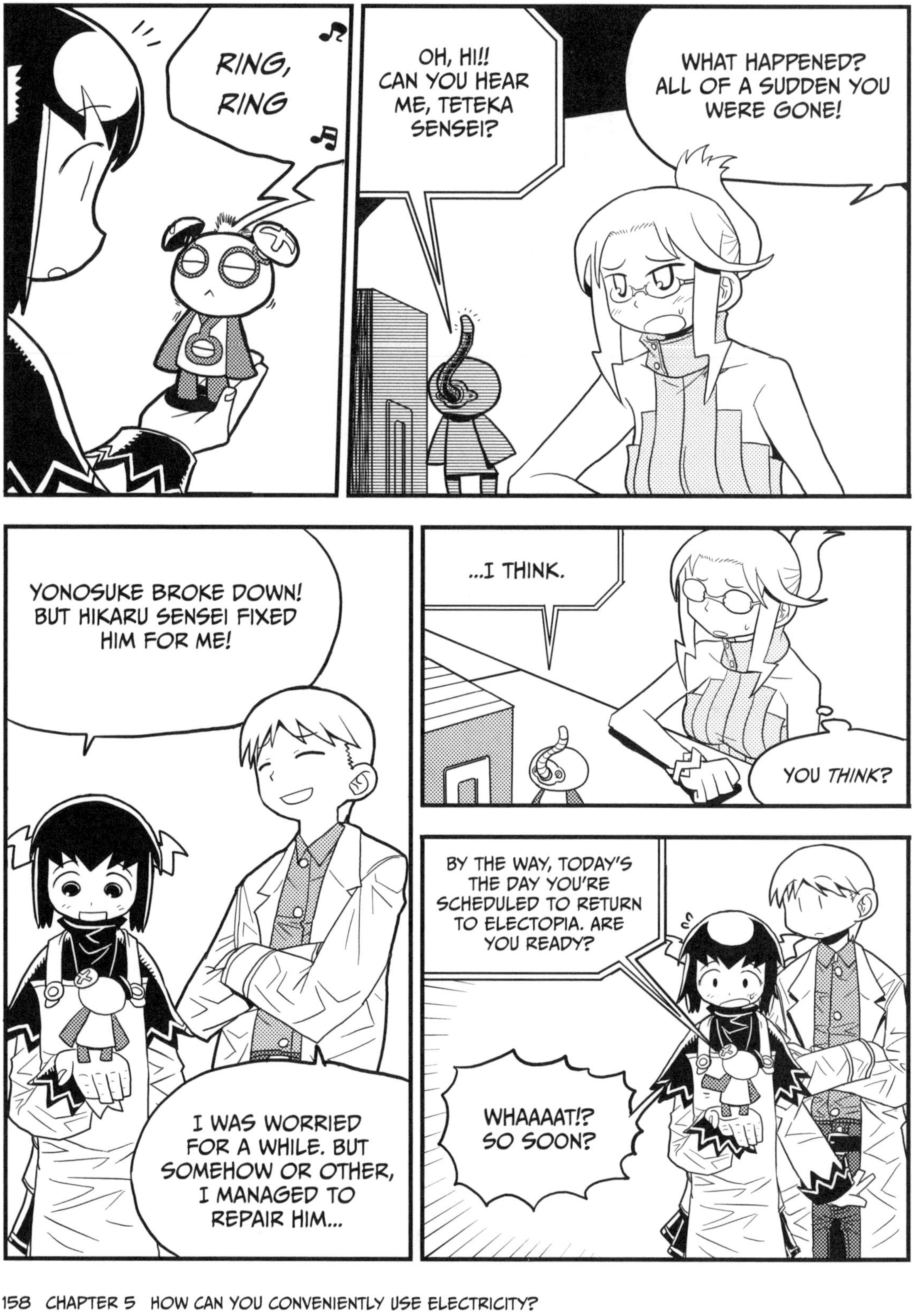
RING, RING
OH, HI!! CAN YOU HEAR ME, TETEKA SENSEI?
WHAT HAPPENED? ALL OF A SUDDEN YOU WERE GONE!
YONOSUKE BROKE DOWN! BUT HIKARU SENSEI FIXED HIM FOR ME!
I WAS WORRIED FOR A WHILE. BUT SOMEHOW OR OTHER, I MANAGED TO REPAIR HIM...
...I THINK.
YOU THINK?
BY THE WAY, TODAY'S THE DAY YOU'RE SCHEDULED TO RETURN TO ELECTOPIA. ARE YOU READY?
WHAAAAT!? SO SOON?

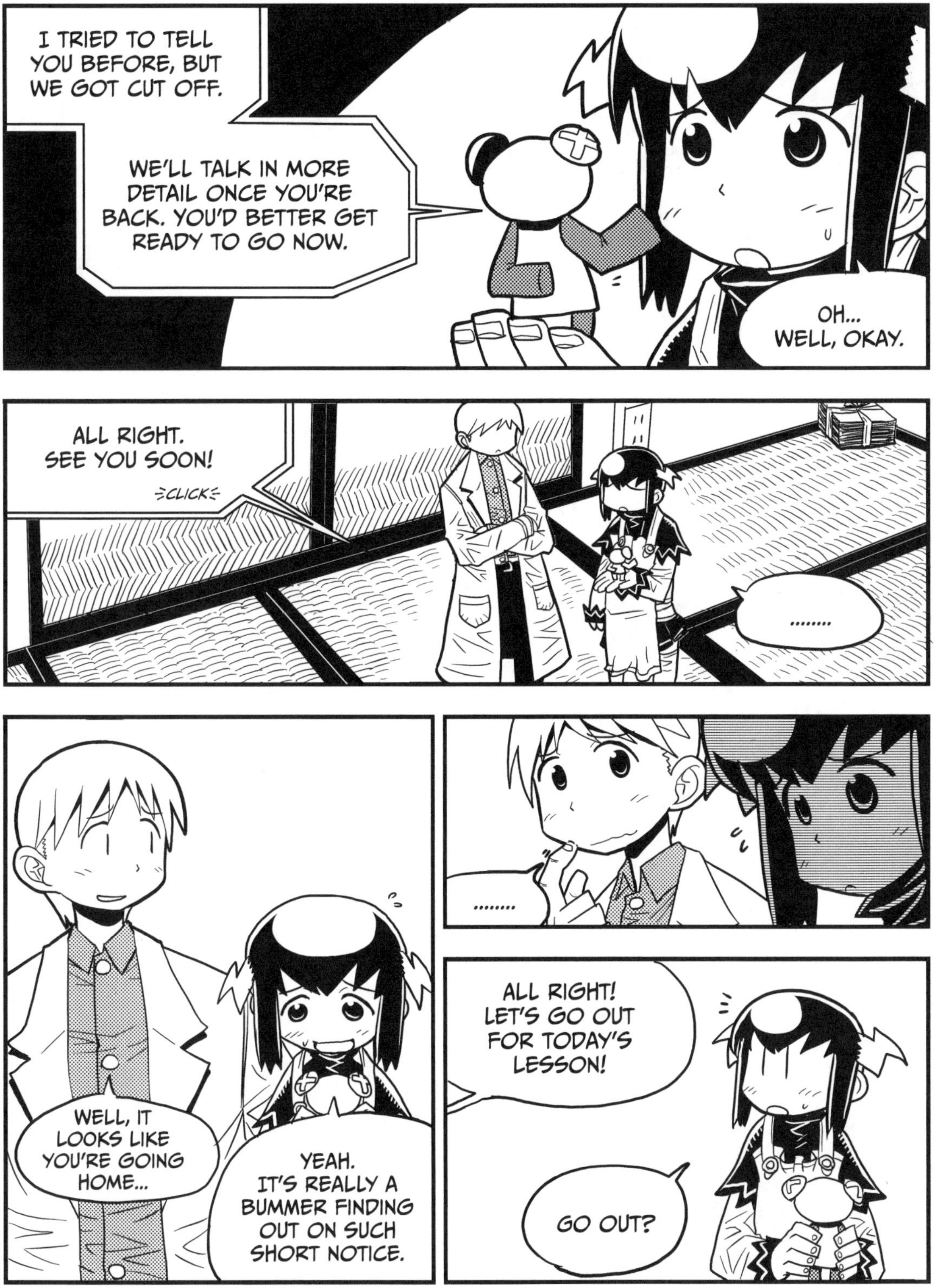
I TRIED TO TELL YOU BEFORE, BUT WE GOT CUT OFF.
WE'LL TALK IN MORE DETAIL ONCE YOU'RE BACK. YOU'D BETTER GET READY TO GO NOW.
OH... WELL, OKAY.
ALL RIGHT. SEE YOU SOON!
CLICK
....
WELL, IT LOOKS LIKE YOU'RE GOING HOME...
YEAH. IT'S REALLY A BUMMER FINDING OUT ON SUCH SHORT NOTICE.
....
ALL RIGHT! LET'S GO OUT FOR TODAY'S LESSON!
GO OUT?

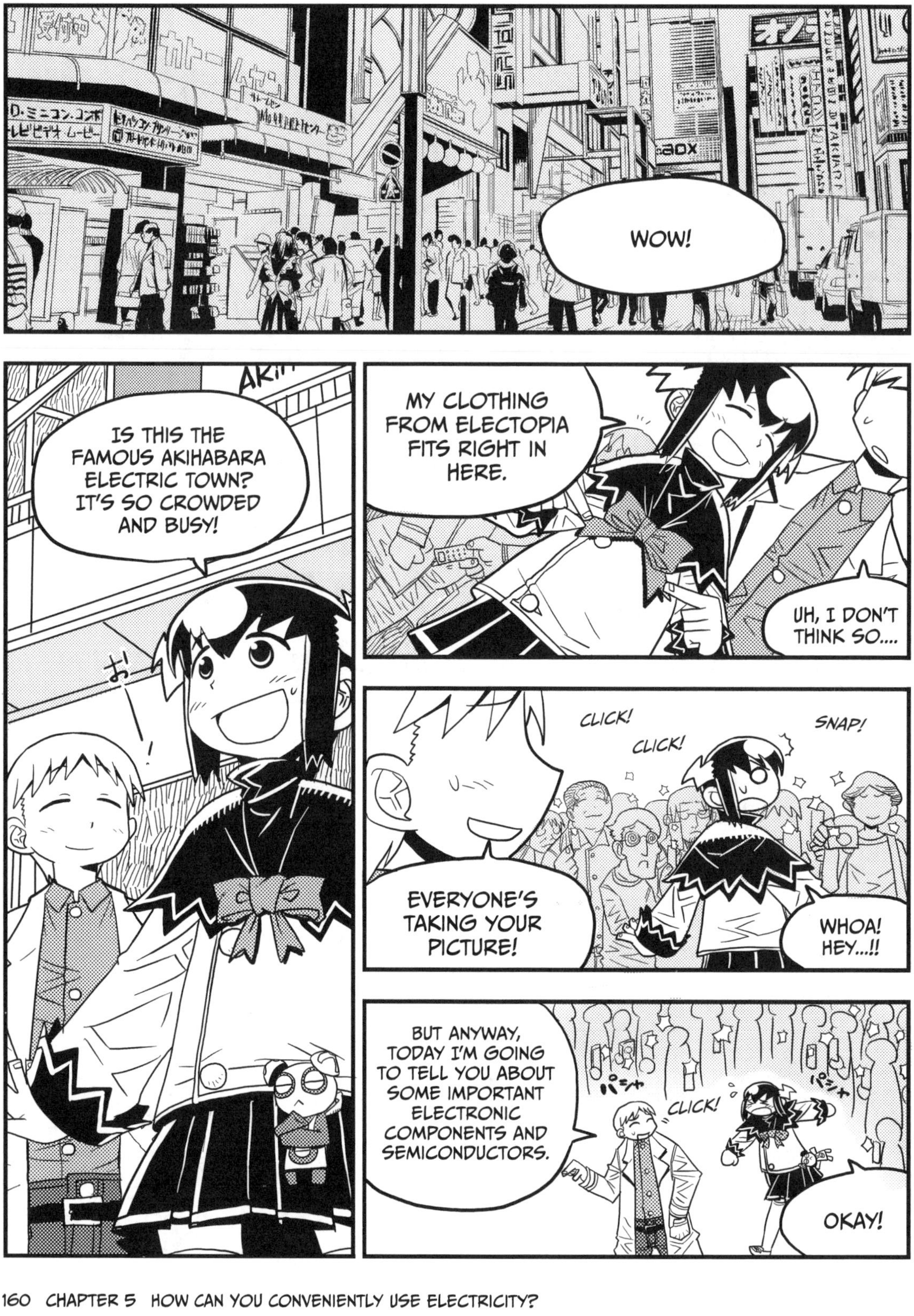
WOW!
IS THIS THE FAMOUS AKIHABARA ELECTRIC TOWN? IT'S SO CROWDED AND BUSY!
お!
MY CLOTHING FROM ELECTOPIA FITS RIGHT IN HERE.
UH, I DON'T THINK SO....
CLICK!
CLICK!
SNAP!
EVERYONE'S TAKING YOUR PICTURE!
WHOA! HEY...!!
BUT ANYWAY, TODAY I'M GOING TO TELL YOU ABOUT SOME IMPORTANT ELECTRONIC COMPONENTS AND SEMICONDUCTORS.
パシャ
CLICK!
パシャ
OKAY!

WHAT ARE SEMICONDUCTORS?

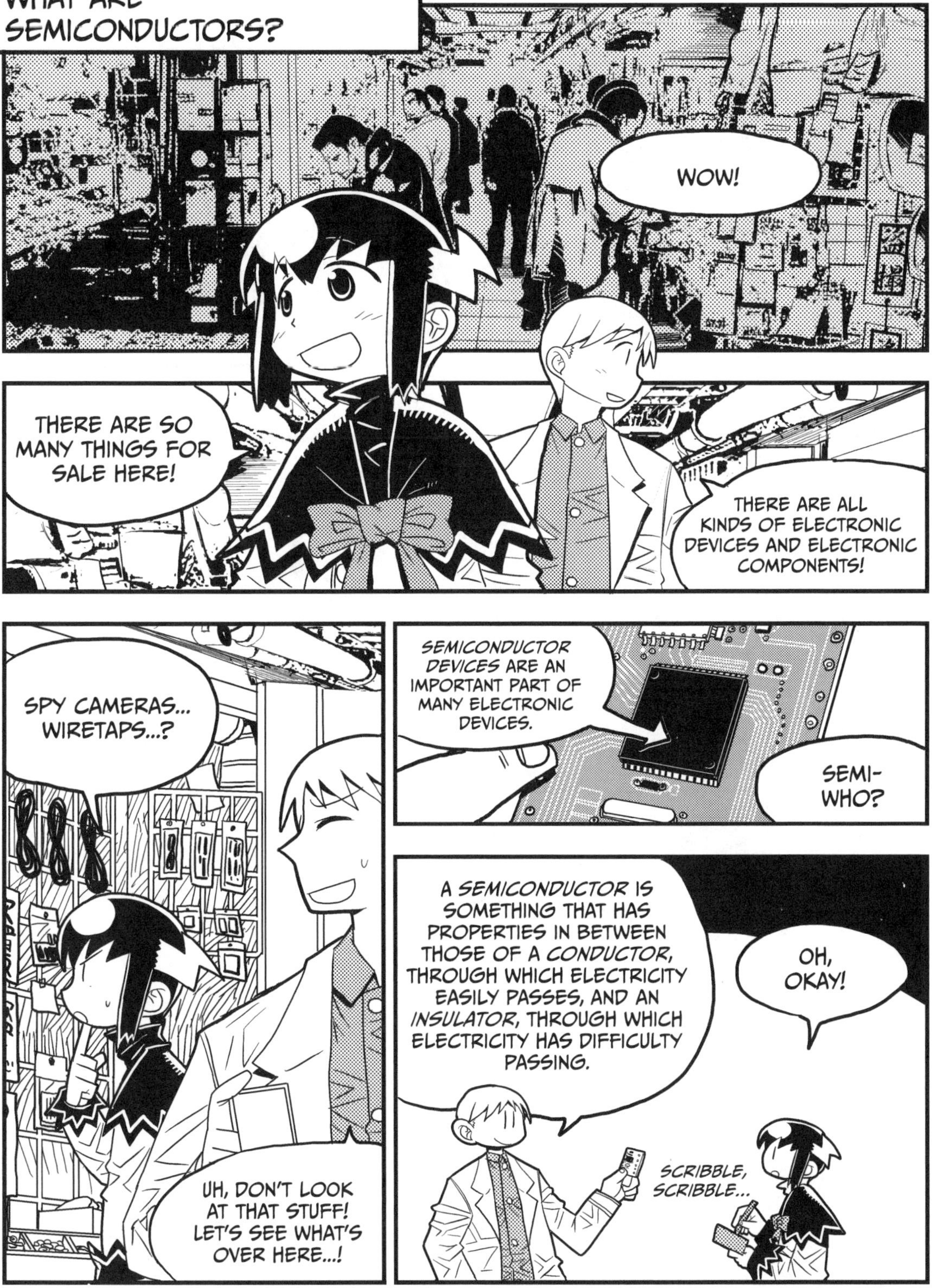
WOW!
THERE ARE SO MANY THINGS FOR SALE HERE!
THERE ARE ALL KINDS OF ELECTRONIC DEVICES AND ELECTRONIC COMPONENTS!
SPY CAMERAS... WIRETAPS...?
UH, DON'T LOOK AT THAT STUFF! LET'S SEE WHAT'S OVER HERE...!
SEMICONDUCTOR DEVICES ARE AN IMPORTANT PART OF MANY ELECTRONIC DEVICES.
SEMI-WHO?
A SEMICONDUCTOR IS SOMETHING THAT HAS PROPERTIES IN BETWEEN THOSE OF A CONDUCTOR, THROUGH WHICH ELECTRICITY EASILY PASSES, AND AN INSULATOR, THROUGH WHICH ELECTRICITY HAS DIFFICULTY PASSING.
OH, OKAY!
SCRIBBLE, SCRIBBLE...

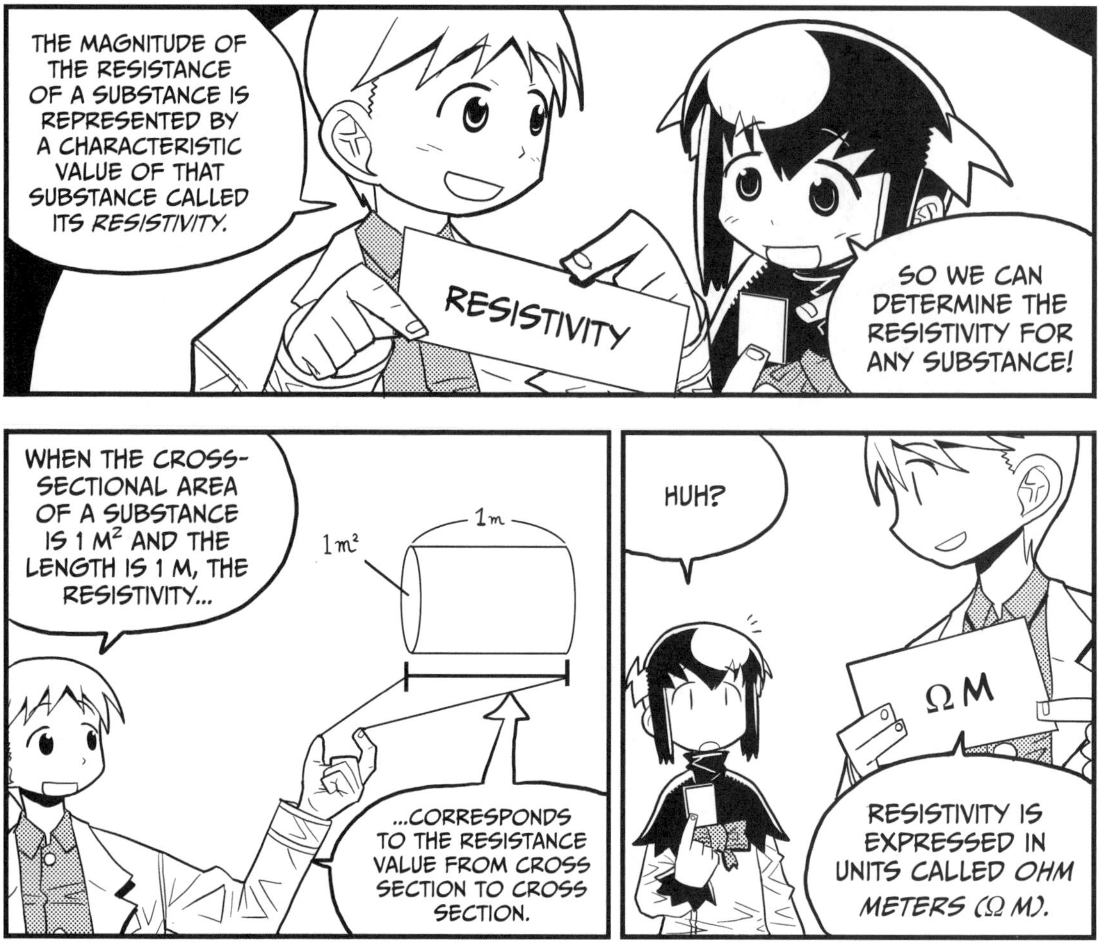
THE MAGNITUDE OF THE RESISTANCE OF A SUBSTANCE IS REPRESENTED BY A CHARACTERISTIC VALUE OF THAT SUBSTANCE CALLED ITS *RESISTIVITY*.
RESISTIVITY
SO WE CAN DETERMINE THE RESISTIVITY FOR ANY SUBSTANCE!
WHEN THE CROSS-SECTIONAL AREA OF A SUBSTANCE IS 1 M² AND THE LENGTH IS 1 M, THE RESISTIVITY...
1m
1m²
...CORRESPONDS TO THE RESISTANCE VALUE FROM CROSS SECTION TO CROSS SECTION.
HUH?
ΩM
RESISTIVITY IS EXPRESSED IN UNITS CALLED *OHM METERS (Ω M)*.

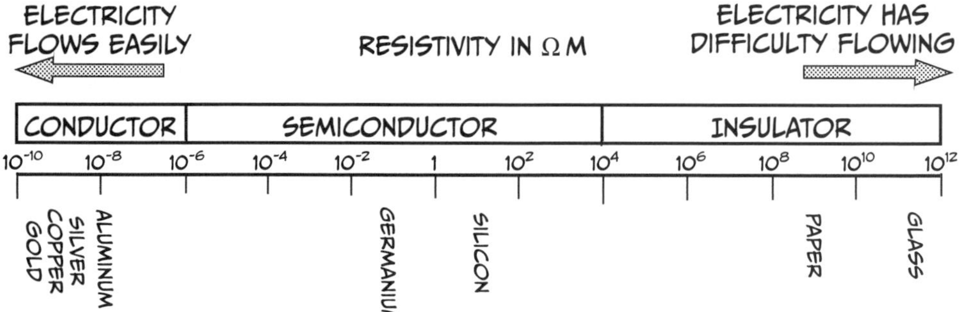
ELECTRICITY FLOWS EASILY
RESISTIVITY IN Ω M
ELECTRICITY HAS DIFFICULTY FLOWING
CONDUCTOR
SEMICONDUCTOR
INSULATOR
10⁻¹⁰ 10⁻⁸ 10⁻⁶ 10⁻⁴ 10⁻² 1 10² 10⁴ 10⁶ 10⁸ 10¹⁰ 10¹²
GOLD
COPPER
SILVER
ALUMINUM
GERMANIUM
SILICON
PAPER
GLASS

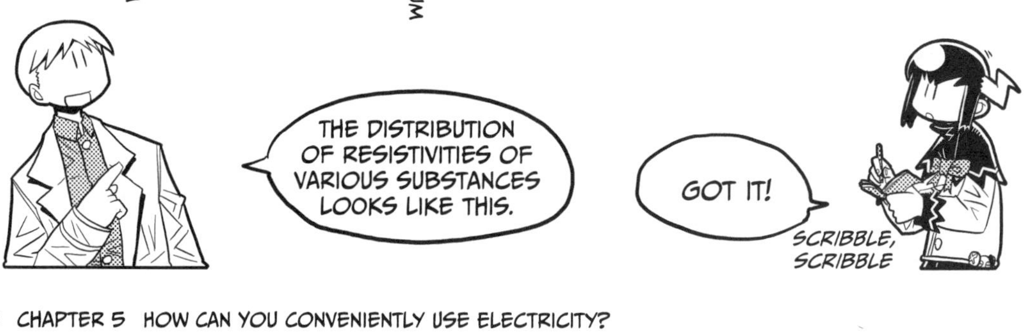
THE DISTRIBUTION OF RESISTIVITIES OF VARIOUS SUBSTANCES LOOKS LIKE THIS.
GOT IT!
SCRIBBLE, SCRIBBLE

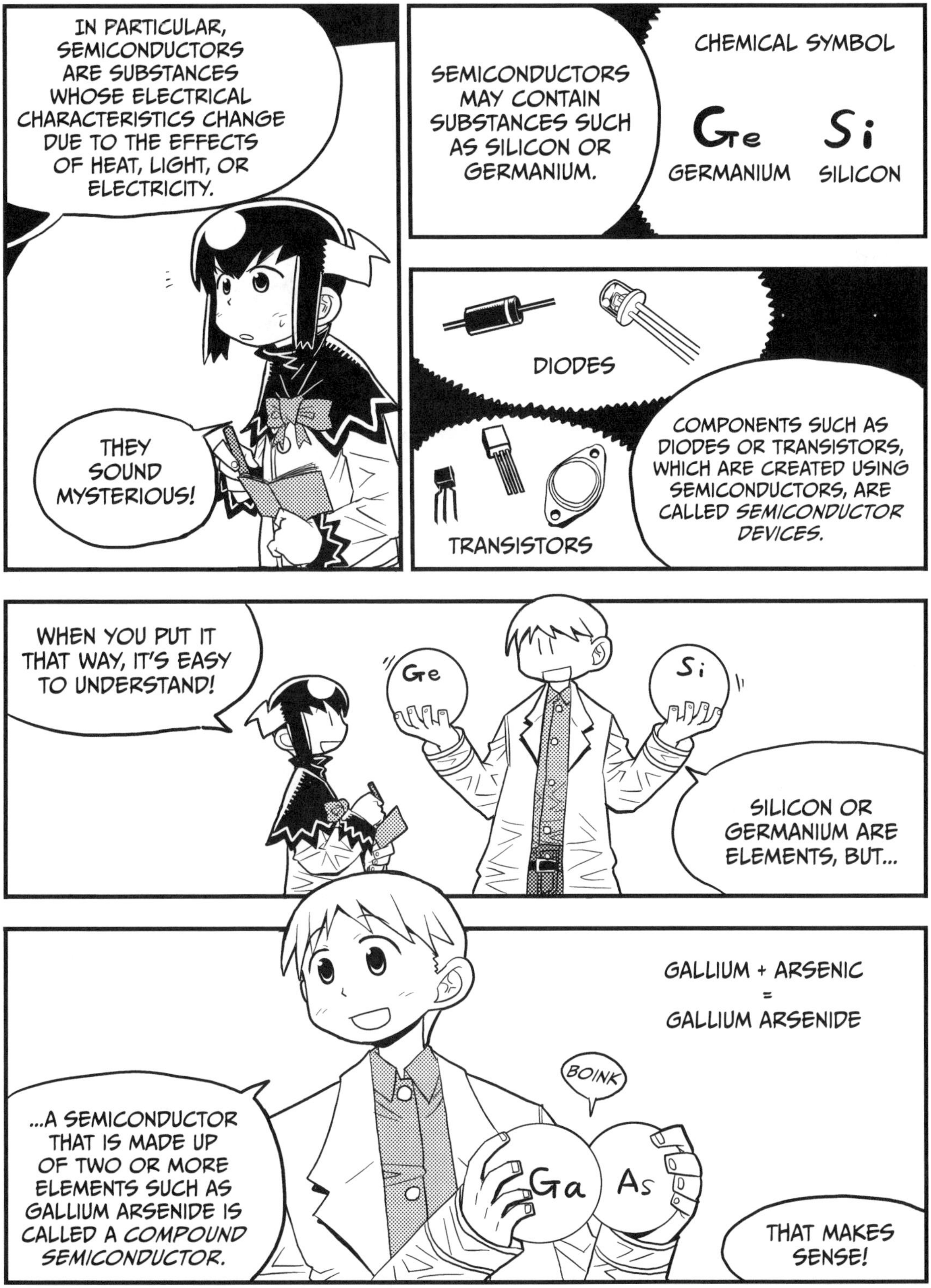

IN PARTICULAR, SEMICONDUCTORS ARE SUBSTANCES WHOSE ELECTRICAL CHARACTERISTICS CHANGE DUE TO THE EFFECTS OF HEAT, LIGHT, OR ELECTRICITY.
THEY SOUND MYSTERIOUS!
SEMICONDUCTORS MAY CONTAIN SUBSTANCES SUCH AS SILICON OR GERMANIUM.
CHEMICAL SYMBOL
Ge
GERMANIUM
Si
SILICON
DIODES
TRANSISTORS
COMPONENTS SUCH AS DIODES OR TRANSISTORS, WHICH ARE CREATED USING SEMICONDUCTORS, ARE CALLED *SEMICONDUCTOR DEVICES*.
WHEN YOU PUT IT THAT WAY, IT'S EASY TO UNDERSTAND!
Ge
Si
SILICON OR GERMANIUM ARE ELEMENTS, BUT...
GALLIUM + ARSENIC = GALLIUM ARSENIDE
...A SEMICONDUCTOR THAT IS MADE UP OF TWO OR MORE ELEMENTS SUCH AS GALLIUM ARSENIDE IS CALLED A *COMPOUND SEMICONDUCTOR*.
BOINK
Ga
As
THAT MAKES SENSE!

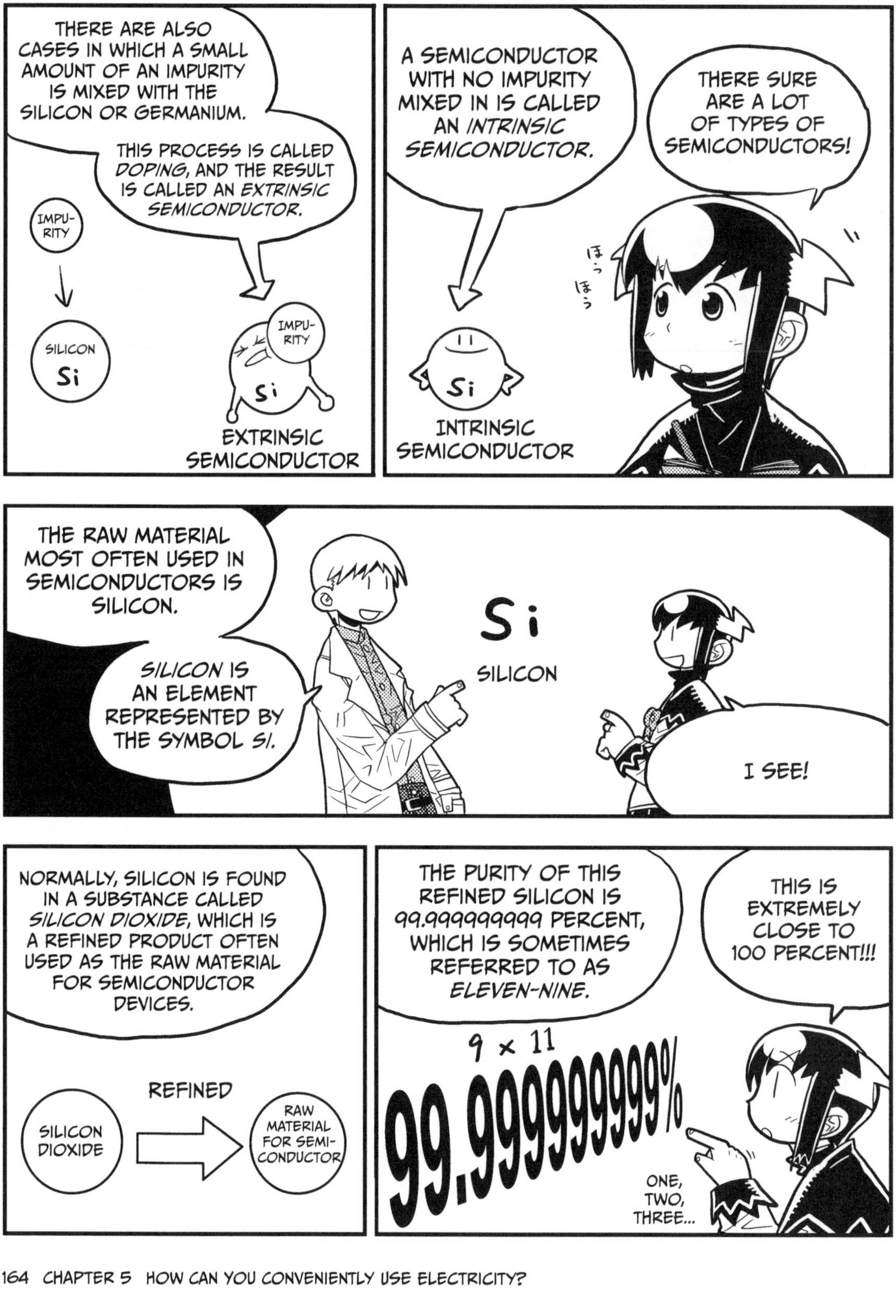
THERE ARE ALSO CASES IN WHICH A SMALL AMOUNT OF AN IMPURITY IS MIXED WITH THE SILICON OR GERMANIUM.
THIS PROCESS IS CALLED *DOPING*, AND THE RESULT IS CALLED AN *EXTRINSIC SEMICONDUCTOR*.
IMPURITY
SILICON
Si
IMPURITY
Si
EXTRINSIC SEMICONDUCTOR
A SEMICONDUCTOR WITH NO IMPURITY MIXED IN IS CALLED AN *INTRINSIC SEMICONDUCTOR*.
Si
INTRINSIC SEMICONDUCTOR
THERE SURE ARE A LOT OF TYPES OF SEMICONDUCTORS!
ほう ほう
THE RAW MATERIAL MOST OFTEN USED IN SEMICONDUCTORS IS SILICON.
SILICON IS AN ELEMENT REPRESENTED BY THE SYMBOL *Si*.
Si
SILICON
I SEE!
NORMALLY, SILICON IS FOUND IN A SUBSTANCE CALLED *SILICON DIOXIDE*, WHICH IS A REFINED PRODUCT OFTEN USED AS THE RAW MATERIAL FOR SEMICONDUCTOR DEVICES.
SILICON DIOXIDE
REFINED
RAW MATERIAL FOR SEMICONDUCTOR
THE PURITY OF THIS REFINED SILICON IS 99.999999999 PERCENT, WHICH IS SOMETIMES REFERRED TO AS *ELEVEN-NINE*.
9 × 11
99.999999999%
ONE, TWO, THREE...
THIS IS EXTREMELY CLOSE TO 100 PERCENT!!!

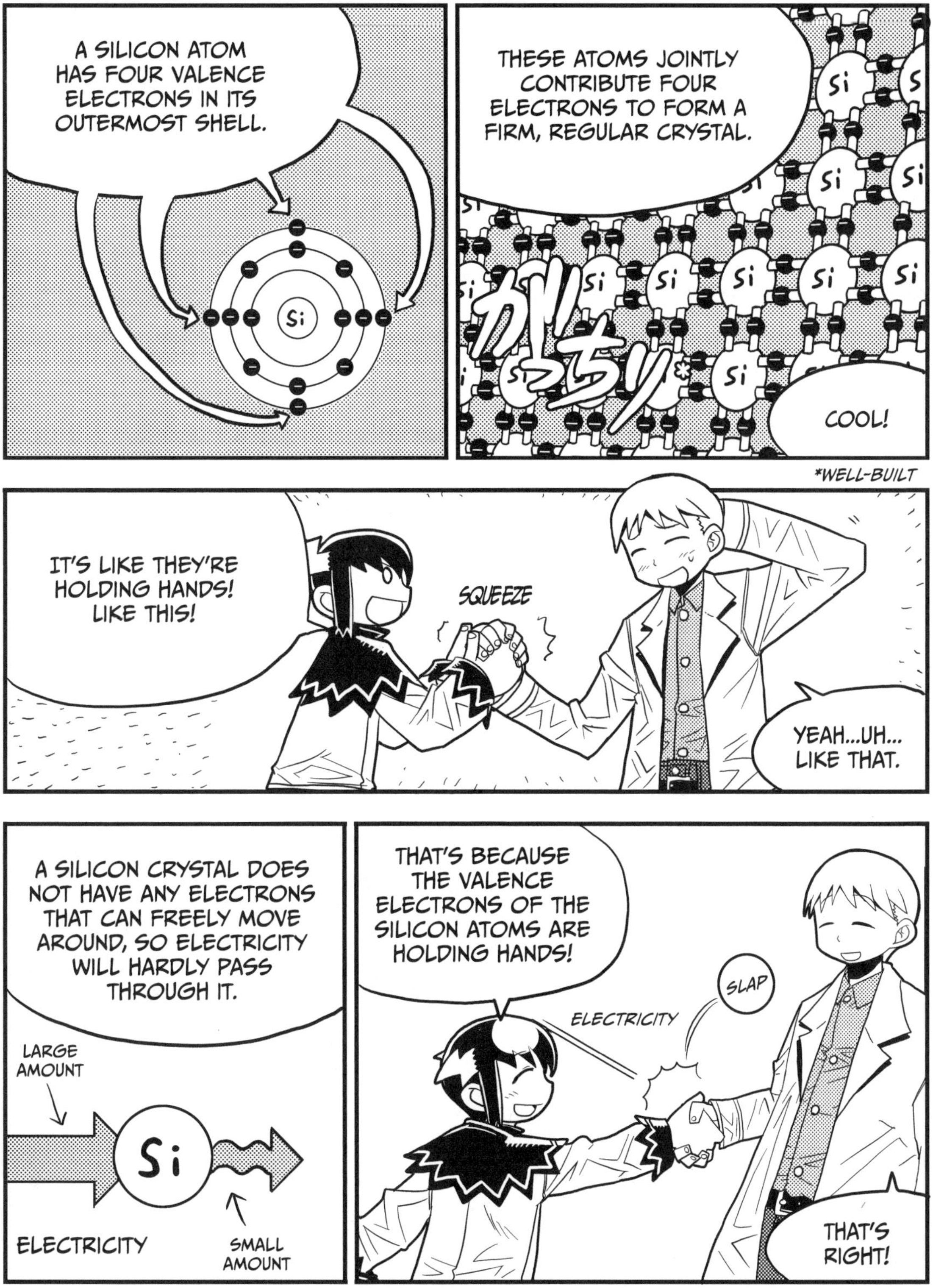
A SILICON ATOM HAS FOUR VALENCE ELECTRONS IN ITS OUTERMOST SHELL.
Si
THESE ATOMS JOINTLY CONTRIBUTE FOUR ELECTRONS TO FORM A FIRM, REGULAR CRYSTAL.
がっちり*
COOL!
*WELL-BUILT
IT'S LIKE THEY'RE HOLDING HANDS! LIKE THIS!
SQUEEZE
YEAH...UH... LIKE THAT.
A SILICON CRYSTAL DOES NOT HAVE ANY ELECTRONS THAT CAN FREELY MOVE AROUND, SO ELECTRICITY WILL HARDLY PASS THROUGH IT.
LARGE AMOUNT
Si
ELECTRICITY
SMALL AMOUNT
THAT'S BECAUSE THE VALENCE ELECTRONS OF THE SILICON ATOMS ARE HOLDING HANDS!
ELECTRICITY
SLAP
THAT'S RIGHT!

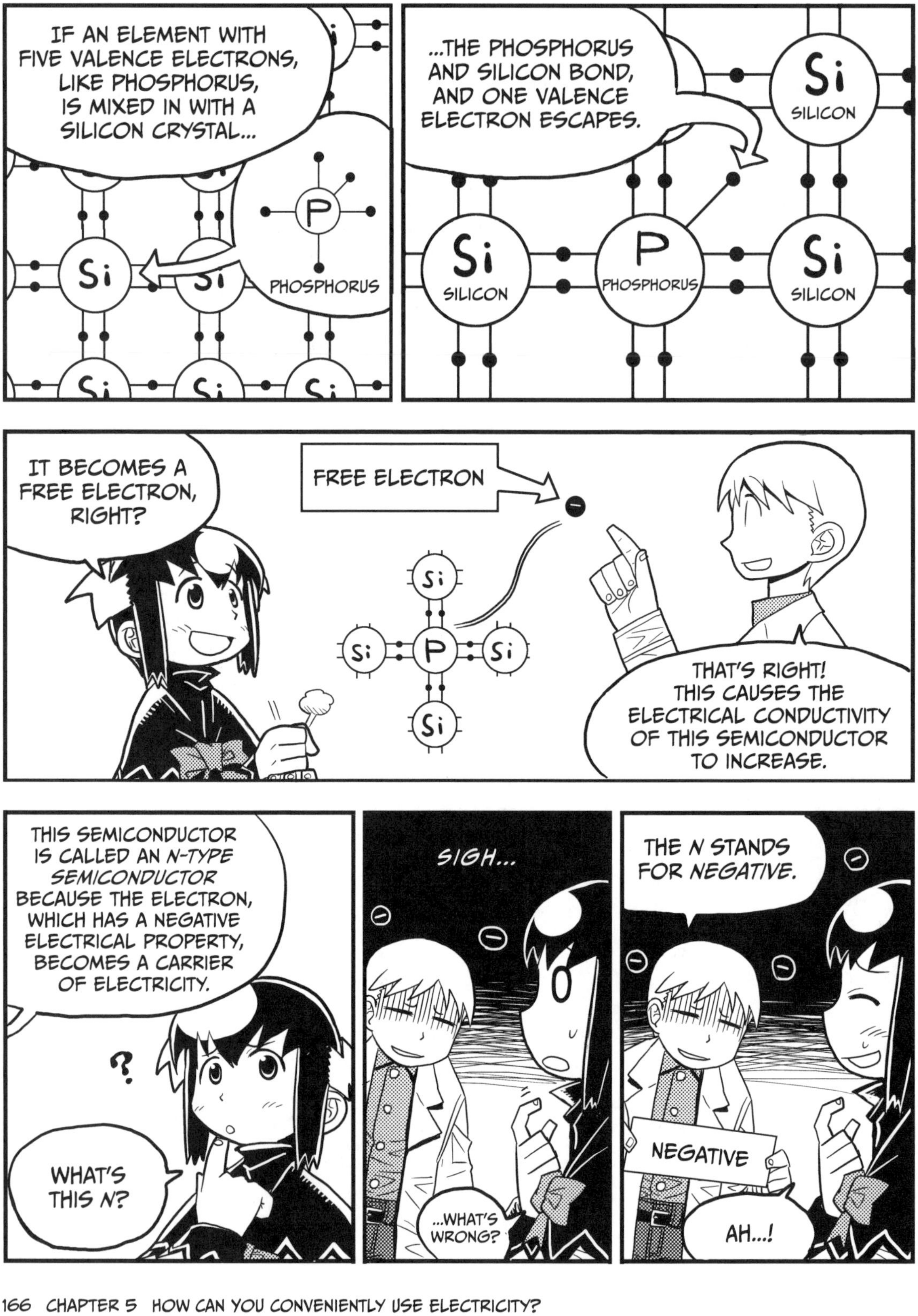
IF AN ELEMENT WITH FIVE VALENCE ELECTRONS, LIKE PHOSPHORUS, IS MIXED IN WITH A SILICON CRYSTAL...
Si
P
PHOSPHORUS
...THE PHOSPHORUS AND SILICON BOND, AND ONE VALENCE ELECTRON ESCAPES.
Si
SILICON
P
PHOSPHORUS
IT BECOMES A FREE ELECTRON, RIGHT?
FREE ELECTRON
THAT'S RIGHT! THIS CAUSES THE ELECTRICAL CONDUCTIVITY OF THIS SEMICONDUCTOR TO INCREASE.
THIS SEMICONDUCTOR IS CALLED AN *N-TYPE SEMICONDUCTOR* BECAUSE THE ELECTRON, WHICH HAS A NEGATIVE ELECTRICAL PROPERTY, BECOMES A CARRIER OF ELECTRICITY.
?
WHAT'S THIS *N*?
SIGH...
...WHAT'S WRONG?
THE *N* STANDS FOR *NEGATIVE*.
NEGATIVE
AH...!

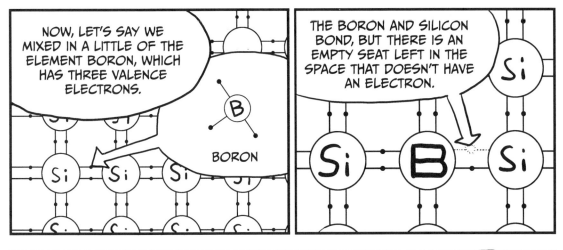

NOW, LET'S SAY WE MIXED IN A LITTLE OF THE ELEMENT BORON, WHICH HAS THREE VALENCE ELECTRONS.

BORON

THE BORON AND SILICON BOND, BUT THERE IS AN EMPTY SEAT LEFT IN THE SPACE THAT DOESN'T HAVE AN ELECTRON.

EMPTY SEAT?

HOLE

THIS EMPTY SEAT IS CALLED A *HOLE*.

HOLES ARE PLACES ELECTRONS ARE VACANT IN COVALENT BONDS.

HOLE

HOLE

DRIFT

DRIFT

HOLE

YOU CAN THINK OF THE HOLE AS A FREE ELECTRON WITH A POSITIVE CHARGE.

OKAY...

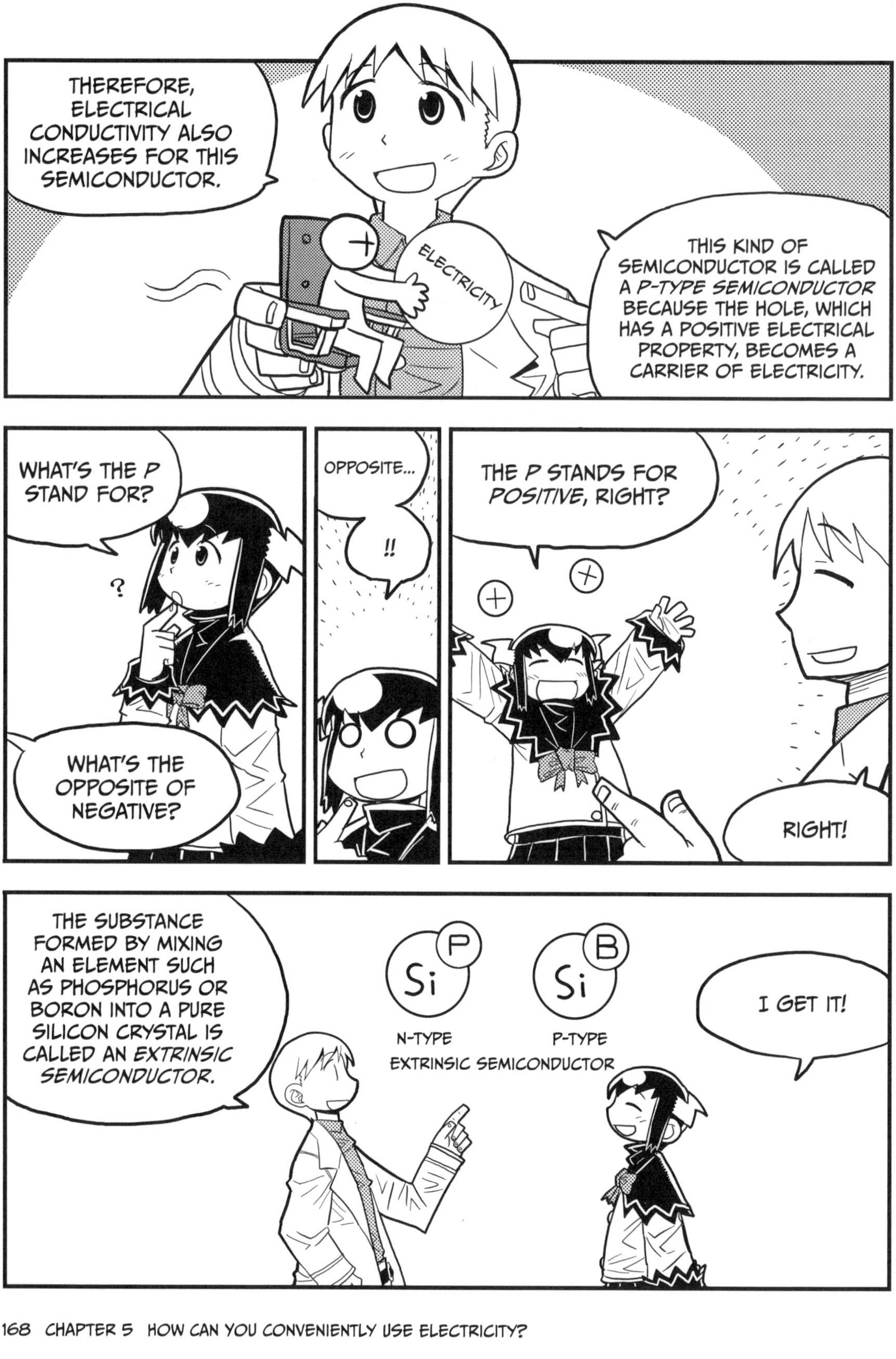
THEREFORE, ELECTRICAL CONDUCTIVITY ALSO INCREASES FOR THIS SEMICONDUCTOR.
ELECTRICITY
THIS KIND OF SEMICONDUCTOR IS CALLED A P-TYPE SEMICONDUCTOR BECAUSE THE HOLE, WHICH HAS A POSITIVE ELECTRICAL PROPERTY, BECOMES A CARRIER OF ELECTRICITY.
WHAT'S THE P STAND FOR?
?
WHAT'S THE OPPOSITE OF NEGATIVE?
OPPOSITE...
!!
THE P STANDS FOR POSITIVE, RIGHT?
RIGHT!
THE SUBSTANCE FORMED BY MIXING AN ELEMENT SUCH AS PHOSPHORUS OR BORON INTO A PURE SILICON CRYSTAL IS CALLED AN EXTRINSIC SEMICONDUCTOR.
Si P
Si B
N-TYPE
P-TYPE
EXTRINSIC SEMICONDUCTOR
I GET IT!

DIODES AND TRANSISTORS

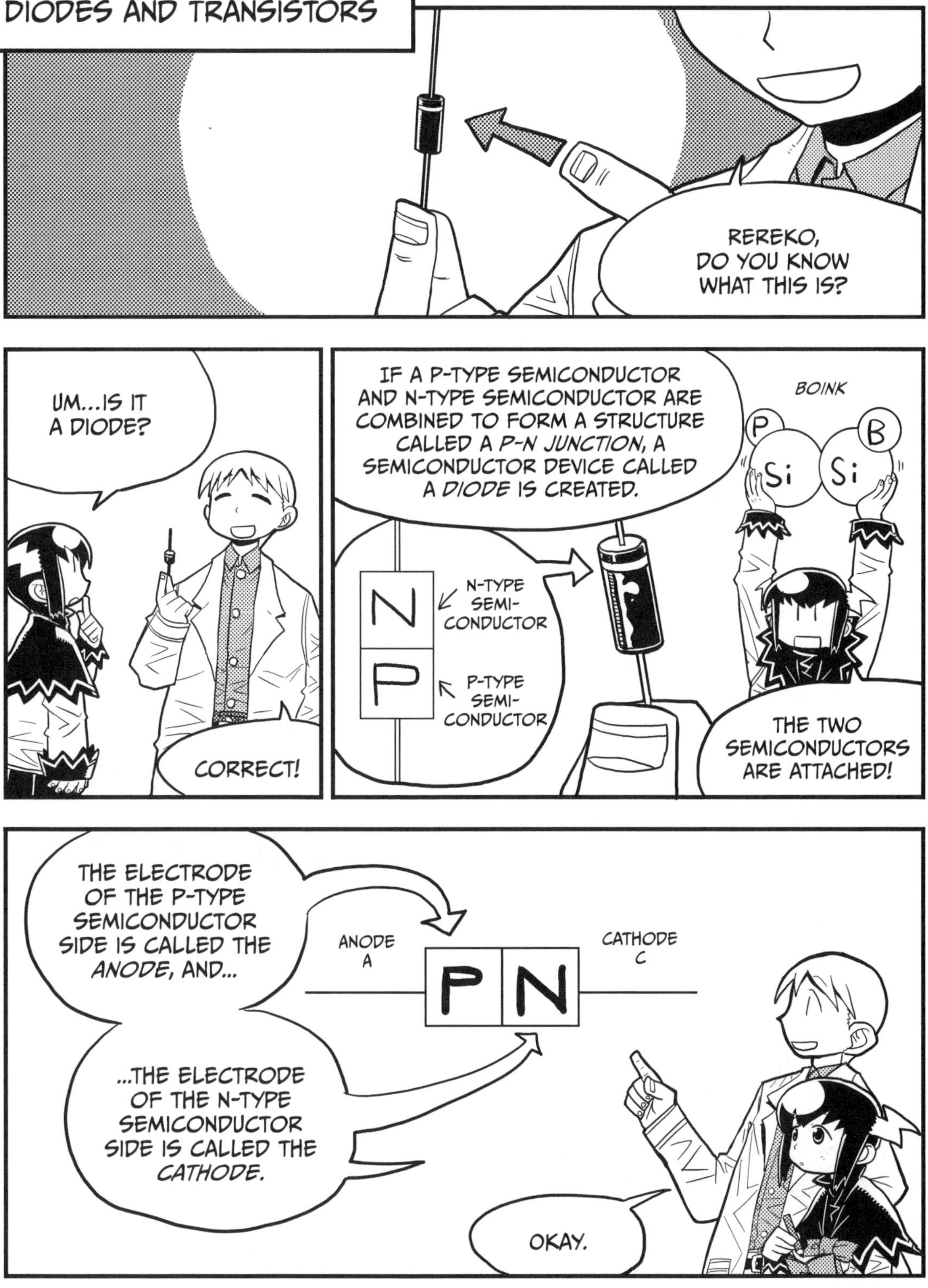
REREKO, DO YOU KNOW WHAT THIS IS?
UM...IS IT A DIODE?
CORRECT!
IF A P-TYPE SEMICONDUCTOR AND N-TYPE SEMICONDUCTOR ARE COMBINED TO FORM A STRUCTURE CALLED A *P-N JUNCTION*, A SEMICONDUCTOR DEVICE CALLED A *DIODE* IS CREATED.
BOINK
P
Si
B
Si
N
N-TYPE SEMI-CONDUCTOR
P
P-TYPE SEMI-CONDUCTOR
THE TWO SEMICONDUCTORS ARE ATTACHED!
THE ELECTRODE OF THE P-TYPE SEMICONDUCTOR SIDE IS CALLED THE *ANODE*, AND...
ANODE A
P N
CATHODE C
...THE ELECTRODE OF THE N-TYPE SEMICONDUCTOR SIDE IS CALLED THE *CATHODE*.
OKAY.

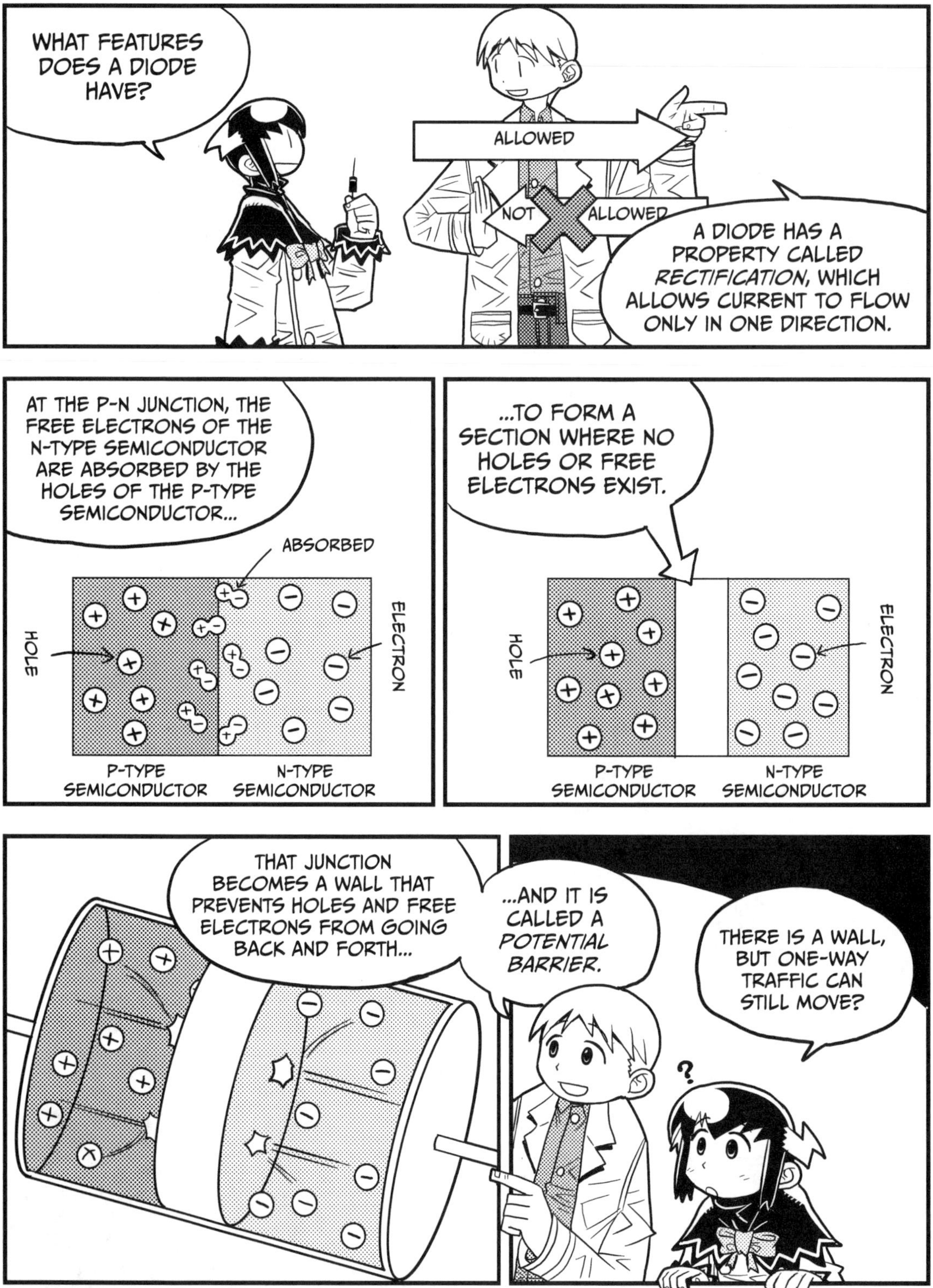

WHAT FEATURES DOES A DIODE HAVE?
ALLOWED
NOT ALLOWED
A DIODE HAS A PROPERTY CALLED *RECTIFICATION*, WHICH ALLOWS CURRENT TO FLOW ONLY IN ONE DIRECTION.
AT THE P-N JUNCTION, THE FREE ELECTRONS OF THE N-TYPE SEMICONDUCTOR ARE ABSORBED BY THE HOLES OF THE P-TYPE SEMICONDUCTOR...
ABSORBED
HOLE
ELECTRON
P-TYPE SEMICONDUCTOR
N-TYPE SEMICONDUCTOR
...TO FORM A SECTION WHERE NO HOLES OR FREE ELECTRONS EXIST.
HOLE
ELECTRON
P-TYPE SEMICONDUCTOR
N-TYPE SEMICONDUCTOR
THAT JUNCTION BECOMES A WALL THAT PREVENTS HOLES AND FREE ELECTRONS FROM GOING BACK AND FORTH...
...AND IT IS CALLED A *POTENTIAL BARRIER*.
THERE IS A WALL, BUT ONE-WAY TRAFFIC CAN STILL MOVE?

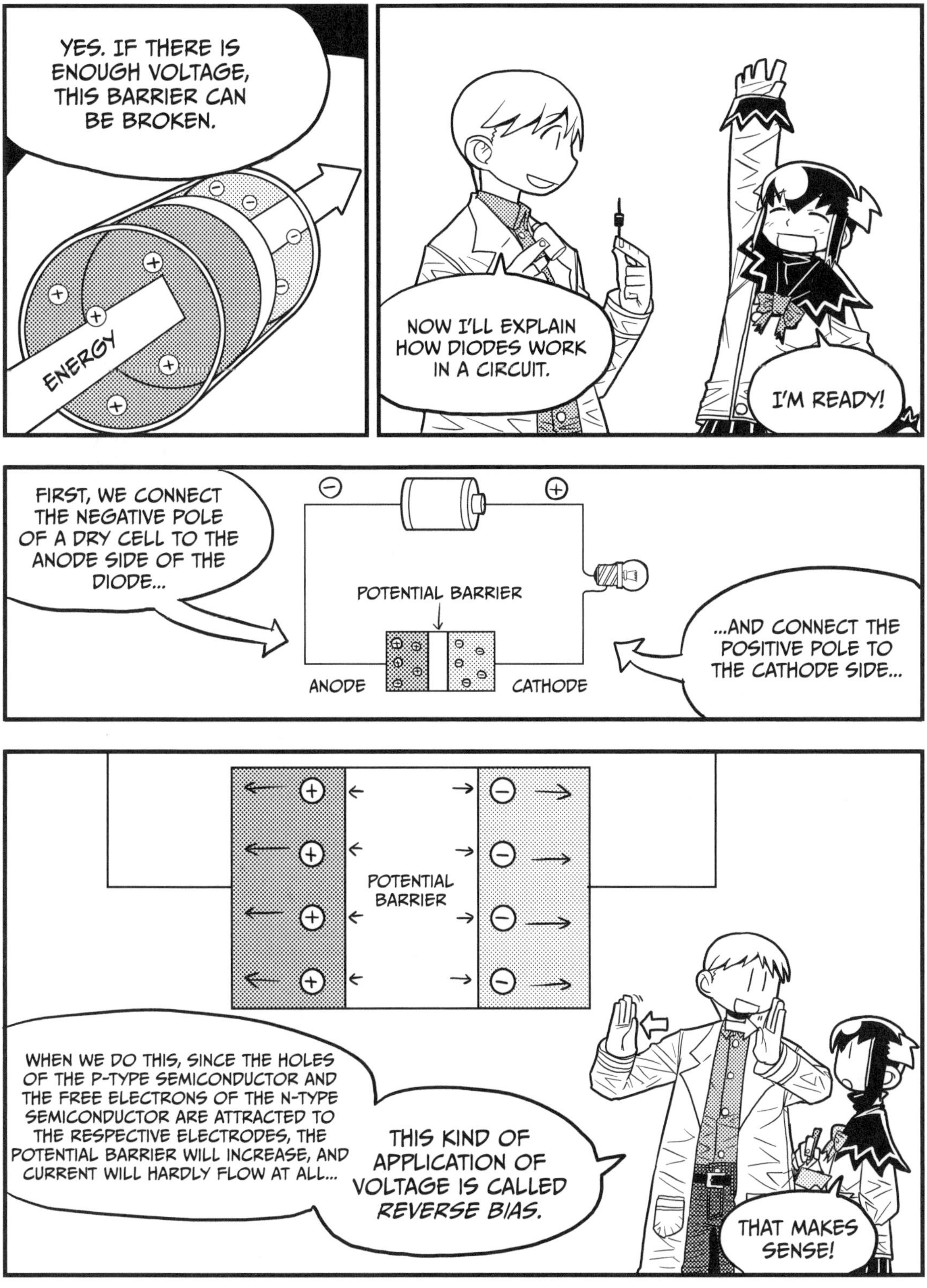
YES. IF THERE IS ENOUGH VOLTAGE, THIS BARRIER CAN BE BROKEN.
ENERGY
NOW I'LL EXPLAIN HOW DIODES WORK IN A CIRCUIT.
I'M READY!
FIRST, WE CONNECT THE NEGATIVE POLE OF A DRY CELL TO THE ANODE SIDE OF THE DIODE...
POTENTIAL BARRIER
ANODE
CATHODE
...AND CONNECT THE POSITIVE POLE TO THE CATHODE SIDE...
POTENTIAL BARRIER
WHEN WE DO THIS, SINCE THE HOLES OF THE P-TYPE SEMICONDUCTOR AND THE FREE ELECTRONS OF THE N-TYPE SEMICONDUCTOR ARE ATTRACTED TO THE RESPECTIVE ELECTRODES, THE POTENTIAL BARRIER WILL INCREASE, AND CURRENT WILL HARDLY FLOW AT ALL...
THIS KIND OF APPLICATION OF VOLTAGE IS CALLED *REVERSE BIAS*.
THAT MAKES SENSE!

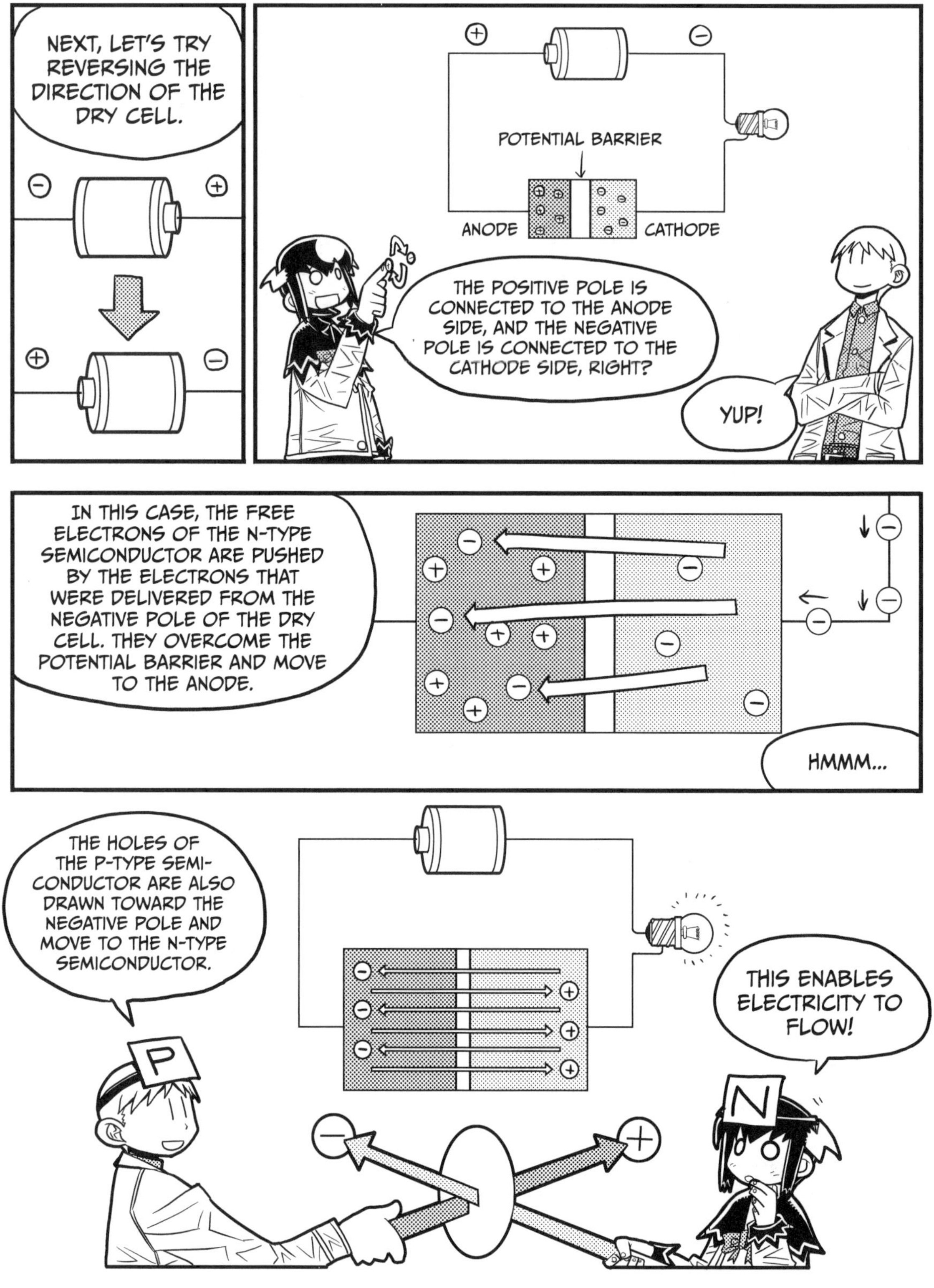
NEXT, LET'S TRY REVERSING THE DIRECTION OF THE DRY CELL.
POTENTIAL BARRIER
ANODE
CATHODE
THE POSITIVE POLE IS CONNECTED TO THE ANODE SIDE, AND THE NEGATIVE POLE IS CONNECTED TO THE CATHODE SIDE, RIGHT?
YUP!
IN THIS CASE, THE FREE ELECTRONS OF THE N-TYPE SEMICONDUCTOR ARE PUSHED BY THE ELECTRONS THAT WERE DELIVERED FROM THE NEGATIVE POLE OF THE DRY CELL. THEY OVERCOME THE POTENTIAL BARRIER AND MOVE TO THE ANODE.
HMMM...
THE HOLES OF THE P-TYPE SEMI-CONDUCTOR ARE ALSO DRAWN TOWARD THE NEGATIVE POLE AND MOVE TO THE N-TYPE SEMICONDUCTOR.
THIS ENABLES ELECTRICITY TO FLOW!
P
N

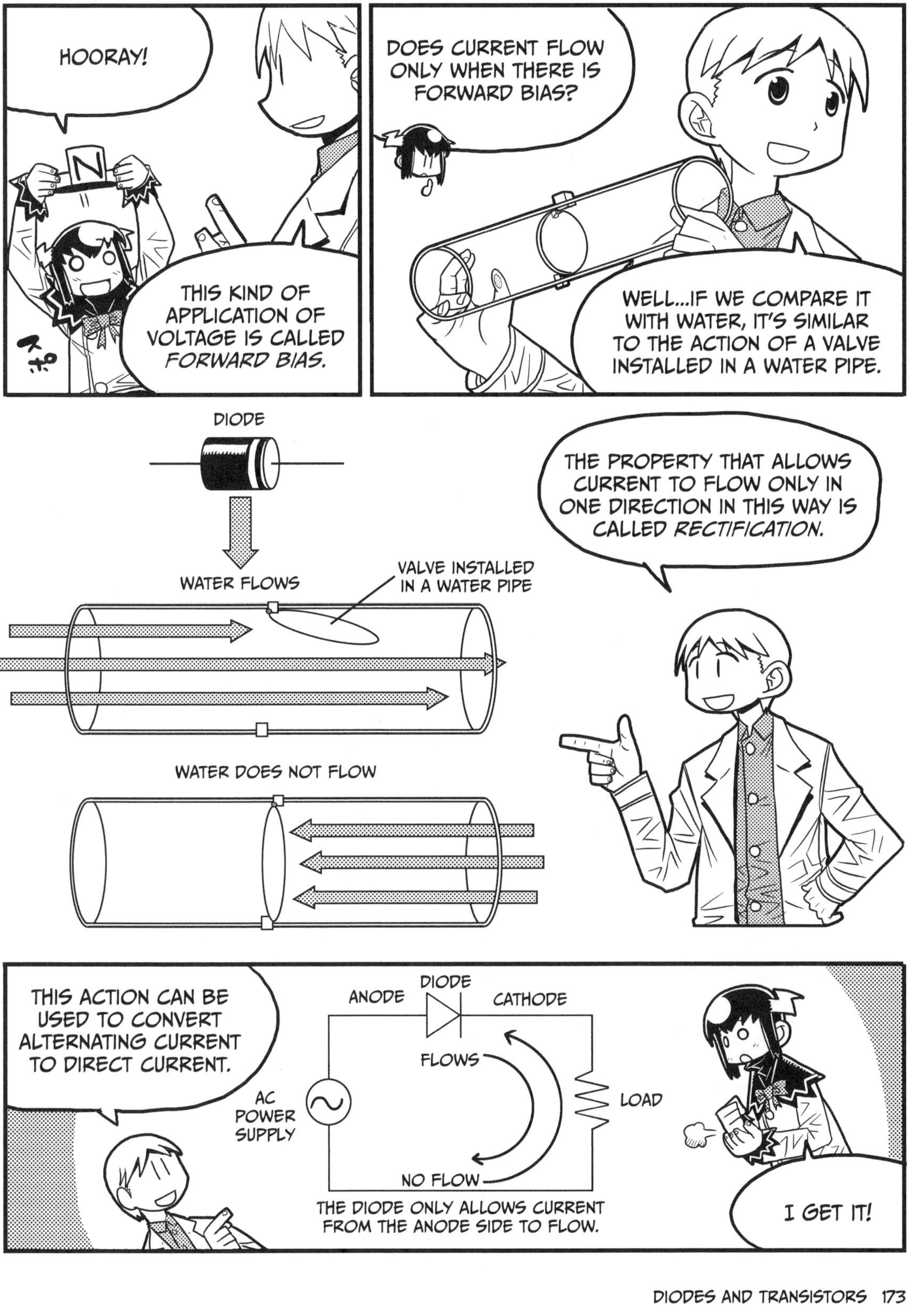

HOORAY!
THIS KIND OF APPLICATION OF VOLTAGE IS CALLED *FORWARD BIAS.*
スポ
DOES CURRENT FLOW ONLY WHEN THERE IS FORWARD BIAS?
WELL...IF WE COMPARE IT WITH WATER, IT'S SIMILAR TO THE ACTION OF A VALVE INSTALLED IN A WATER PIPE.
DIODE
THE PROPERTY THAT ALLOWS CURRENT TO FLOW ONLY IN ONE DIRECTION IN THIS WAY IS CALLED *RECTIFICATION.*
WATER FLOWS
VALVE INSTALLED IN A WATER PIPE
WATER DOES NOT FLOW
THIS ACTION CAN BE USED TO CONVERT ALTERNATING CURRENT TO DIRECT CURRENT.
DIODE
ANODE
CATHODE
FLOWS
AC POWER SUPPLY
LOAD
NO FLOW
THE DIODE ONLY ALLOWS CURRENT FROM THE ANODE SIDE TO FLOW.
I GET IT!

DIODES THAT EMIT LIGHT

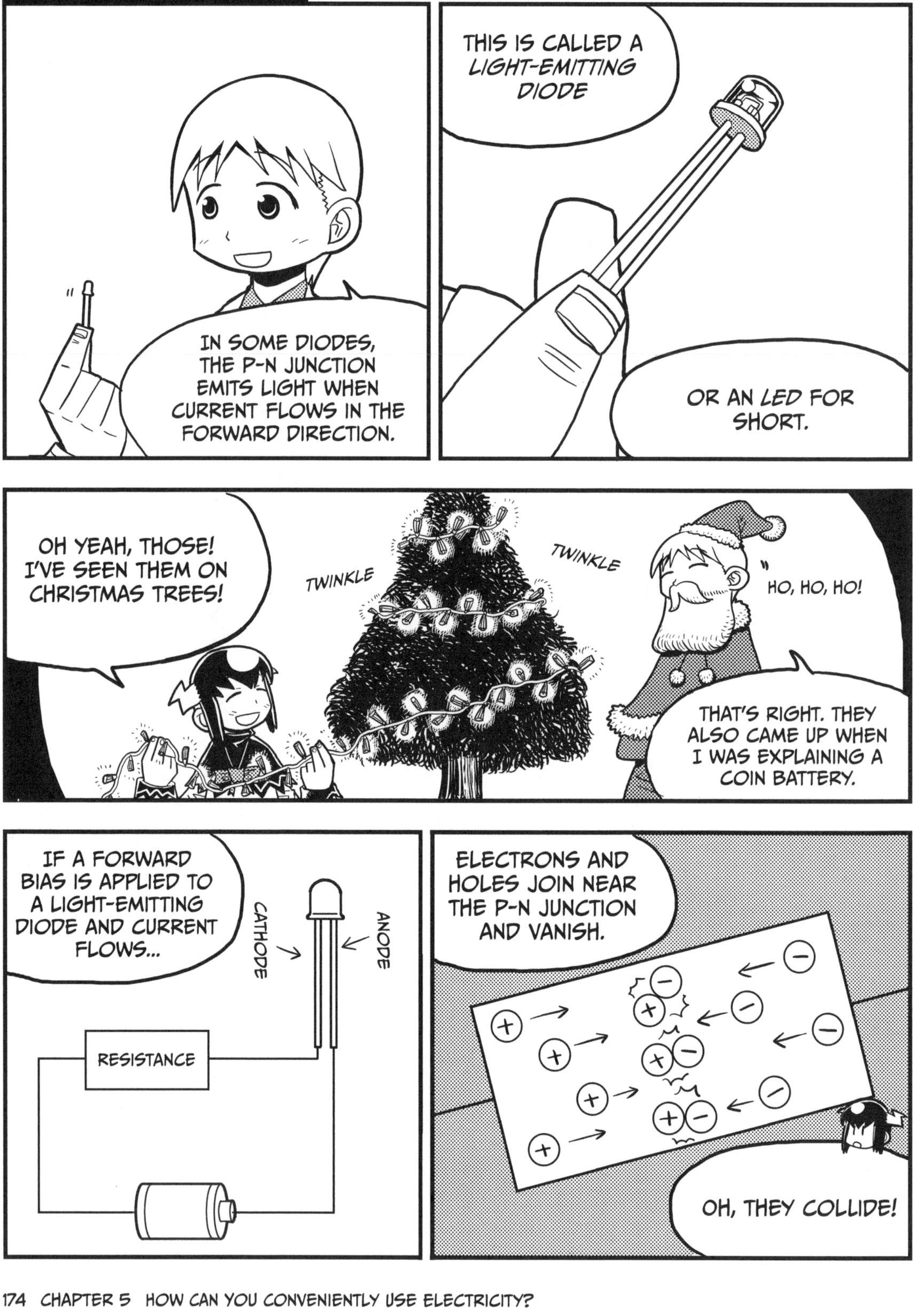
IN SOME DIODES, THE P-N JUNCTION EMITS LIGHT WHEN CURRENT FLOWS IN THE FORWARD DIRECTION.
THIS IS CALLED A *LIGHT-EMITTING DIODE*
OR AN *LED* FOR SHORT.
OH YEAH, THOSE! I'VE SEEN THEM ON CHRISTMAS TREES!
TWINKLE
TWINKLE
HO, HO, HO!
THAT'S RIGHT. THEY ALSO CAME UP WHEN I WAS EXPLAINING A COIN BATTERY.
IF A FORWARD BIAS IS APPLIED TO A LIGHT-EMITTING DIODE AND CURRENT FLOWS...
CATHODE
ANODE
RESISTANCE
ELECTRONS AND HOLES JOIN NEAR THE P-N JUNCTION AND VANISH.
OH, THEY COLLIDE!

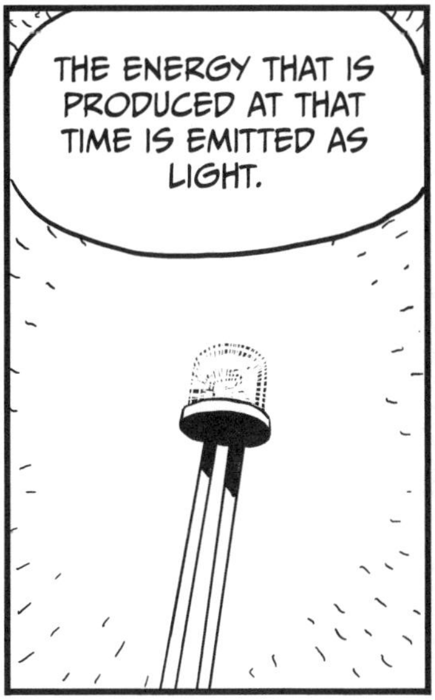
THE ENERGY THAT IS PRODUCED AT THAT TIME IS EMITTED AS LIGHT.

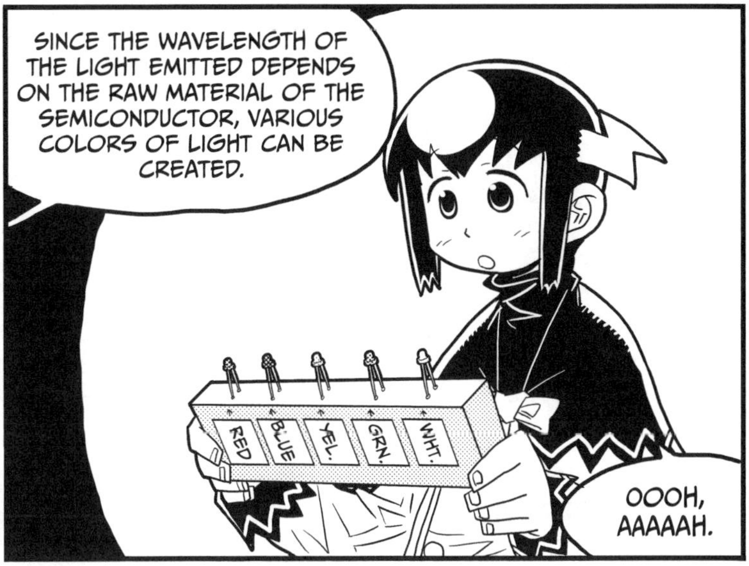
SINCE THE WAVELENGTH OF THE LIGHT EMITTED DEPENDS ON THE RAW MATERIAL OF THE SEMICONDUCTOR, VARIOUS COLORS OF LIGHT CAN BE CREATED.
RED
BLUE
YEL.
GRN.
WHT.
OOOH, AAAAAH.

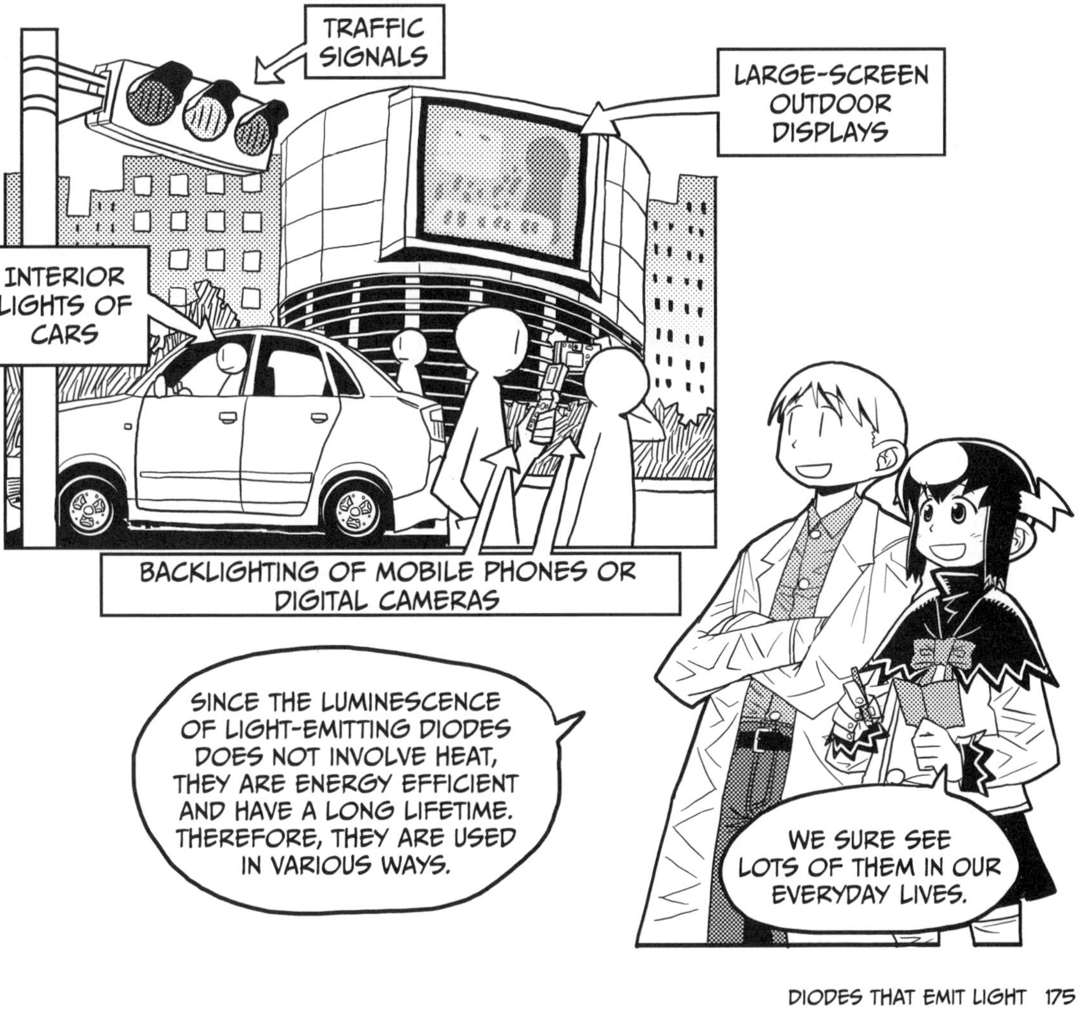
TRAFFIC SIGNALS
LARGE-SCREEN OUTDOOR DISPLAYS
INTERIOR LIGHTS OF CARS
BACKLIGHTING OF MOBILE PHONES OR DIGITAL CAMERAS
SINCE THE LUMINESCENCE OF LIGHT-EMITTING DIODES DOES NOT INVOLVE HEAT, THEY ARE ENERGY EFFICIENT AND HAVE A LONG LIFETIME. THEREFORE, THEY ARE USED IN VARIOUS WAYS.
WE SURE SEE LOTS OF THEM IN OUR EVERYDAY LIVES.

TRANSISTORS

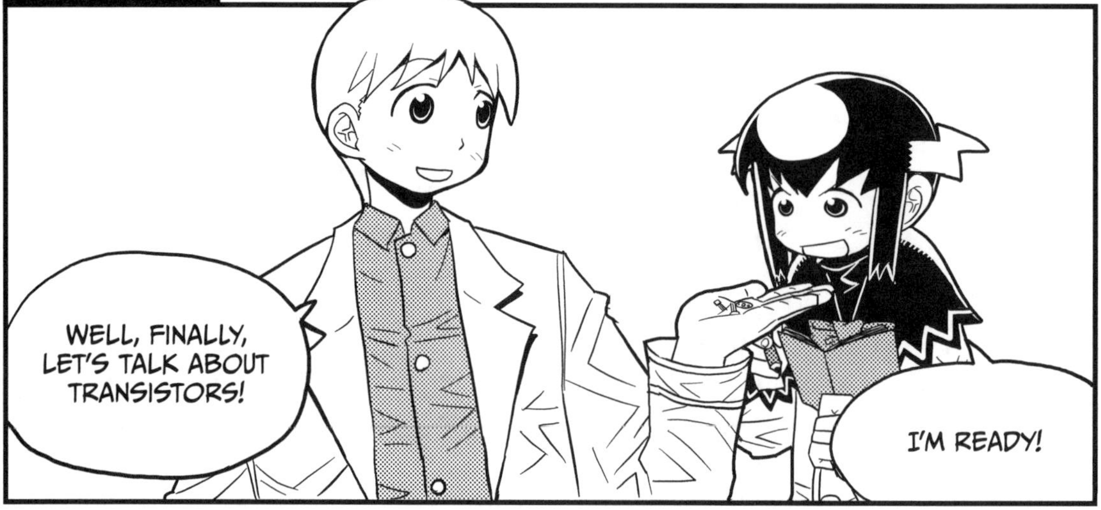
WELL, FINALLY, LET'S TALK ABOUT TRANSISTORS!
I'M READY!

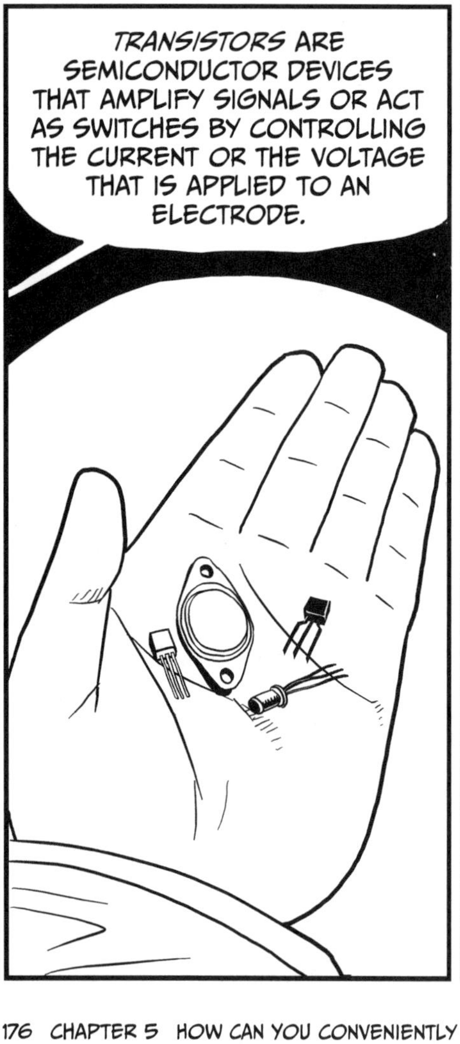
TRANSISTORS ARE SEMICONDUCTOR DEVICES THAT AMPLIFY SIGNALS OR ACT AS SWITCHES BY CONTROLLING THE CURRENT OR THE VOLTAGE THAT IS APPLIED TO AN ELECTRODE.

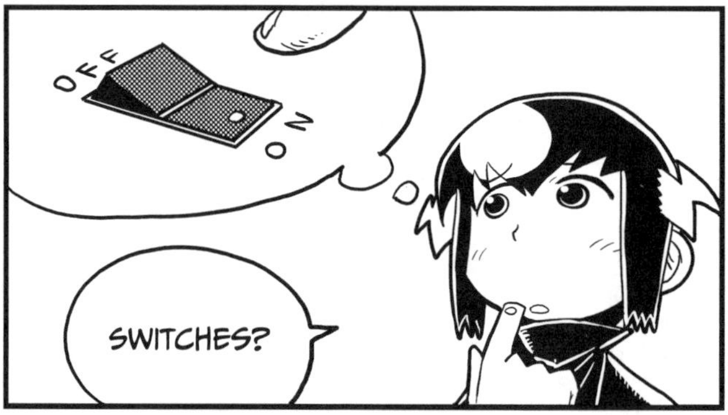
OFF
ON
SWITCHES?

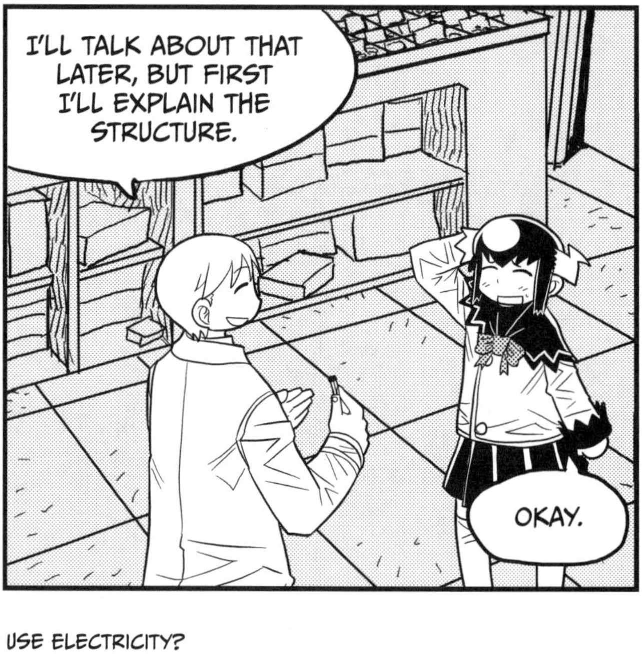
I'LL TALK ABOUT THAT LATER, BUT FIRST I'LL EXPLAIN THE STRUCTURE.
OKAY.

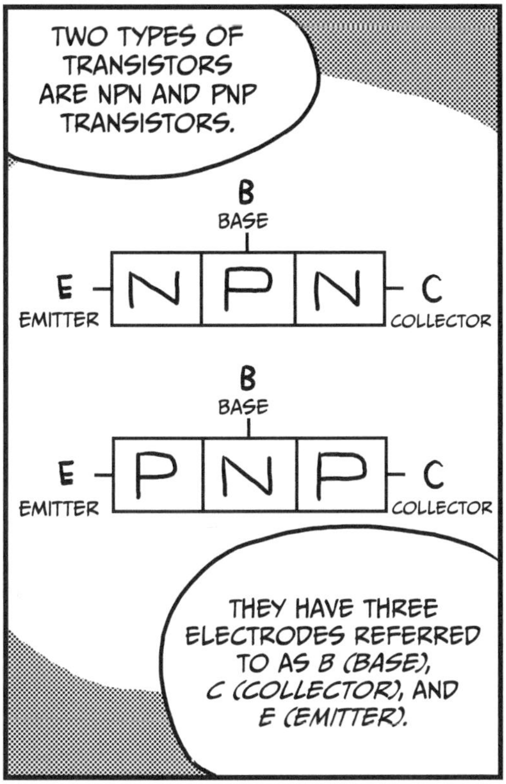
TWO TYPES OF TRANSISTORS ARE NPN AND PNP TRANSISTORS.
B
BASE
E
EMITTER
N P N
C
COLLECTOR
B
BASE
E
EMITTER
P N P
C
COLLECTOR
THEY HAVE THREE ELECTRODES REFERRED TO AS B (BASE), C (COLLECTOR), AND E (EMITTER).

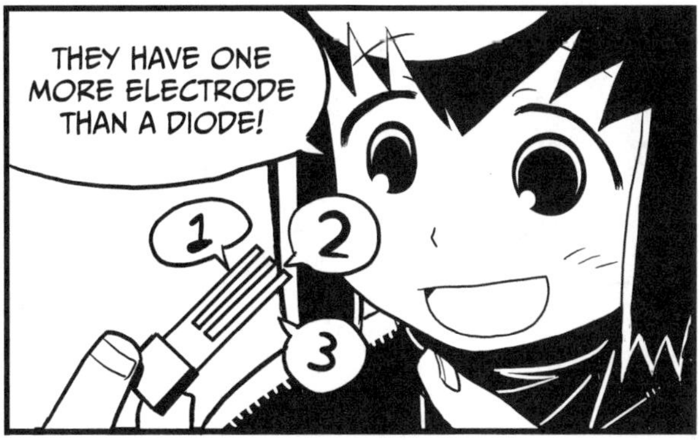
THEY HAVE ONE MORE ELECTRODE THAN A DIODE!
1
2
3

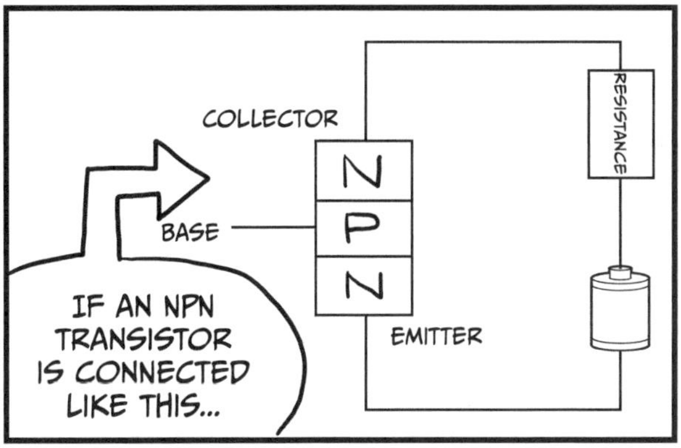
COLLECTOR
N
BASE
P
N
EMITTER
RESISTANCE
IF AN NPN TRANSISTOR IS CONNECTED LIKE THIS...

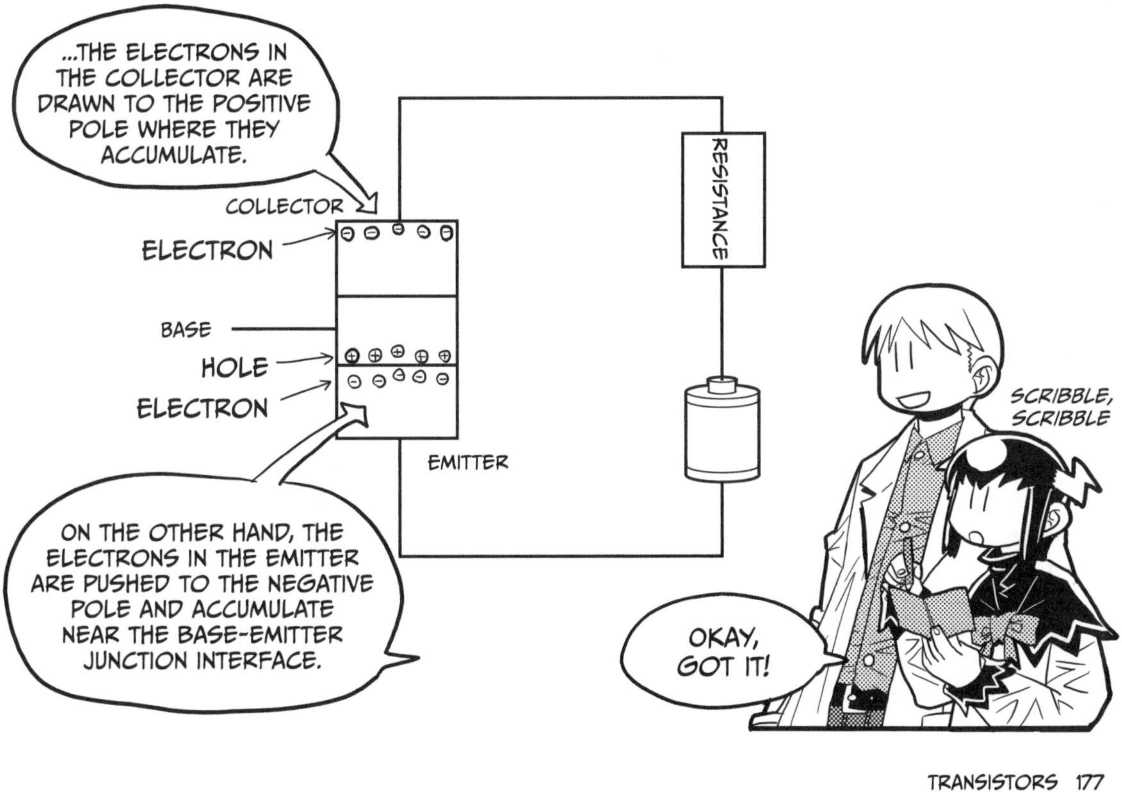
...THE ELECTRONS IN THE COLLECTOR ARE DRAWN TO THE POSITIVE POLE WHERE THEY ACCUMULATE.
COLLECTOR
ELECTRON
BASE
HOLE
ELECTRON
EMITTER
RESISTANCE
ON THE OTHER HAND, THE ELECTRONS IN THE EMITTER ARE PUSHED TO THE NEGATIVE POLE AND ACCUMULATE NEAR THE BASE-EMITTER JUNCTION INTERFACE.
SCRIBBLE, SCRIBBLE
OKAY, GOT IT!

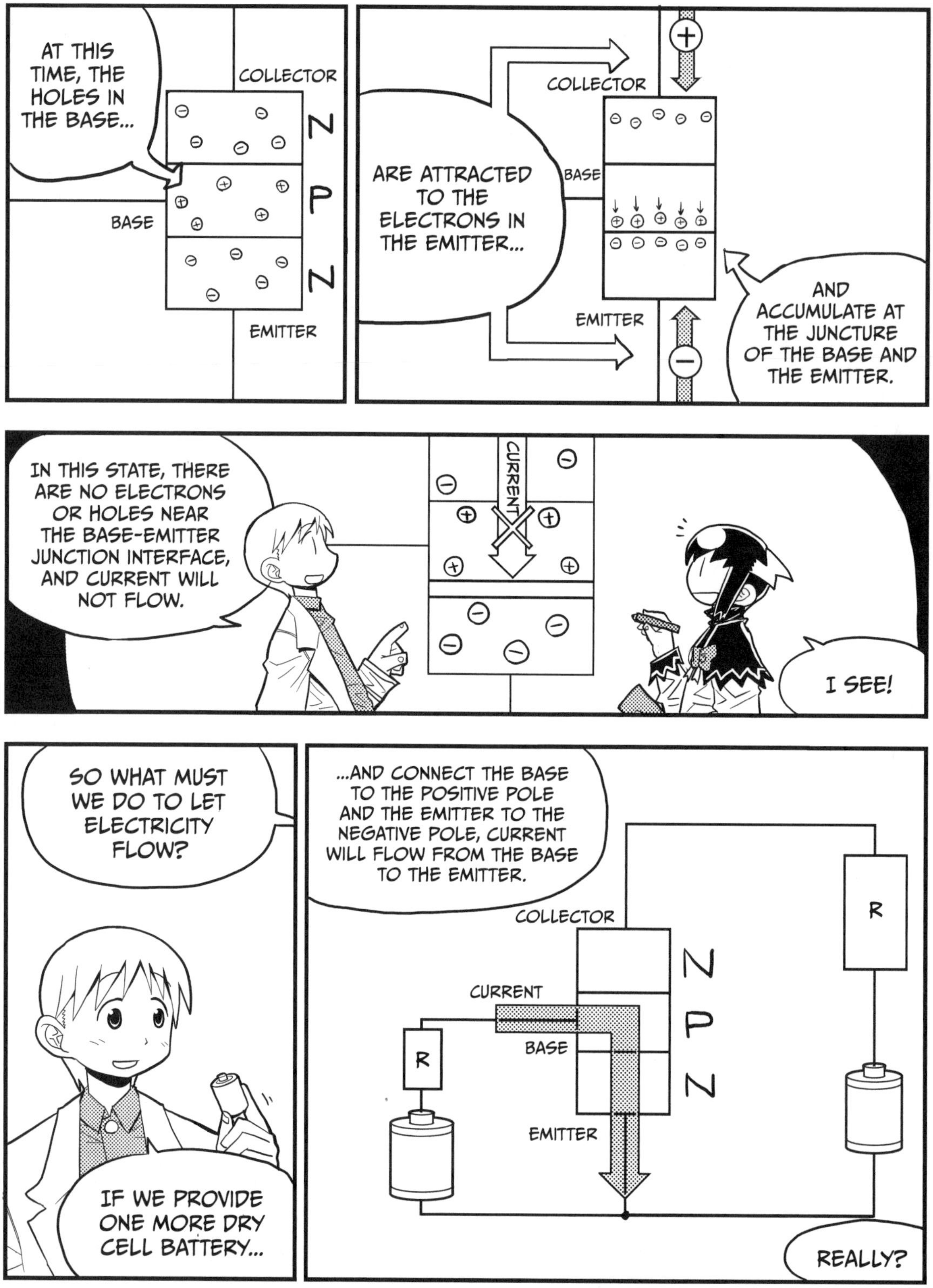
AT THIS TIME, THE HOLES IN THE BASE...
COLLECTOR
N
P
N
BASE
EMITTER
ARE ATTRACTED TO THE ELECTRONS IN THE EMITTER...
COLLECTOR
BASE
EMITTER
AND ACCUMULATE AT THE JUNCTURE OF THE BASE AND THE EMITTER.
IN THIS STATE, THERE ARE NO ELECTRONS OR HOLES NEAR THE BASE-EMITTER JUNCTION INTERFACE, AND CURRENT WILL NOT FLOW.
CURRENT
I SEE!
SO WHAT MUST WE DO TO LET ELECTRICITY FLOW?
IF WE PROVIDE ONE MORE DRY CELL BATTERY...
...AND CONNECT THE BASE TO THE POSITIVE POLE AND THE EMITTER TO THE NEGATIVE POLE, CURRENT WILL FLOW FROM THE BASE TO THE EMITTER.
COLLECTOR
N
P
N
CURRENT
BASE
EMITTER
R
R
REALLY?

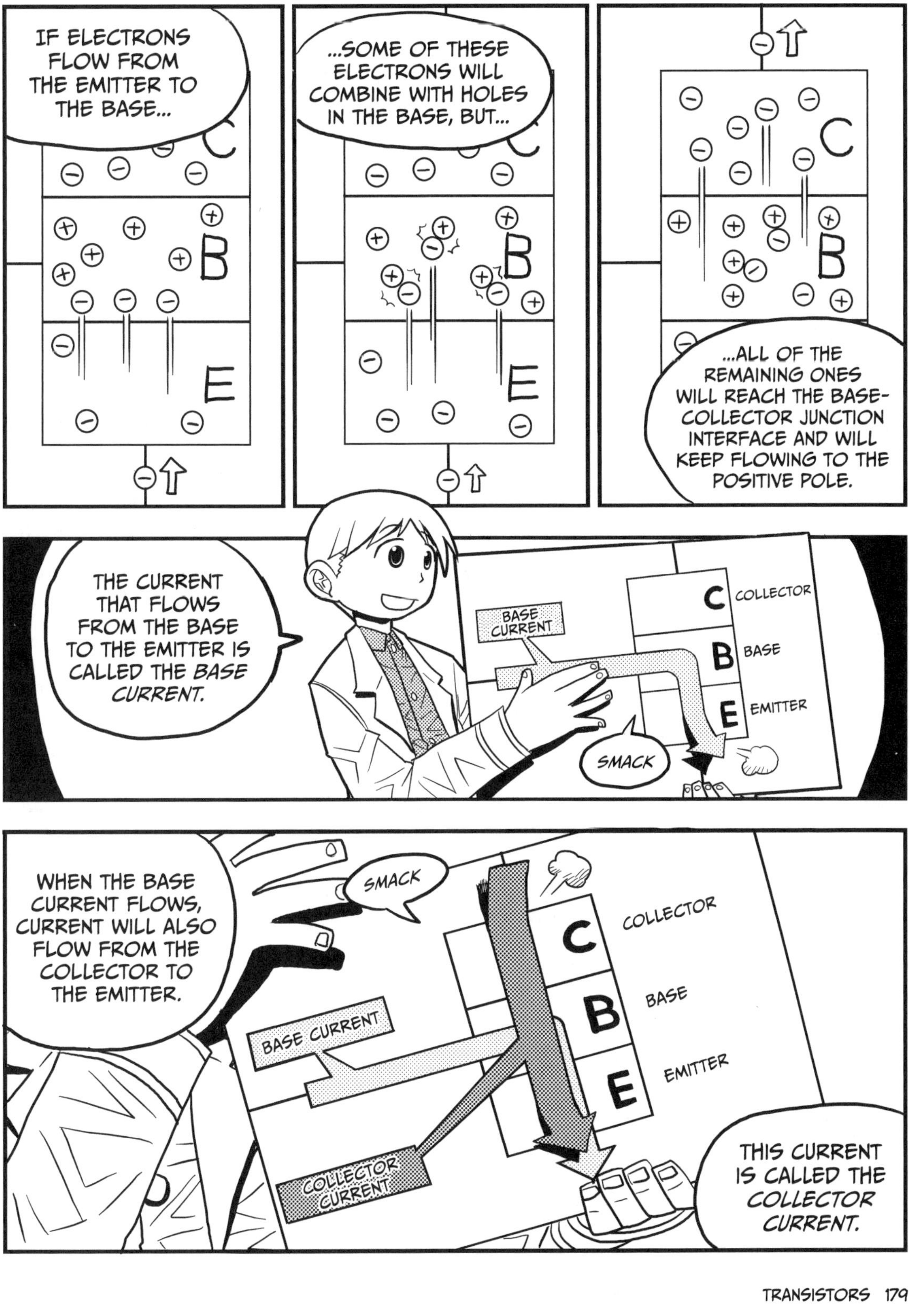
IF ELECTRONS FLOW FROM THE EMITTER TO THE BASE...
C
B
E
...SOME OF THESE ELECTRONS WILL COMBINE WITH HOLES IN THE BASE, BUT...
C
B
E
C
B
...ALL OF THE REMAINING ONES WILL REACH THE BASE-COLLECTOR JUNCTION INTERFACE AND WILL KEEP FLOWING TO THE POSITIVE POLE.
THE CURRENT THAT FLOWS FROM THE BASE TO THE EMITTER IS CALLED THE *BASE CURRENT.*
BASE CURRENT
C
COLLECTOR
B
BASE
E
EMITTER
SMACK
WHEN THE BASE CURRENT FLOWS, CURRENT WILL ALSO FLOW FROM THE COLLECTOR TO THE EMITTER.
SMACK
C
COLLECTOR
B
BASE
E
EMITTER
BASE CURRENT
COLLECTOR CURRENT
THIS CURRENT IS CALLED THE *COLLECTOR CURRENT.*

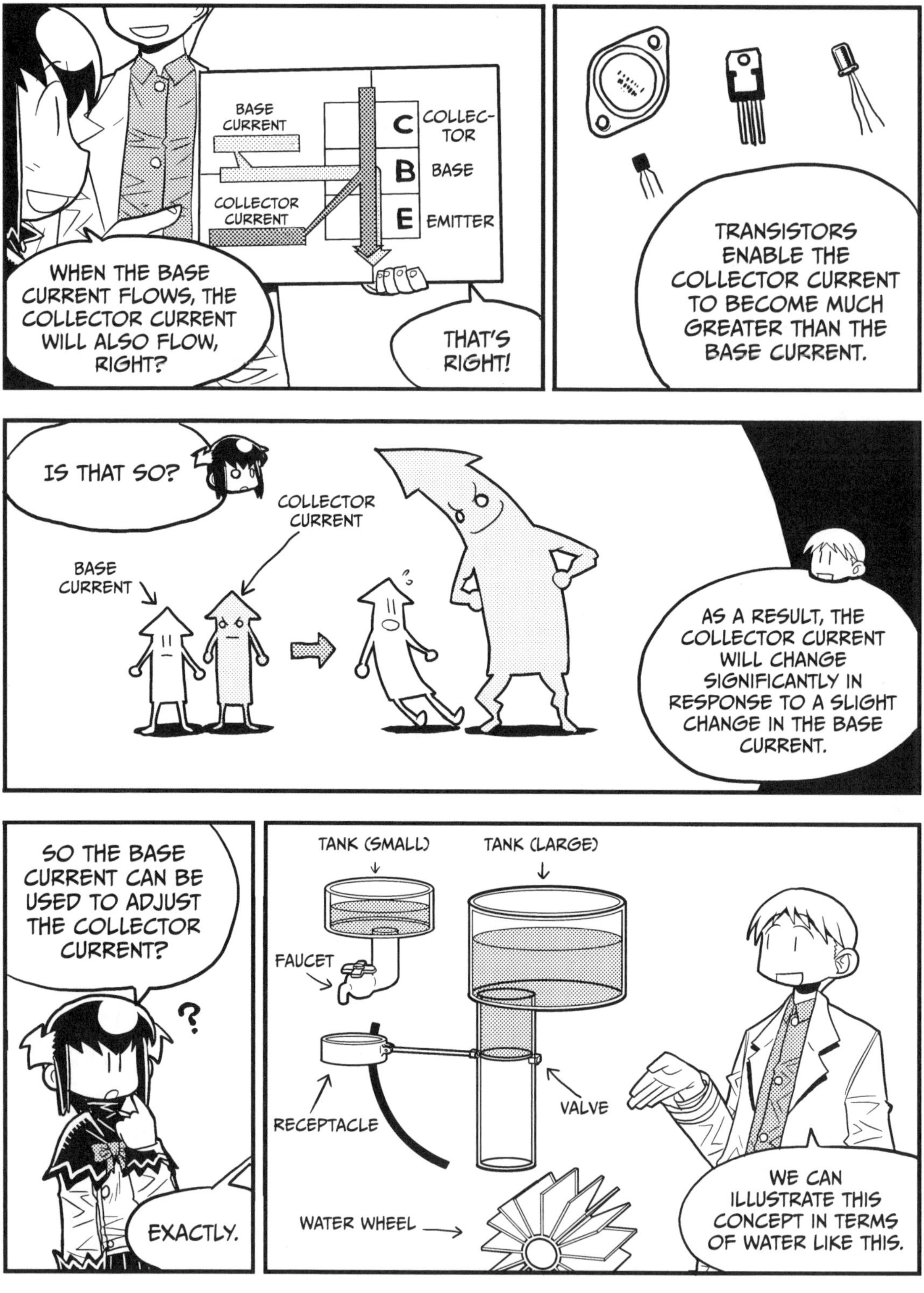
BASE CURRENT
COLLECTOR CURRENT
C COLLEC-TOR
B BASE
E EMITTER
WHEN THE BASE CURRENT FLOWS, THE COLLECTOR CURRENT WILL ALSO FLOW, RIGHT?
THAT'S RIGHT!
TRANSISTORS ENABLE THE COLLECTOR CURRENT TO BECOME MUCH GREATER THAN THE BASE CURRENT.
IS THAT SO?
BASE CURRENT
COLLECTOR CURRENT
AS A RESULT, THE COLLECTOR CURRENT WILL CHANGE SIGNIFICANTLY IN RESPONSE TO A SLIGHT CHANGE IN THE BASE CURRENT.
SO THE BASE CURRENT CAN BE USED TO ADJUST THE COLLECTOR CURRENT?
EXACTLY.
TANK (SMALL)
TANK (LARGE)
FAUCET
RECEPTACLE
VALVE
WATER WHEEL
WE CAN ILLUSTRATE THIS CONCEPT IN TERMS OF WATER LIKE THIS.

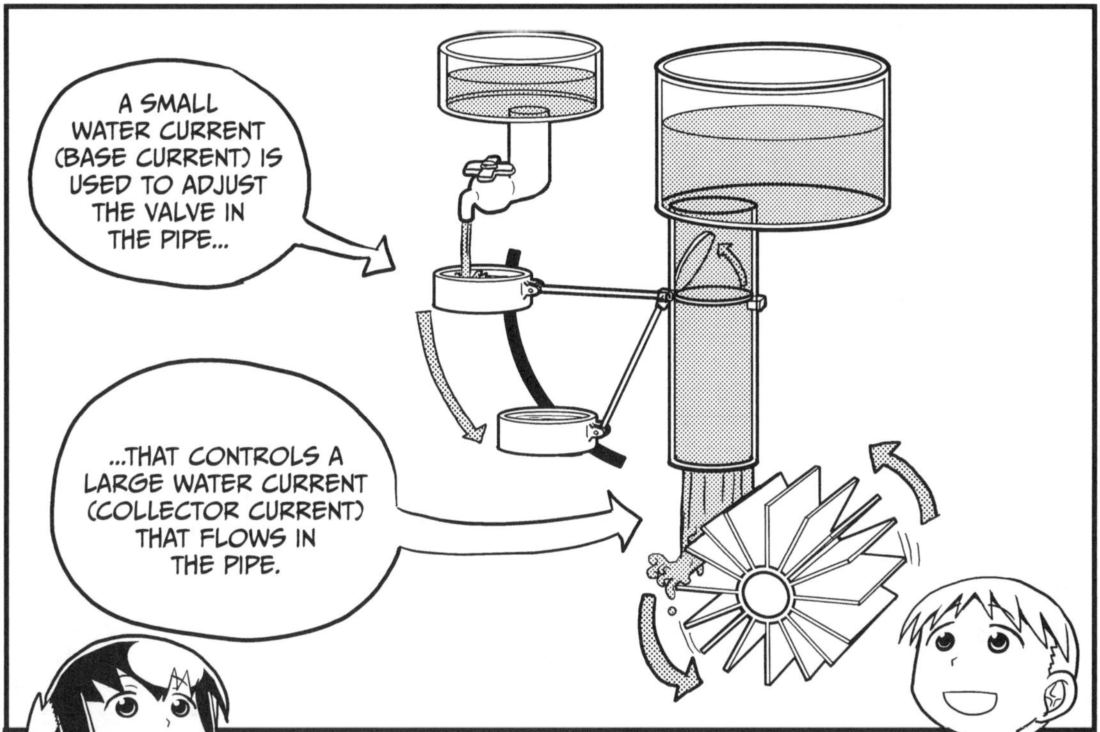
A SMALL WATER CURRENT (BASE CURRENT) IS USED TO ADJUST THE VALVE IN THE PIPE...
...THAT CONTROLS A LARGE WATER CURRENT (COLLECTOR CURRENT) THAT FLOWS IN THE PIPE.

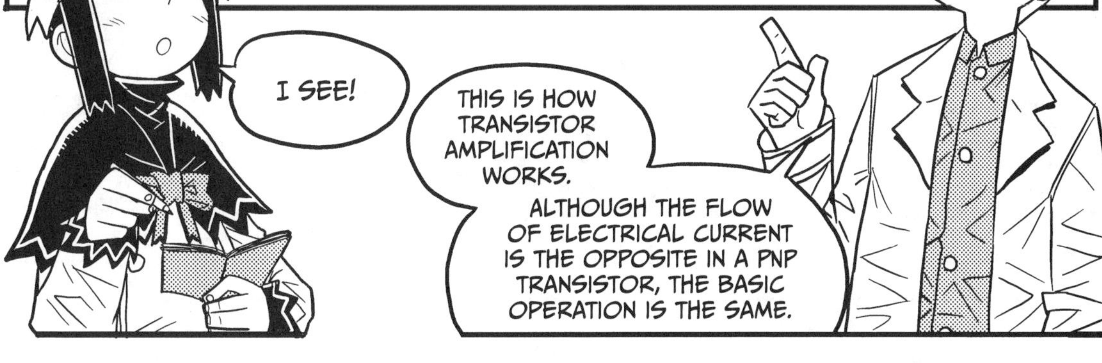
I SEE!
THIS IS HOW TRANSISTOR AMPLIFICATION WORKS.
ALTHOUGH THE FLOW OF ELECTRICAL CURRENT IS THE OPPOSITE IN A PNP TRANSISTOR, THE BASIC OPERATION IS THE SAME.

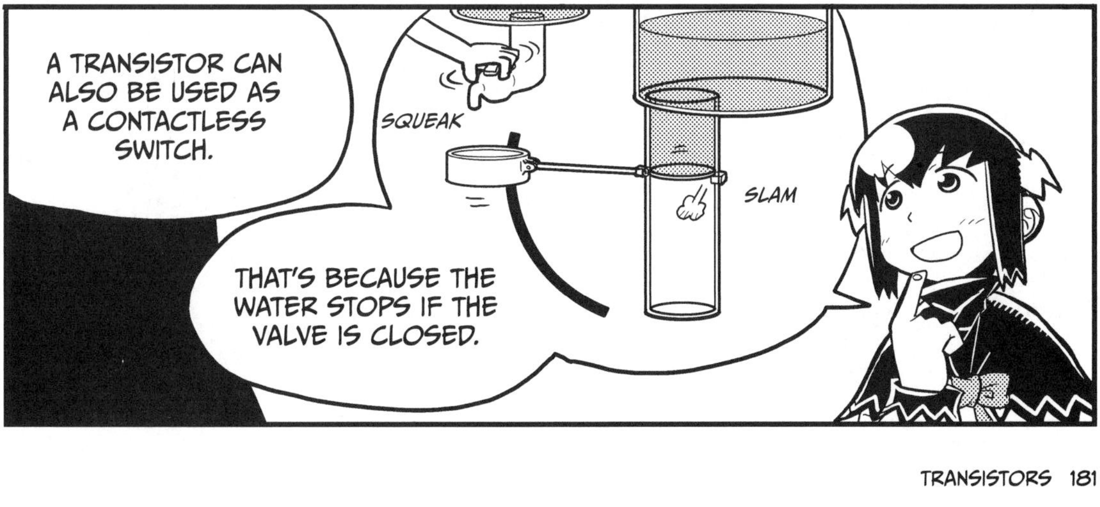
A TRANSISTOR CAN ALSO BE USED AS A CONTACTLESS SWITCH.
SQUEAK
SLAM
THAT'S BECAUSE THE WATER STOPS IF THE VALVE IS CLOSED.

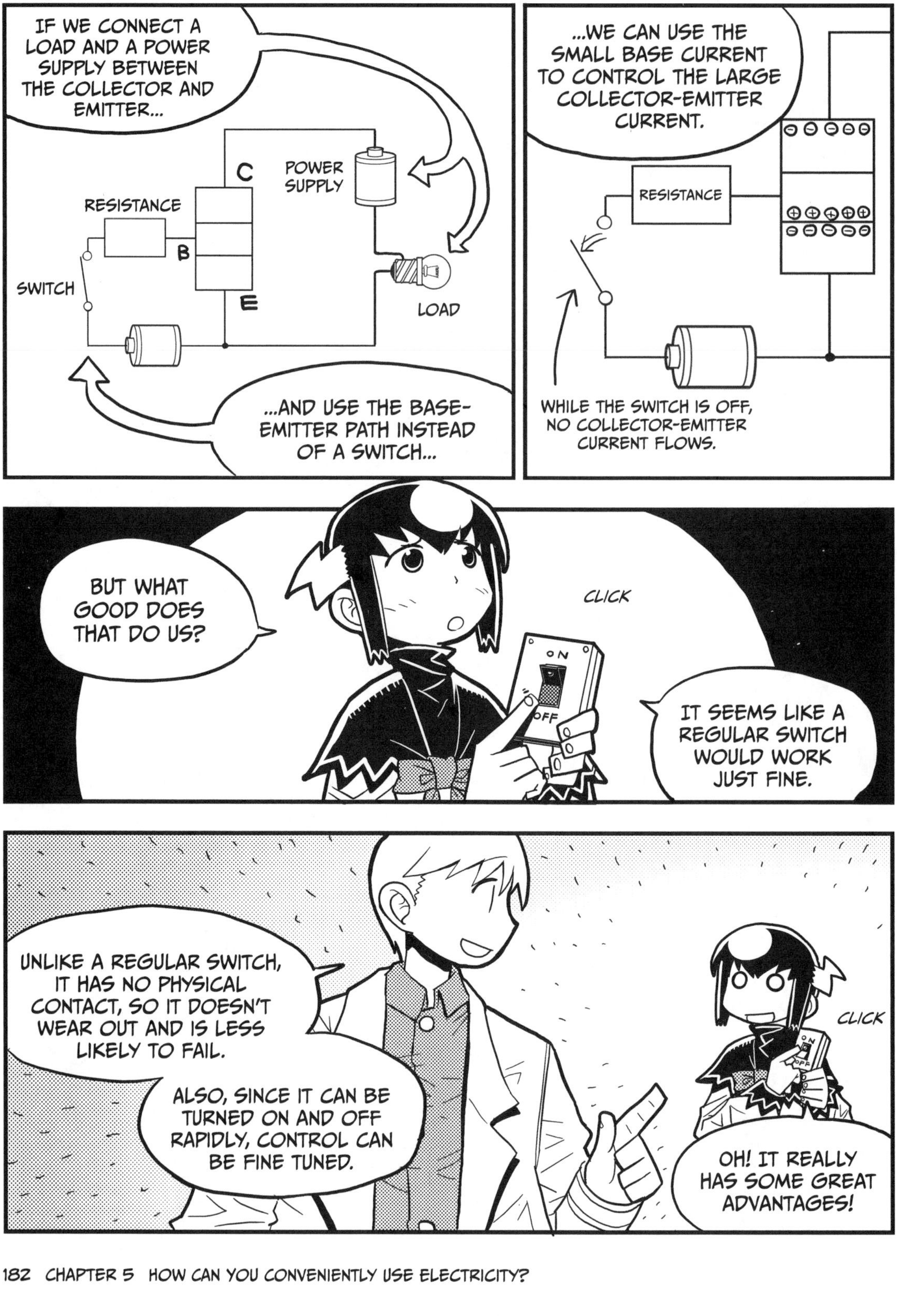
IF WE CONNECT A LOAD AND A POWER SUPPLY BETWEEN THE COLLECTOR AND EMITTER...
C
POWER SUPPLY
RESISTANCE
B
SWITCH
E
LOAD
...AND USE THE BASE-EMITTER PATH INSTEAD OF A SWITCH...
...WE CAN USE THE SMALL BASE CURRENT TO CONTROL THE LARGE COLLECTOR-EMITTER CURRENT.
RESISTANCE
WHILE THE SWITCH IS OFF, NO COLLECTOR-EMITTER CURRENT FLOWS.
BUT WHAT GOOD DOES THAT DO US?
CLICK
ON
OFF
IT SEEMS LIKE A REGULAR SWITCH WOULD WORK JUST FINE.
UNLIKE A REGULAR SWITCH, IT HAS NO PHYSICAL CONTACT, SO IT DOESN'T WEAR OUT AND IS LESS LIKELY TO FAIL.
ALSO, SINCE IT CAN BE TURNED ON AND OFF RAPIDLY, CONTROL CAN BE FINE TUNED.
CLICK
OH! IT REALLY HAS SOME GREAT ADVANTAGES!

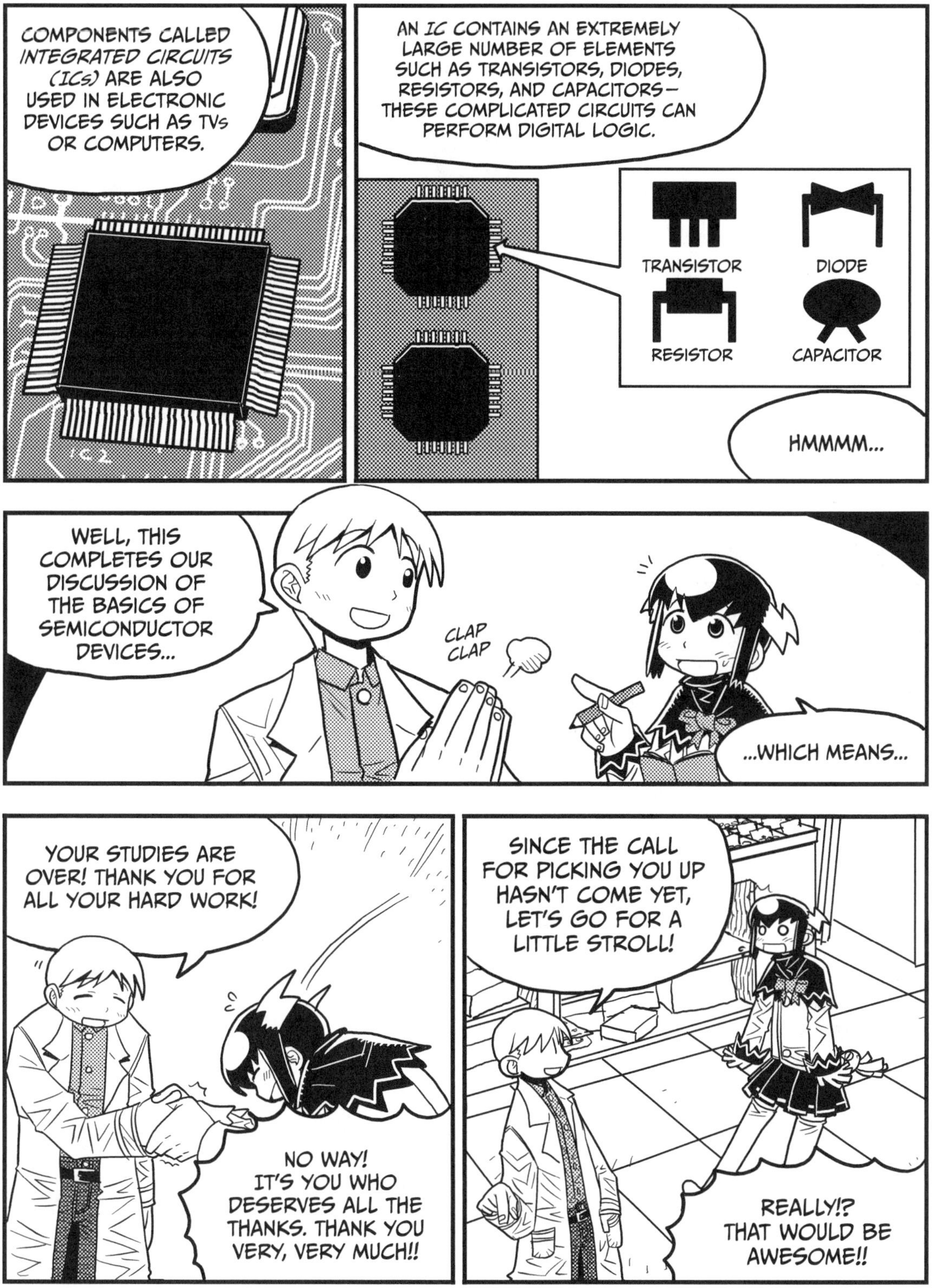
COMPONENTS CALLED *INTEGRATED CIRCUITS* (*ICs*) ARE ALSO USED IN ELECTRONIC DEVICES SUCH AS TVs OR COMPUTERS.
AN *IC* CONTAINS AN EXTREMELY LARGE NUMBER OF ELEMENTS SUCH AS TRANSISTORS, DIODES, RESISTORS, AND CAPACITORS— THESE COMPLICATED CIRCUITS CAN PERFORM DIGITAL LOGIC.
TRANSISTOR
DIODE
RESISTOR
CAPACITOR
HMMMM...
WELL, THIS COMPLETES OUR DISCUSSION OF THE BASICS OF SEMICONDUCTOR DEVICES...
CLAP CLAP
...WHICH MEANS...
YOUR STUDIES ARE OVER! THANK YOU FOR ALL YOUR HARD WORK!
NO WAY! IT'S YOU WHO DESERVES ALL THE THANKS. THANK YOU VERY, VERY MUCH!!
SINCE THE CALL FOR PICKING YOU UP HASN'T COME YET, LET'S GO FOR A LITTLE STROLL!
REALLY!? THAT WOULD BE AWESOME!!

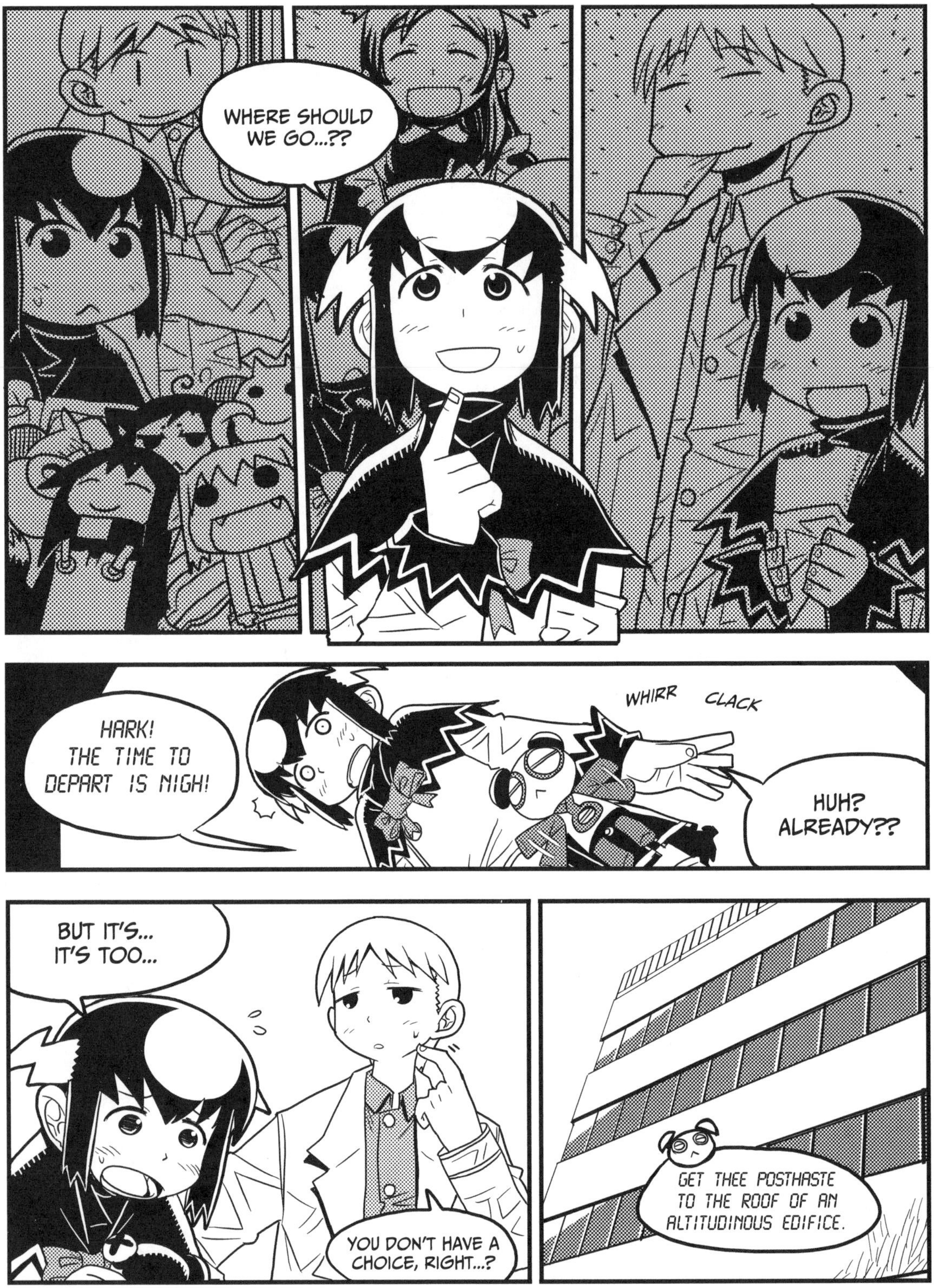
WHERE SHOULD WE GO...??
HARK! THE TIME TO DEPART IS NIGH!
WHIRR
CLACK
HUH? ALREADY??
BUT IT'S... IT'S TOO...
YOU DON'T HAVE A CHOICE, RIGHT...?
GET THEE POSTHASTE TO THE ROOF OF AN ALTITUDINOUS EDIFICE.

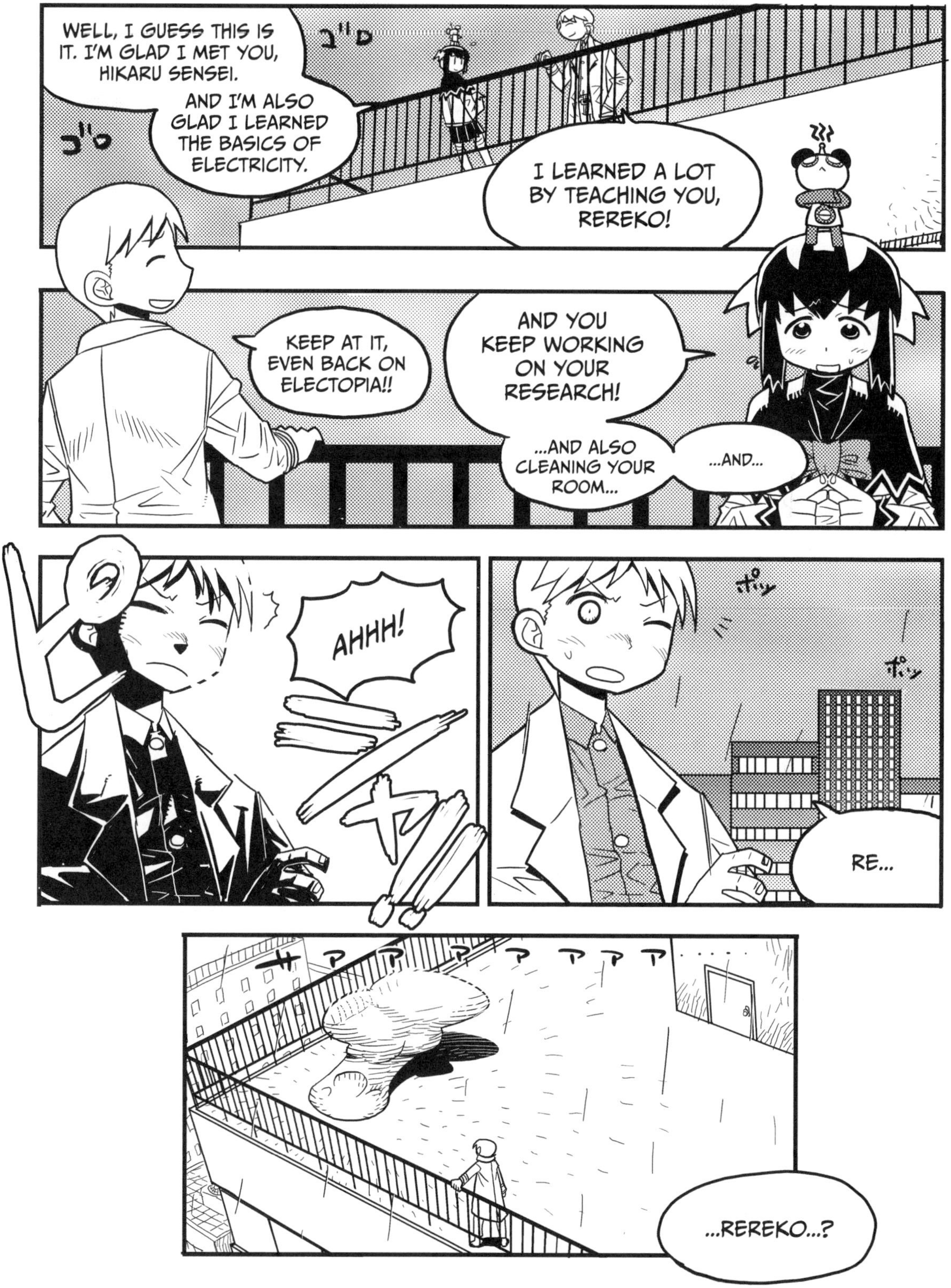
WELL, I GUESS THIS IS IT. I'M GLAD I MET YOU, HIKARU SENSEI.
AND I'M ALSO GLAD I LEARNED THE BASICS OF ELECTRICITY.
I LEARNED A LOT BY TEACHING YOU, REREKO!
KEEP AT IT, EVEN BACK ON ELECTOPIA!!
AND YOU KEEP WORKING ON YOUR RESEARCH!
...AND ALSO CLEANING YOUR ROOM...
...AND...
AHHH!
RE...
...REREKO...?

DIODES

When a single diode is connected to an AC power supply, current flows to the load for only one direction of the AC power supply due to rectification. Rectification that only allows a half-cycle of the alternating current to flow is called *half-wave rectification*, and the current that flows to the load in this process flows in only one direction, just like a direct current power supply. But since only a half-cycle of the AC waveform flows, this kind of rectification is inefficient.

DIODE
AC POWER SUPPLY
CURRENT
LOAD
DIODE SYMBOL
HALF-WAVE RECTIFICATION
CURRENT THAT FLOWS TO THE LOAD

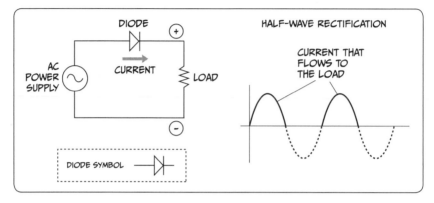

Half-wave rectification

When four diodes are arranged in a bridge configuration and an AC power supply is connected, the current of the entire cycle becomes positive and flows to the load. This kind of rectification is called *full-wave rectification*, and diodes that are connected in this way are called a *diode bridge*. Full-wave rectification enables current from the entire cycle of the AC power supply to be used as direct current.

DIODE BRIDGE
AC POWER SUPPLY
LOAD
FULL-WAVE RECTIFICATION
CURRENT THAT FLOWS TO THE LOAD

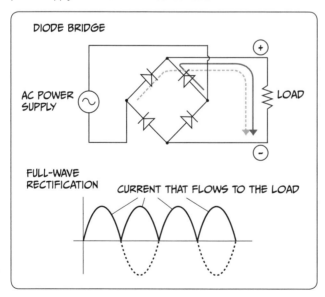

Full-wave rectification

Although this kind of full-wave rectification is more efficient than half-wave rectification, the waveform exhibits large pulsations. However, if an electrolytic capacitor is connected to the output, the charging and discharging of the capacitor can change the pulsations in the waveform into a flat, smooth direct current. A capacitor that is used to change a pulsating flow to a flat waveform in this way is called a *smoothing capacitor*.

AC POWER SUPPLY
CHARGE
DISCHARGE
LOAD
+
-
SMOOTHING CAPACITOR
THE CHARGING AND DISCHARGING OF THE SMOOTHING CAPACITOR SMOOTHS THE PULSATIONS.

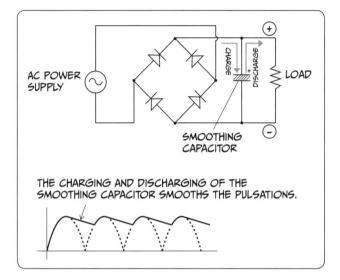

Smoothing capacitor

If a reverse-direction voltage is applied to a *Zener diode* (or *constant-voltage diode*) and the value of the voltage is steadily increased, current will flow once a certain voltage is reached. This phenomenon is called *breakdown*, and when the circuit voltage rises more than necessary, current can flow from the cathode to the anode to suppress the rise in voltage. This characteristic of a Zener diode is used in a constant voltage circuit that maintains a fixed voltage.

ZENER DIODE
CURRENT I
K
A
VOLTAGE V
AS THE VOLTAGE V INCREASES, THE CURRENT SUDDENLY FLOWS.
CURRENT
FORWARD DIRECTION CURRENT
FORWARD DIRECTION REGION
ZENER VOLTAGE
VOLTAGE
REVERSE DIRECTION VOLTAGE
FORWARD DIRECTION VOLTAGE
REVERSE DIRECTION CURRENT

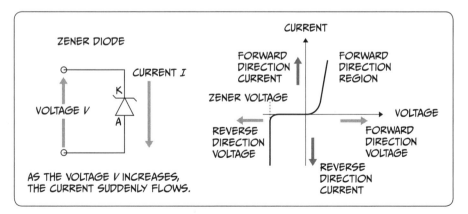

Characteristics of a Zener diode

If an ordinary diode is used as a Zener diode, it will be damaged because breakdown and the Zener current will be concentrated locally within the diode.

TRANSISTORS

A *transistor* is a semiconductor device that amplifies signals or acts as a switch by controlling the current or the voltage that is applied to an electrode.

When a large amount of electric power is controlled by a transistor that is used as a switch, the transistor is called a *power transistor*. Generally, an NPN-type transistor is used in this way.

NPN-TYPE TRANSISTOR
PNP-TYPE TRANSISTOR
C
B
E
C
B
E

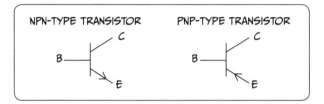

Transistor symbols

A switch that uses a transistor has no contact that will wear out, reducing the occurrence of failures, allowing it to be turned on and off rapidly, and allowing users to finely tune control of the device.

THE COLLECTOR-EMITTER PATH DOES THE WORK OF A CONTACT.
C
B
E
+
-
+
-
+
-
(CONTACTLESS)
(WITH A CONTACT)

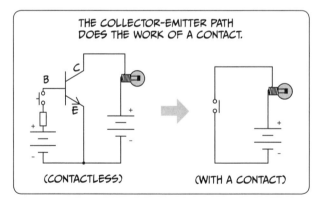

Transistor that does the work of a contact

FIELD-EFFECT TRANSISTOR

A transistor in which the collector current is controlled by the change of current that is input to the base is called a *bipolar transistor (junction transistor)*. In contrast, a transistor that is controlled by changing the voltage that is input, rather than the current, is called a *field-effect transistor (FET)*.

The merits of a field-effect transistor are that power consumption is low and response speed is extremely fast because current does not flow to the input. A field-effect transistor has three terminals that are referred to as *G (gate)*, *D (drain)*, and *S (source)*, which correspond to the base, collector, and emitter of a bipolar transistor, respectively. A field-effect transistor controls the drain current according to changes in the voltage that is input to the gate.

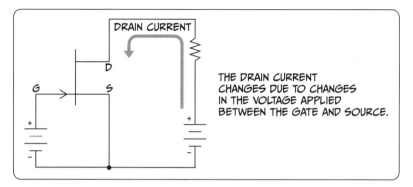

DRAIN CURRENT
D
G
S
+
-
+
-
THE DRAIN CURRENT CHANGES DUE TO CHANGES IN THE VOLTAGE APPLIED BETWEEN THE GATE AND SOURCE.

Field-effect transistor (N-channel type)

An *integrated circuit (IC)* is a device in which an extremely large number of elements such as transistors or resistors are placed on one component; ICs are used in more complex electronic devices such as TVs and computers. An amplifier called a *MOSFET (metal-oxide semiconductor field-effect transistor)*, in which the input gate is insulated by a thin film of silicon dioxide, is used in ICs.

CONVERTERS AND INVERTERS

A device that uses a diode to convert alternating current to direct current is called a *converter*, and a device that converts direct current to alternating current is called an *inverter*.

CONVERTER
ALTERNATING CURRENT
DIRECT CURRENT
INVERTER
DIRECT CURRENT
ALTERNATING CURRENT

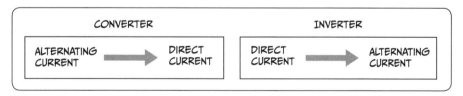

Converter and inverter

An inverter uses a semiconductor switching device such as a transistor to do the work of a switch. Single-phase alternating current can be produced by connecting four semiconductor switching devices and alternately turning on and off A, D, B, and C, as shown in the next figure. The frequency of the single-phase alternating current can be changed at will by varying the switching speed of the semiconductor switching devices.

CURRENT THAT FLOWS
WHEN A AND D ARE ON
DC
POWER
SUPPLY
A
B
C
D
LOAD
0
CURRENT THAT FLOWS
WHEN B AND C ARE ON
SINGLE-PHASE ALTERNATING
CURRENT FLOWS TO THE LOAD

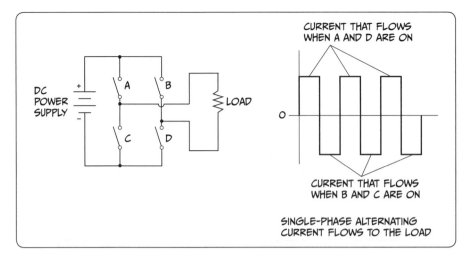

Single-phase alternating current created by an inverter

The rotational speed of an *induction motor* is directly proportional to the power supply frequency. If the supply frequency is constant, the rotational speed will also be constant.

For an air conditioner to cool the air, a motor must turn a compressor to compress the refrigerant gas. If the rotational speed of the motor is constant, a large capacity will be output even when a small capacity is required, and electrical power will be wasted.

Therefore, energy-saving operation with no waste can be achieved by using an inverter to create alternating current with the frequency required to continuously vary the rotational speed of the motor according to the required capacity.

A DC motor that is rotated by a DC power supply is used in the newest inverter air conditioners. In order to vary the rotational speed of the DC motor, the voltage must also vary, so a semiconductor switching device is used for this purpose.

In addition to air conditioners, inverters are also widely used in other familiar electrical appliances such as lighting or refrigerators and even in railroad cars.

SENSORS

Various sensors are used in electrical appliances in place of the perceptions of our eyes or skin. For example, an electric thermostat uses a temperature sensor to detect the temperature and turn a heater on and off, so we need not repeatedly turn the switch on and off ourselves.

Since sensors convert physical information such as light or heat to electrical information, if they are incorporated into an electric circuit, they can allow an electrical appliance to operate automatically. There are also sensors that can detect magnetism, which cannot be perceived by humans, or infrared rays, which cannot be seen by the naked eye.

TEMPERATURE SENSORS

A *temperature sensor* is a device that opens or closes a contact or varies electrical resistance according to the temperature it detects. Temperature sensors include contact-type sensors, which perceive the temperature by directly touching the substance whose temperature they are trying to measure, and non-contact-type sensors, which perceive emitted thermal energy without directly touching the substance whose temperature they are trying to measure.

There are many types of contact-type temperature sensors such as thermostats, thermistors, and thermocouples. Non-contact-type temperature sensors include infrared sensors.

TEMPERATURE SENSORS
CONTACT-TYPE
NON-CONTACT-TYPE
THERMISTOR
PTC (POSITIVE TEMPERATURE COEFFICIENT)
NTC (NEGATIVE TEMPERATURE COEFFICIENT)
THERMOCOUPLE (SEEBECK EFFECT)
PYROMETER (PERCEIVES INFRARED RAYS)

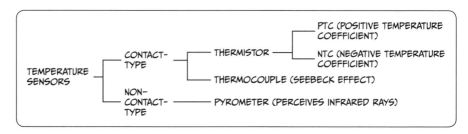

Classification of temperature sensors

A bi-metal thermostat is the simplest temperature sensor. It uses a bi-metal strip consisting of two types of metal with different thermal expansion rates, which curves in response to a temperature change. Although a thermostat is used in an appliance like an electric blanket, since the heater is turned on and off directly by a contact, the thermostat can only control large temperature fluctuations. A temperature sensor using a bi-metal strip is also used for the overcurrent action of a circuit breaker.

TEMPERATURE SENSOR
BI-METAL STRIP
LARGE THERMAL EXPANSION RATE
SMALL THERMAL EXPANSION RATE
HEATING
BENDS
COOLING
WHEN THE TEMPERATURE RETURNS TO THE ORIGINAL TEMPERATURE, THE SHAPE ALSO RETURNS.
THERMOSTAT
BI-METAL STRIP
CONTACT
POWER SUPPLY
HEATER
SET TEMPERATURE
SINCE IT TAKES TIME FOR THE BI-METAL STRIP TO BEND, THE TEMPERATURE FLUCTUATIONS ARE LARGE.

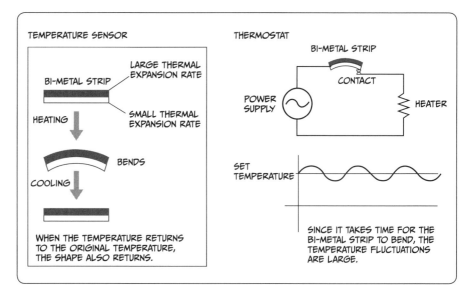

Temperature control by a bi-metal thermostat

A *thermistor* is a temperature sensor whose electrical resistance varies according to a temperature change. Generally speaking, electrical resistance also varies with temperature for any metal. However, thermistors' resistance changes significantly, even in response to a small temperature change. Since a large current does not flow directly to a thermistor, it is used in combination with an electrical circuit to control temperature.

Thermistors are classified into *positive temperature coefficient (PTC) thermistors*, those whose resistance value rises when the temperature rises, and *negative temperature coefficient (NTC) thermistors*, those whose resistance value falls when the temperature rises.

The newest air conditioners and electric refrigerators use thermistors for temperature sensors, combined with electrical circuits that use semiconductor devices to enable temperature control to be finely tuned.

NEGATIVE TEMPERATURE COEFFICIENT THERMISTOR (NTC THERMISTOR)
RESISTANCE VALUE
TEMPERATURE
WHEN THE TEMPERATURE RISES, THE RESISTANCE VALUE FALLS.
POSITIVE TEMPERATURE COEFFICIENT THERMISTOR (PTC THERMISTOR)
RESISTANCE VALUE
TEMPERATURE
WHEN THE TEMPERATURE RISES, THE RESISTANCE VALUE RISES.

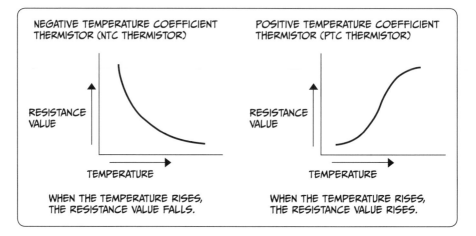

Temperature characteristics of thermistors

OPTICAL SENSORS

An *optical sensor* perceives light like our eyes do. These sensors are frequently used to automatically turn on street lights when it gets dark, and they function as the receiver of an infrared remote controller on an electrical appliance.

An optical sensor converts light energy to electrical signals. The phenomenon in which a substance such as a metal absorbs light energy and emits electrons as a result is called the *photoelectric effect*.

ELECTRONS
LIGHT
ABSORBS LIGHT ENERGY AND EMITS ELECTRONS
METAL

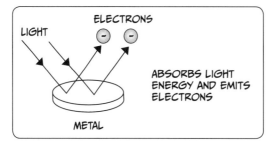

Photoelectric effect

The phenomenon that describes how voltage appears at the junction of a semiconductor due to the photoelectric effect is called the *photovoltaic effect*. Optical sensors that use the photovoltaic effect include *photodiodes* and *phototransistors*. A solar cell that is used for photovoltaic power generation also uses the photovoltaic effect to create electricity.

A *solar cell* generates an electromotive force when light energy strikes the p-n junction surface, causing the electrons and holes to move to the negative and positive poles, respectively. When a load is connected to a solar cell, current flows.

LIGHT
NEGATIVE POLE
N-TYPE SEMI-CONDUCTOR
ELECTRON
P-N JUNCTION SURFACE
CURRENT
HOLE
P-TYPE SEMI-CONDUCTOR
POSITIVE POLE
WHEN LIGHT STRIKES THE P-N JUNCTION SURFACE, AN ELECTROMOTIVE FORCE IS GENERATED AND CURRENT FLOWS.

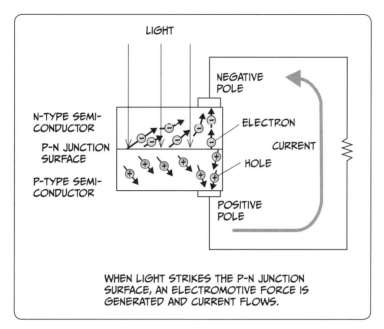

Photovoltaic effect of a solar cell

The effect in which a carrier of electricity such as an electron is generated by the photoelectic effect, thus causing the internal resistance value of a substance to change, is called *photoconductivity*. A *cadmium sulphide (CdS) cell* is a solar cell that functions using photoconductivity.

A *photodiode* is a semiconductor device in which current flows from the cathode to the anode due to the photovoltaic effect when light or infrared rays are received. The current that flows at this time varies according to the intensity of the light, and the photodiode measures this current.

The current when the light is received is extremely small. It is generally used by applying a reverse-bias voltage.

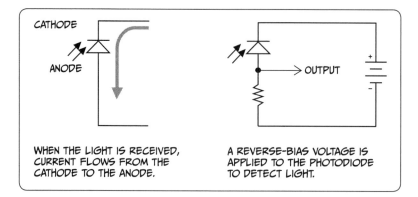

CATHODE

ANODE

WHEN THE LIGHT IS RECEIVED,
CURRENT FLOWS FROM THE
CATHODE TO THE ANODE.

OUTPUT

A REVERSE-BIAS VOLTAGE IS
APPLIED TO THE PHOTODIODE
TO DETECT LIGHT.

Photodiode

A photodiode combined with a transistor is called a *phototransistor*. Although a photo-transistor has no base, a collector current flows when light is received in a manner similar to how base current flows in a transistor. The current in the collector varies according to the intensity of the light.

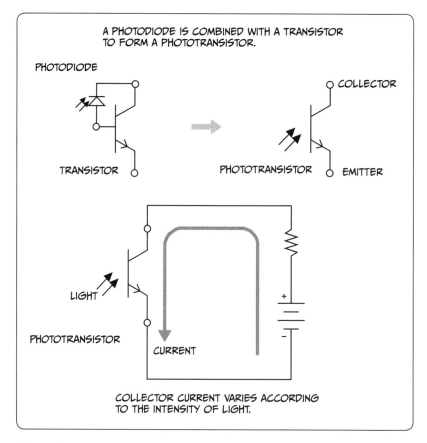

A PHOTODIODE IS COMBINED WITH A TRANSISTOR
TO FORM A PHOTOTRANSISTOR.

PHOTODIODE

COLLECTOR

TRANSISTOR

PHOTOTRANSISTOR

EMITTER

LIGHT

PHOTOTRANSISTOR

CURRENT

COLLECTOR CURRENT VARIES ACCORDING
TO THE INTENSITY OF LIGHT.

Phototransistor

An optical sensor like a phototransistor can be used to determine the position or existence of a target object without touching it.

PHOTOVOLTAIC EFFECT
PHOTOELECTRIC EFFECT
PHOTOTRANSISTOR
PHOTO IC
PHOTODIODE
PHOTOCONDUCTIVITY
CADMIUM SULFIDE CELL
PHOTOEMISSION EFFECT
PHOTOELECTRIC TUBE
PHOTOMULTIPLIER TUBE

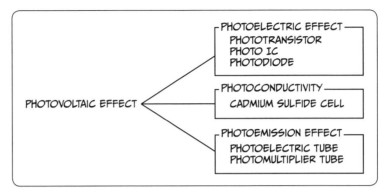

Photoelectric effect and optical sensors

Optical sensors are widely used for various purposes such as detecting brightness and turning on or dimming lights; an optical sensor can also be used in a security system as a photoelectric eye that detects the changes in light—that is, movement.

STREET LIGHT
AN OPTICAL SENSOR DETECTS BRIGHTNESS AND TURNS ON THE LIGHT WHEN IT GETS DARK.
TV
AIR CONDITIONER
RECEIVES THE INFRARED SIGNAL OF THE REMOTE CONTROL
REMOTE CONTROL RECEIVER
BEAM EMITTER
INFRARED BEAM
SECURITY SYSTEM
RECEIVER
OPTICAL SENSOR

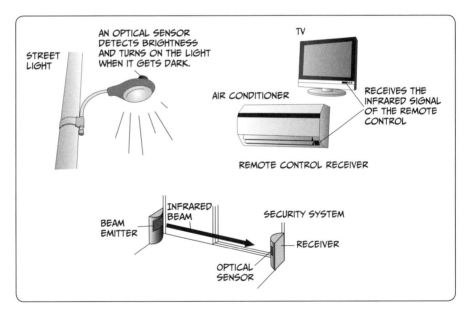

Uses of optical sensor

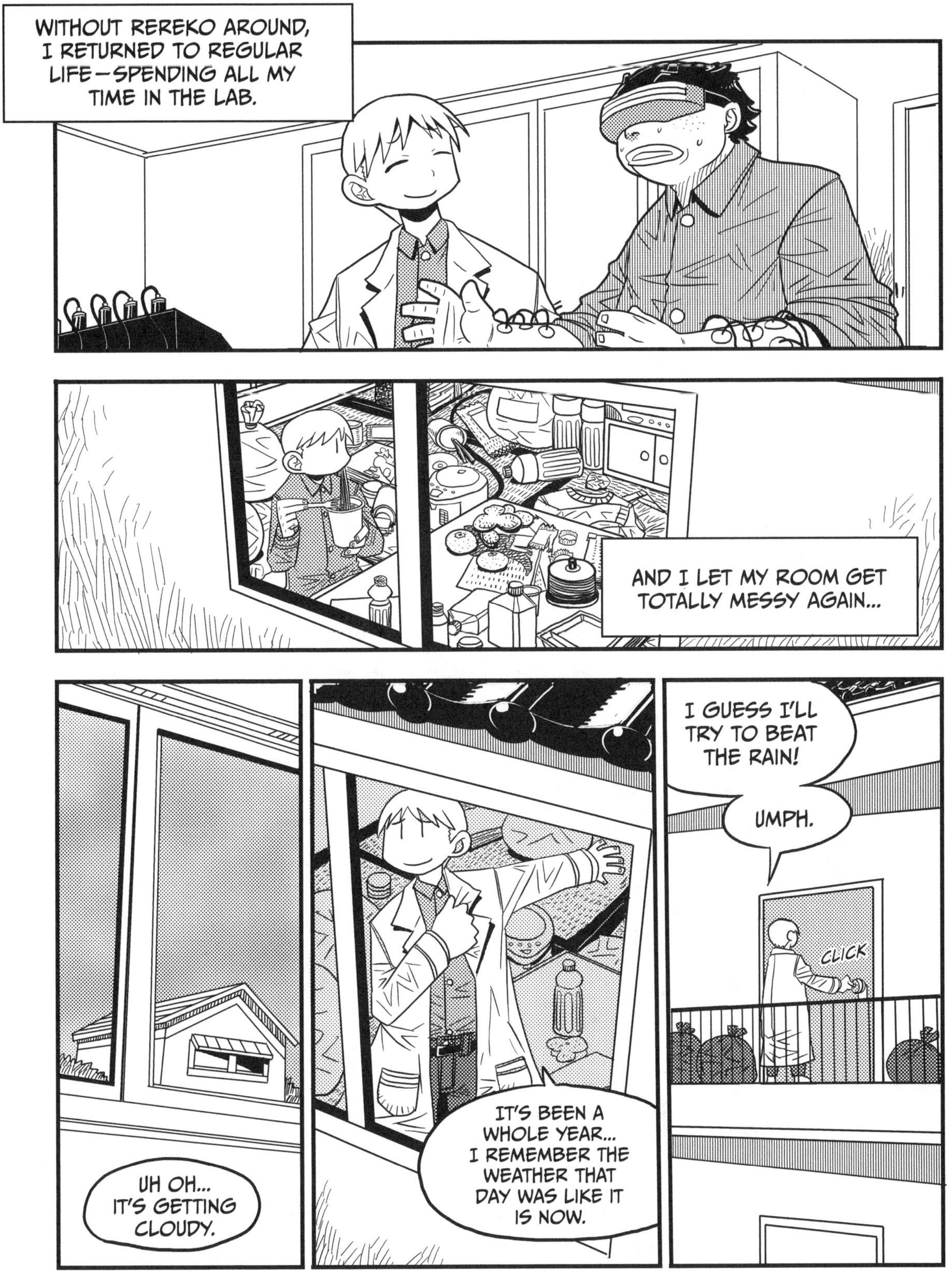
WITHOUT REREKO AROUND, I RETURNED TO REGULAR LIFE—SPENDING ALL MY TIME IN THE LAB.
AND I LET MY ROOM GET TOTALLY MESSY AGAIN...
UH OH... IT'S GETTING CLOUDY.
IT'S BEEN A WHOLE YEAR... I REMEMBER THE WEATHER THAT DAY WAS LIKE IT IS NOW.
I GUESS I'LL TRY TO BEAT THE RAIN!
UMPH.
CLICK

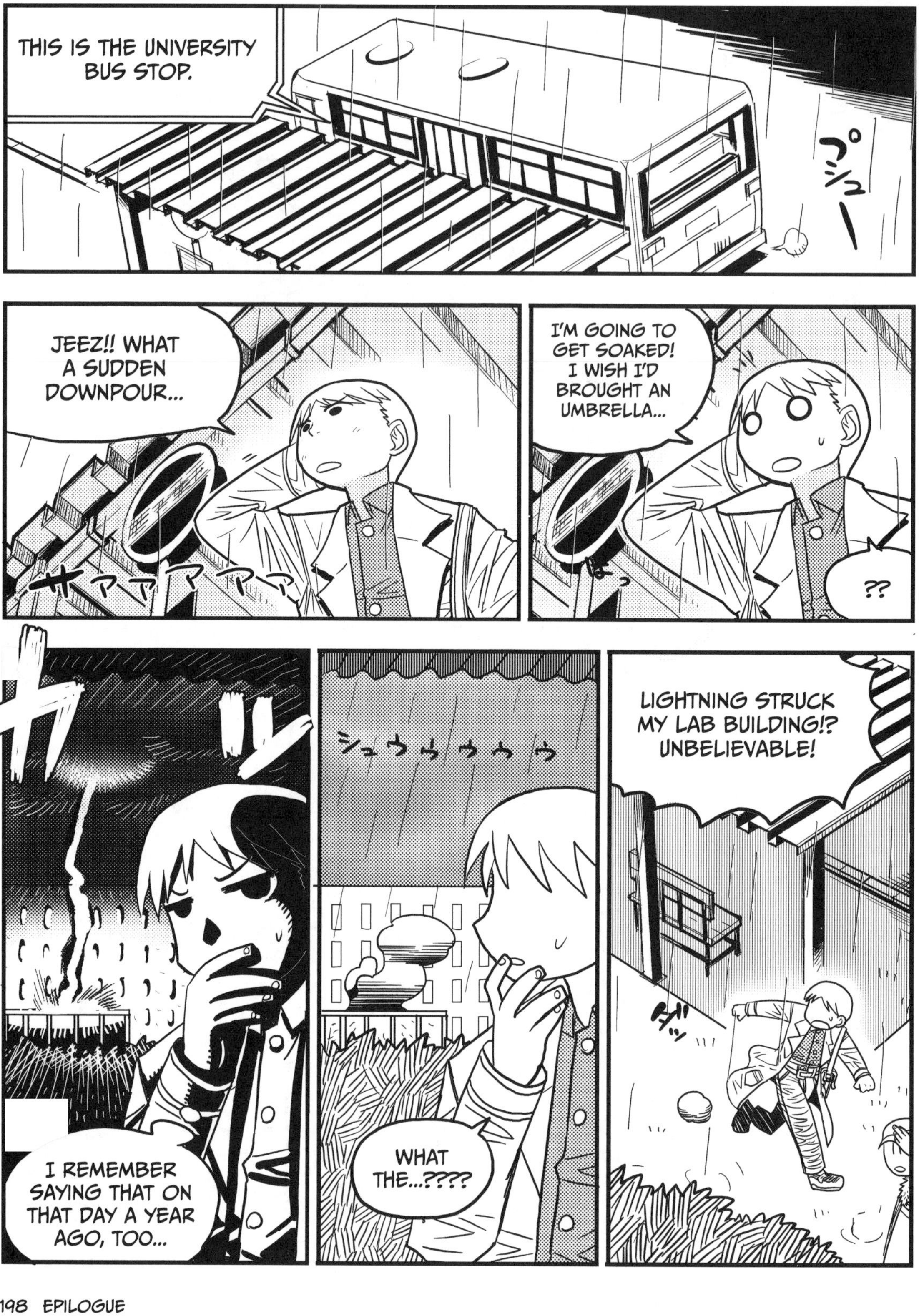
THIS IS THE UNIVERSITY BUS STOP.
JEEZ!! WHAT A SUDDEN DOWNPOUR...
I'M GOING TO GET SOAKED! I WISH I'D BROUGHT AN UMBRELLA...
??
I REMEMBER SAYING THAT ON THAT DAY A YEAR AGO, TOO...
WHAT THE...????
LIGHTNING STRUCK MY LAB BUILDING!? UNBELIEVABLE!

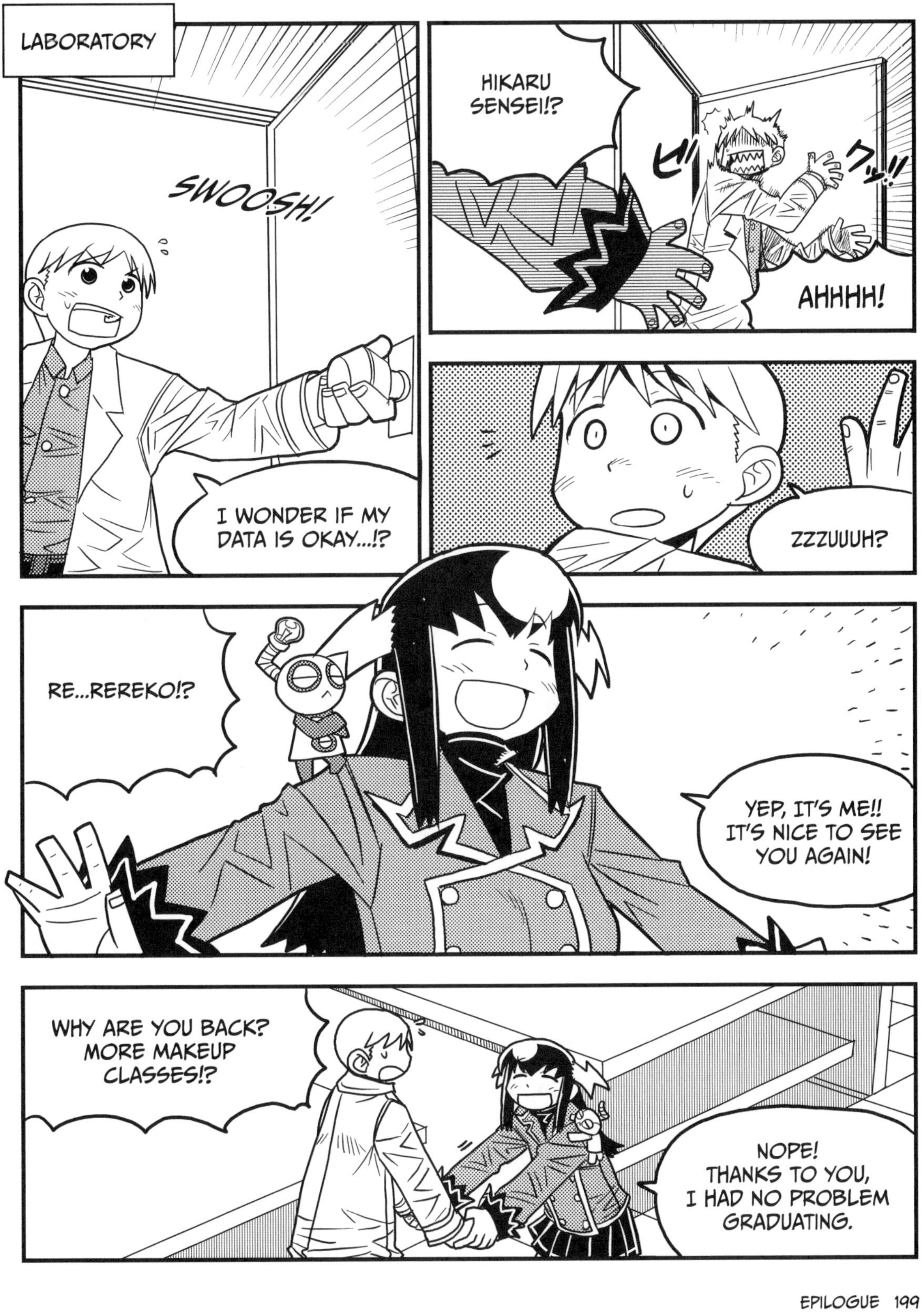
LABORATORY
SWOOSH!
I WONDER IF MY DATA IS OKAY...!?
HIKARU SENSEI!?
ビクッ!!
AHHHH!
ZZZUUUH?
RE...REREKO!?
YEP, IT'S ME!! IT'S NICE TO SEE YOU AGAIN!
WHY ARE YOU BACK? MORE MAKEUP CLASSES!?
NOPE! THANKS TO YOU, I HAD NO PROBLEM GRADUATING.

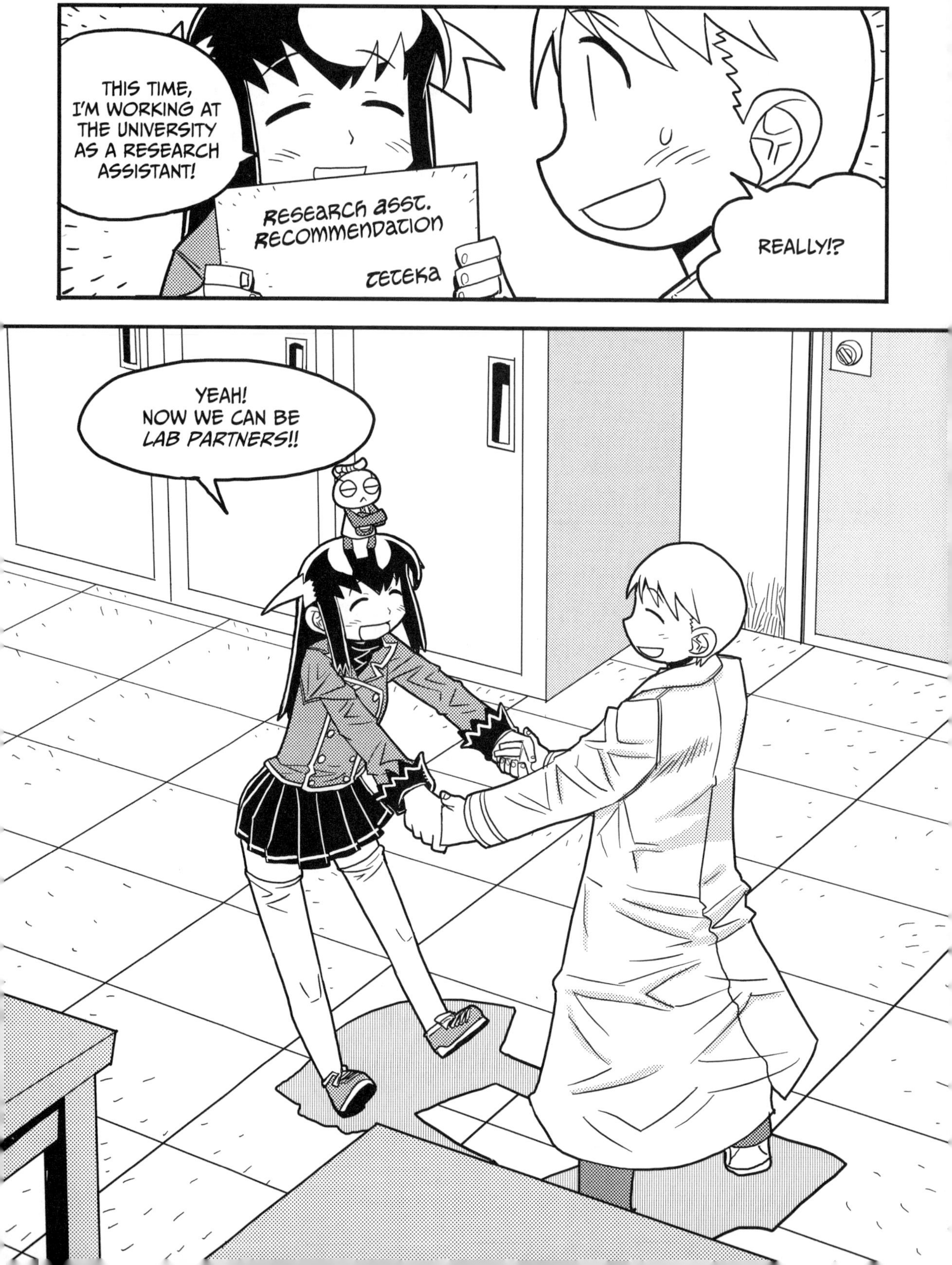
THIS TIME, I'M WORKING AT THE UNIVERSITY AS A RESEARCH ASSISTANT!
RESEARCH ASST. RECOMMENDATION
TETEKA
REALLY!?
YEAH! NOW WE CAN BE LAB PARTNERS!!

INDEX

triboelectricity, 38
triboelectric series, 40–43, 51–53
turbine engines, 147, 149,
         152, 153

U

UHF waves, 91
ultraviolet rays, 91, 93, 106, 107
United States, voltage in, 45
utility pole transformers, 20

V

valence electrons, 34–35, 48,
         165–166, 167
valence shells, 48
VHF waves, 91
vinyl (PVC), 37, 39, 41–42, 51
violet light, 91, 106
visible light, 91, 93, 106
Volta, Alessandro, 45, 126
voltage (*V*)
         100V, 14, 45
         120V, 22, 45, 72, 75, 76, 80,
             123, 149
         170V, 75, 76, 149
         240V, 45
         changes in, 20
         defined, 15, 45, 46–47
         effective, 75–76, 123, 149
         in electrical appliances, 19,
             24, 45, 46, 71–72, 96,
             123, 190
         forward bias, 173, 174
         instantaneous, 75–76
         international differences, 45
         peak, 75–76, 123, 149
         positive and negative poles,
             46–47, 61, 73, 132, 171,
             177–178, 179
         potential and, 46–47
         power supply, 60–61, 62,
             73–74, 78, 115
         reverse-bias, 171, 193

reverse-direction, 187
         supply, 46–47
voltaic cells, 125
volts (V), 15, 17–18, 45, 68

W

water
      cells, 135–137
      current, 181
      levels, 15–16
      turbine, 147, 152, 153–154
      wheels, 17, 59–60
Watt, James, 45
watt seconds (Ws), 46, 104
watts (W), 14, 17, 19, 45
waveforms, 187
wavelength, 91, 106
waves
      electromagnetic, 91, 106–107
      radio, 91, 106
      UHF, 91
      VHF, 91
wind power generation, 154
wires
      coils, 108, 111–115
      copper, 31, 104, 128
      electric, 74, 77, 88, 96, 107,
          108, 148
wood, 41–42, 51
wool, 41–42, 51

X

X rays, 91

Z

Zener diodes, 187–188
zinc
      chloride, 132
      ions, 129
      plate, 126, 127, 128
      powder, 133

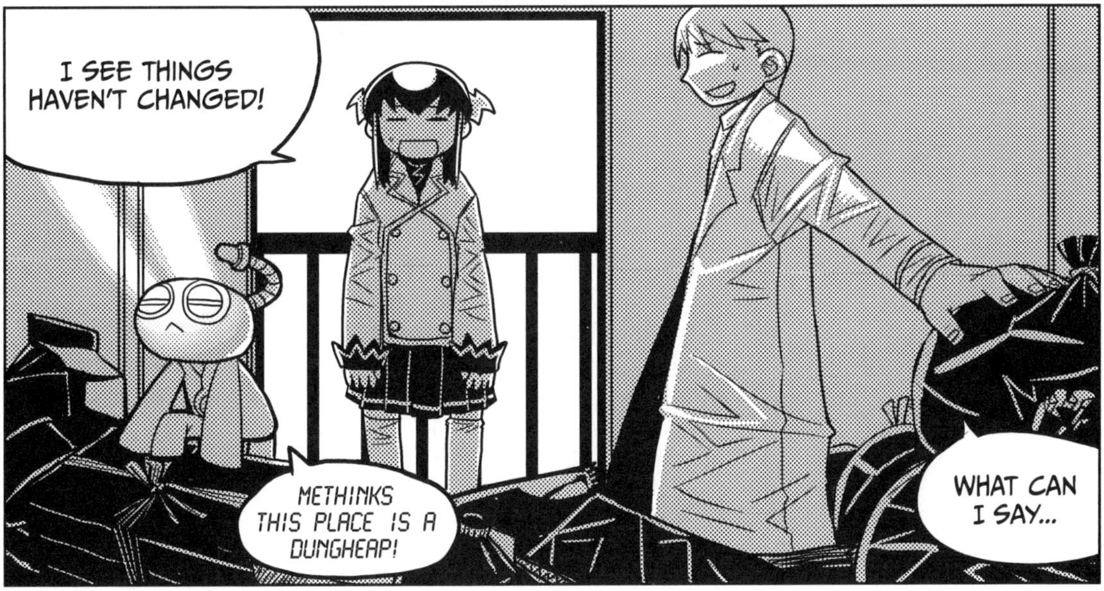
I SEE THINGS HAVEN'T CHANGED!
METHINKS THIS PLACE IS A DUNGHEAP!
WHAT CAN I SAY...

NOTES

NOTES

ABOUT THE AUTHOR

Kazuhiro Fujitaki is a lecturer at the Tokyo Metropolitan Vocational Skills Development Center. He has written a number of books on electrical engineering and runs a website offering information about Japan's qualifying examinations for electrical technicians. The site (in Japanese) is at *http://www10.ocn.ne.jp/~denkou/*.

PROJECT TEAM FOR THE JAPANESE EDITION

Production: TREND-PRO Co., Ltd.

Founded in 1988, TREND-PRO produces newspaper and magazine advertisements incorporating manga for a wide range of clients from government agencies to major corporations and associations. Currently, TREND-PRO is actively participating in advertisement and publishing projects using digital content. Some results of past creations are publicly available on the company's website, *http://www.ad-manga.com/*.

Ikeden Bldg., 3F, 2-12-5 Shinbashi, Minato-ku, Tokyo, Japan

Telephone: 03-3519-6769; Fax: 03-3519-6110

Scenario writer: re_akino

Artist: Matsuda

DTP: Mackey Soft Corporation

MORE MANGA GUIDES

The *Manga Guide* series is a co-publication of No Starch Press and Ohmsha, Ltd. of Tokyo, Japan, one of Japan's oldest and most respected scientific and technical book publishers. Each title in the best-selling *Manga Guide* series is the product of the combined work of a manga illustrator, scenario writer, and expert scientist or mathematician. Once each title is translated into English, we rewrite and edit the translation as necessary and have an expert review each volume for technical correctness. The result is the English version you hold in your hands.

Find more *Manga Guides* at your favorite bookstore, and learn more about the series at *http://www.edumanga.me/*.

THE MANGA GUIDE TO
STATISTICS
SHIN TAKAHASHI
TREND-PRO, CO., LTD.

THE MANGA GUIDE TO
DATABASES
MANA TAKAHASHI
SHOKO AZUMA
TREND-PRO CO., LTD.

THE MANGA GUIDE TO
PHYSICS
HIDEO NITTA
KEITA TAKATSU
TREND-PRO CO., LTD.

THE MANGA GUIDE TO
CALCULUS
HIROYUKI KOJIMA
SHIN TOGAMI
BECOM CO., LTD.

THE MANGA GUIDE TO
MOLECULAR BIOLOGY
MASAHARU TAKEMURA
SAKURA
BECOM CO., LTD

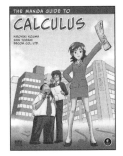

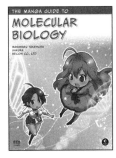

COLOPHON

The Manga Guide to Electricity was laid out in Adobe InDesign. The fonts are CCMeanwhile and Chevin.

The book was printed and bound at Malloy Incorporated in Ann Arbor, Michigan. The paper is Glatfelter Spring Forge 60# Eggshell, which is certified by the Sustainable Forestry Initiative (SFI).

UPDATES

Visit *http://www.nostarch.com/mg_electricity.htm* for updates, errata, and other information.